SCHAUM'S OUTLINE OF

THEORY AND PROBLEMS

OF

MATRICES
SI (METRIC) EDITION

BY

FRANK AYRES, JR., Ph.D.

Formerly Professor and Head,
Department of Mathematics
Dickinson College

McGraw-Hill International Book Company, New York
co-published with
McGraw-Hill Book Company (UK) Limited, London
McGraw-Hill Book Company GmbH, Düsseldorf
McGraw-Hill Book Company Australia Pty, Limited
McGraw-Hill Book Company (SA) (Pty) Limited, Johannesburg

07 084379 1

CONTENTS

		Page
Chapter *1*	MATRICES	1
	Matrices. Equal matrices. Sums of matrices. Products of matrices. Products by partitioning.	
Chapter *2*	SOME TYPES OF MATRICES	10
	Triangular matrices. Scalar matrices. Diagonal matrices. The identity matrix. Inverse of a matrix. Transpose of a matrix. Symmetric matrices. Skew-symmetric matrices. Conjugate of a matrix. Hermitian matrices. Skew-Hermitian matrices. Direct sums.	
Chapter *3*	DETERMINANT OF A SQUARE MATRIX	20
	Determinants of orders 2 and 3. Properties of determinants. Minors and cofactors. Algebraic complements.	
Chapter *4*	EVALUATION OF DETERMINANTS	32
	Expansion along a row or column. The Laplace expansion. Expansion along the first row and column. Determinant of a product. Derivative of a determinant.	
Chapter *5*	EQUIVALENCE	39
	Rank of a matrix. Non-singular and singular matrices. Elementary transformations. Inverse of an elementary transformation. Equivalent matrices. Row canonical form. Normal form. Elementary matrices. Canonical sets under equivalence. Rank of a product.	
Chapter *6*	THE ADJOINT OF A SQUARE MATRIX	49
	The adjoint. The adjoint of a product. Minor of an adjoint.	
Chapter *7*	THE INVERSE OF A MATRIX	55
	Inverse of a diagonal matrix. Inverse from the adjoint. Inverse from elementary matrices. Inverse by partitioning. Inverse of symmetric matrices. Right and left inverses of $m \times n$ matrices.	
Chapter *8*	FIELDS	64
	Number fields. General fields. Sub-fields. Matrices over a field.	

CONTENTS

Page

Chapter 9 LINEAR DEPENDENCE OF VECTORS AND FORMS........ 67

Vectors. Linear dependence of vectors, linear forms, polynomials, and matrices.

Chapter 10 LINEAR EQUATIONS 75

System of non-homogeneous equations. Solution using matrices. Cramer's rule. Systems of homogeneous equations.

Chapter 11 VECTOR SPACES 85

Vector spaces. Sub-spaces. Basis and dimension. Sum space. Intersection space. Null space of a matrix. Sylvester's laws of nullity. Bases and coordinates.

Chapter 12 LINEAR TRANSFORMATIONS 94

Singular and non-singular transformations. Change of basis. Invariant space. Permutation matrix.

Chapter 13 VECTORS OVER THE REAL FIELD...................... 100

Inner product. Length. Schwarz inequality. Triangle inequality. Orthogonal vectors and spaces. Orthonormal basis. Gram-Schmidt orthogonalization process. The Gramian. Orthogonal matrices. Orthogonal transformations. Vector product.

Chapter 14 VECTORS OVER THE COMPLEX FIELD.................. 110

Complex numbers. Inner product. Length. Schwarz inequality. Triangle inequality. Orthogonal vectors and spaces. Orthonormal basis. Gram-Schmidt orthogonalization process. The Gramian. Unitary matrices. Unitary transformations.

Chapter 15 CONGRUENCE 115

Congruent matrices. Congruent symmetric matrices. Canonical forms of real symmetric, skew-symmetric, Hermitian, skew-Hermitian matrices under congruence.

Chapter 16 BILINEAR FORMS 125

Matrix form. Transformations. Canonical forms. Cogredient transformations. Contragredient transformations. Factorable forms.

Chapter 17 QUADRATIC FORMS 131

Matrix form. Transformations. Canonical forms. Lagrange reduction. Sylvester's law of inertia. Definite and semi-definite forms. Principal minors. Regular form. Kronecker's reduction. Factorable forms.

CONTENTS

		Page
Chapter *18*	**HERMITIAN FORMS**	**146**
	Matrix form. Transformations. Canonical forms. Definite and semi-definite forms.	
Chapter *19*	**THE CHARACTERISTIC EQUATION OF A MATRIX**	**149**
	Characteristic equation and roots. Invariant vectors and spaces.	
Chapter *20*	**SIMILARITY**	**156**
	Similar matrices. Reduction to triangular form. Diagonable matrices.	
Chapter *21*	**SIMILARITY TO A DIAGONAL MATRIX**	**163**
	Real symmetric matrices. Orthogonal similarity. Pairs of real quadratic forms. Hermitian matrices. Unitary similarity. Normal matrices. Spectral decomposition. Field of values.	
Chapter *22*	**POLYNOMIALS OVER A FIELD**	**172**
	Sum, product, quotient of polynomials. Remainder theorem. Greatest common divisor. Least common multiple. Relatively prime polynomials. Unique factorization.	
Chapter *23*	**LAMBDA MATRICES**	**179**
	The λ-matrix or matrix polynomial. Sums, products, and quotients. Remainder theorem. Cayley-Hamilton theorem. Derivative of a matrix.	
Chapter *24*	**SMITH NORMAL FORM**	**188**
	Smith normal form. Invariant factors. Elementary divisors.	
Chapter *25*	**THE MINIMUM POLYNOMIAL OF A MATRIX**	**196**
	Similarity invariants. Minimum polynomial. Derogatory and non-derogatory matrices. Companion matrix.	
Chapter *26*	**CANONICAL FORMS UNDER SIMILARITY**	**203**
	Rational canonical form. A second canonical form. Hypercompanion matrix. Jacobson canonical form. Classical canonical form. A reduction to rational canonical form.	
	INDEX	**215**
	INDEX OF SYMBOLS	**219**

Preface

Elementary matrix algebra has now become an integral part of the mathematical background necessary for such diverse fields as electrical engineering and education, chemistry and sociology, as well as for statistics and pure mathematics. This book, in presenting the more essential material, is designed primarily to serve as a useful supplement to current texts and as a handy reference book for those working in the several fields which require some knowledge of matrix theory. Moreover, the statements of theory and principle are sufficiently complete that the book could be used as a text by itself.

The material has been divided into twenty-six chapters, since the logical arrangement is thereby not disturbed while the usefulness as a reference book is increased. This also permits a separation of the treatment of real matrices, with which the majority of readers will be concerned, from that of matrices with complex elements. Each chapter contains a statement of pertinent definitions, principles, and theorems, fully illustrated by examples. These, in turn, are followed by a carefully selected set of solved problems and a considerable number of supplementary exercises.

The student new to matrix algebra soon finds that the solutions of numerical exercises are disarmingly simple. Difficulties are likely to arise from the constant round of definition, theorem, proof. The trouble here is essentially a matter of lack of mathematical maturity, and normally to be expected, since usually the student's previous work in mathematics has been concerned with the solution of numerical problems while precise statements of principles and proofs of theorems have in large part been deferred for later courses. The aim of the present book is to enable the reader, if he persists through the introductory paragraphs and solved problems in any chapter, to develop a reasonable degree of self-assurance about the material.

The solved problems, in addition to giving more variety to the examples illustrating the theorems, contain most of the proofs of any considerable length together with representative shorter proofs. The supplementary problems call both for the solution of numerical exercises and for proofs. Some of the latter require only proper modifications of proofs given earlier; more important, however, are the many theorems whose proofs require but a few lines. Some are of the type frequently misnamed "obvious" while others will be found to call for considerable ingenuity. None should be treated lightly, however, for it is due precisely to the abundance of such theorems that elementary matrix algebra becomes a natural first course for those seeking to attain a degree of mathematical maturity. While the large number of these problems in any chapter makes it impractical to solve all of them before moving to the next, special attention is directed to the supplementary problems of the first two chapters. A mastery of these will do much to give the reader confidence to stand on his own feet thereafter.

The author wishes to take this opportunity to express his gratitude to the staff of the Schaum Publishing Company for their splendid cooperation.

FRANK AYRES, JR.

Carlisle, Pa.
October, 1962

Chapter 1

Matrices

A RECTANGULAR ARRAY OF NUMBERS enclosed by a pair of brackets, such as

$$(a)\ \begin{bmatrix} 2 & 3 & 7 \\ 1 & -1 & 5 \end{bmatrix} \quad \text{and} \quad (b)\ \begin{bmatrix} 1 & 3 & 1 \\ 2 & 1 & 4 \\ 4 & 7 & 6 \end{bmatrix},$$

and subject to certain rules of operations given below is called a **matrix**. The matrix (a) could be considered as the **coefficient matrix** of the system of homogeneous linear equations $\begin{cases} 2x+3y+7z = 0 \\ x-\ y+5z = 0 \end{cases}$ or as the **augmented** matrix of the system of non-homogeneous linear equations $\begin{cases} 2x+3y = 7 \\ x-\ y = 5 \end{cases}$ Later, we shall see how the matrix may be used to obtain solutions of these systems. The matrix (b) could be given a similar interpretation or we might consider its rows as simply the coordinates of the points $(1,3,1)$, $(2,1,4)$, and $(4,7,6)$ in ordinary space. The matrix will be used later to settle such questions as whether or not the three points lie in the same plane with the origin or on the same line through the origin.

In the matrix

$$\textbf{(1.1)} \qquad \begin{bmatrix} a_{11}\ a_{12}\ a_{13} \cdots\cdots\cdots a_{1n} \\ a_{21}\ a_{22}\ a_{23} \cdots\cdots\cdots a_{2n} \\ \cdots\cdots\cdots\cdots\cdots\cdots\cdots \\ \cdots\cdots\cdots\cdots\cdots\cdots\cdots \\ a_{m1}\ a_{m2}\ a_{m3} \cdots\cdots\cdots a_{mn} \end{bmatrix}$$

the numbers or functions a_{ij} are called its **elements**. In the double subscript notation, the first subscript indicates the row and the second subscript indicates the column in which the element stands. Thus, all elements in the second row have 2 as first subscript and all the elements in the fifth column have 5 as second subscript. A matrix of m rows and n columns is said to be of order "m by n" or $m \times n$.

(In indicating a matrix pairs of parentheses, ( ), and double bars, ‖ ‖, are sometimes used. We shall use the double bracket notation throughout.)

At times the matrix **(1.1)** will be called "the $m\times n$ matrix $[a_{ij}]$" or "the $m\times n$ matrix $A = [a_{ij}]$". When the order has been established, we shall write simply "the matrix A".

SQUARE MATRICES. When $m = n$, **(1.1)** is square and will be called a **square matrix** of order n or an n-square matrix.

In a square matrix, the elements $a_{11}, a_{22}, \ldots, a_{nn}$ are called its **diagonal elements**.

The sum of the diagonal elements of a square matrix A is called the **trace** of A.

EQUAL MATRICES. Two matrices $A = [a_{ij}]$ and $B = [b_{ij}]$ are said to be **equal** $(A = B)$ if and only if they have the same order and each element of one is equal to the corresponding element of the other, that is, if and only if

$$a_{ij} = b_{ij}, \qquad (i = 1, 2, \ldots, m;\ j = 1, 2, \ldots, n)$$

Thus, two matrices are equal if and only if one is a duplicate of the other.

ZERO MATRIX. A matrix, every element of which is zero, is called a **zero matrix**. When A is a zero matrix and there can be no confusion as to its order, we shall write $A = 0$ instead of the $m \times n$ array of zero elements.

SUMS OF MATRICES. If $A = [a_{ij}]$ and $B = [b_{ij}]$ are two $m \times n$ matrices, their sum (difference), $A \pm B$, is defined as the $m \times n$ matrix $C = [c_{ij}]$, where each element of C is the sum (difference) of the corresponding elements of A and B. Thus, $A \pm B = [a_{ij} \pm b_{ij}]$.

Example 1. If $A = \begin{bmatrix} 1 & 2 & 3 \\ 0 & 1 & 4 \end{bmatrix}$ and $B = \begin{bmatrix} 2 & 3 & 0 \\ -1 & 2 & 5 \end{bmatrix}$ then

$$A + B = \begin{bmatrix} 1+2 & 2+3 & 3+0 \\ 0+(-1) & 1+2 & 4+5 \end{bmatrix} = \begin{bmatrix} 3 & 5 & 3 \\ -1 & 3 & 9 \end{bmatrix}$$

and

$$A - B = \begin{bmatrix} 1-2 & 2-3 & 3-0 \\ 0-(-1) & 1-2 & 4-5 \end{bmatrix} = \begin{bmatrix} -1 & -1 & 3 \\ 1 & -1 & -1 \end{bmatrix}$$

Two matrices of the same order are said to be **conformable** for addition or subtraction. *Two matrices of different orders cannot be added or subtracted.* For example, the matrices (a) and (b) above are non-conformable for addition and subtraction.

The sum of k matrices A is a matrix of the same order as A and each of its elements is k times the corresponding element of A. We define: If k is any **scalar** (we call k a scalar to distinguish it from $[k]$ which is a 1×1 matrix) then by $kA = Ak$ is meant the matrix obtained from A by multiplying each of its elements by k.

Example 2. If $A = \begin{bmatrix} 1 & -2 \\ 2 & 3 \end{bmatrix}$ then

$$A + A + A = \begin{bmatrix} 1 & -2 \\ 2 & 3 \end{bmatrix} + \begin{bmatrix} 1 & -2 \\ 2 & 3 \end{bmatrix} + \begin{bmatrix} 1 & -2 \\ 2 & 3 \end{bmatrix} = \begin{bmatrix} 3 & -6 \\ 6 & 9 \end{bmatrix} = 3A = A \cdot 3$$

and

$$-5A = \begin{bmatrix} -5(1) & -5(-2) \\ -5(2) & -5(3) \end{bmatrix} = \begin{bmatrix} -5 & 10 \\ -10 & -15 \end{bmatrix}$$

In particular, by $-A$, called the **negative** of A, is meant the matrix obtained from A by multiplying each of its elements by -1 or by simply changing the sign of **all** of its elements. For every A, we have $A + (-A) = 0$, where 0 indicates the zero matrix of the same order as A.

Assuming that the matrices A, B, C are conformable for addition, we state:

(a) $A + B = B + A$ (commutative law)

(b) $A + (B + C) = (A + B) + C$ (associative law)

(c) $k(A + B) = kA + kB = (A + B)k$, k a scalar

(d) There exists a matrix D such that $A + D = B$.

These laws are a result of the laws of elementary algebra governing the addition of numbers and polynomials. They show, moreover,

1. Conformable matrices obey the same laws of addition as the elements of these matrices.

MULTIPLICATION. By the product AB **in that order** of the $1\times m$ matrix $A = [a_{11}\ a_{12}\ a_{13} \ldots a_{1m}]$ and the $m\times 1$ matrix $B = \begin{bmatrix} b_{11} \\ b_{21} \\ b_{31} \\ \cdot \\ \cdot \\ \cdot \\ b_{m1} \end{bmatrix}$ is meant the 1×1 matrix $C = [a_{11}b_{11} + a_{12}b_{21} + \cdots + a_{1m}b_{m1}]$.

That is,

$$[a_{11}\ a_{12} \ldots a_{1m}] \cdot \begin{bmatrix} b_{11} \\ b_{21} \\ \cdot \\ \cdot \\ \cdot \\ \cdot \\ b_{m1} \end{bmatrix} = [a_{11}b_{11} + a_{12}b_{21} + \cdots + a_{1m}b_{m1}] = \left[\sum_{k=1}^{m} a_{1k}b_{k1}\right].$$

Note that the operation is **row by column**; each element of the row is multiplied into the corresponding element of the column and then the products are summed.

Example 3. (a) $[2\ 3\ 4] \begin{bmatrix} 1 \\ -1 \\ 2 \end{bmatrix} = [2(1)+3(-1)+4(2)] = [7]$

(b) $[3\ -1\ 4] \begin{bmatrix} -2 \\ 6 \\ 3 \end{bmatrix} = [-6-6+12] = 0$

By the product AB **in that order** of the $m\times p$ matrix $A = [a_{ij}]$ and the $p\times n$ matrix $B = [b_{ij}]$ is meant the $m\times n$ matrix $C = [c_{ij}]$ where

$$c_{ij} = a_{i1}b_{1j} + a_{i2}b_{2j} + \cdots + a_{ip}b_{pj} = \sum_{k=1}^{p} a_{ik}b_{kj}, \quad (i = 1, 2, \ldots, m;\ j = 1, 2, \ldots, n).$$

Think of A as consisting of m rows and B as consisting of n columns. In forming $C = AB$ each row of A is multiplied once and only once into each column of B. The element c_{ij} of C is then the product of the ith row of A and the jth column of B.

Example 4.

$$A\ B = \begin{bmatrix} a_{11} & a_{12} \\ a_{21} & a_{22} \\ a_{31} & a_{32} \end{bmatrix} \begin{bmatrix} b_{11} & b_{12} \\ b_{21} & b_{22} \end{bmatrix} = \begin{bmatrix} a_{11}b_{11} + a_{12}b_{21} & a_{11}b_{12} + a_{12}b_{22} \\ a_{21}b_{11} + a_{22}b_{21} & a_{21}b_{12} + a_{22}b_{22} \\ a_{31}b_{11} + a_{32}b_{21} & a_{31}b_{12} + a_{32}b_{22} \end{bmatrix}$$

The product AB is defined or A is **conformable** to B for multiplication only when the number of columns of A is equal to the number of rows of B. If A is conformable to B for multiplication (AB is defined), B is not necessarily conformable to A for multiplication (BA may or may not be defined). See Problems 3-4.

Assuming that A, B, C are conformable for the indicated sums and products, we have

(e) $A(B+C) = AB + AC$ (first distributive law)

(f) $(A+B)C = AC + BC$ (second distributive law)

(g) $A(BC) = (AB)C$ (associative law)

However,

(h) $AB \neq BA$, generally,

(i) $AB = 0$ does not necessarily imply $A = 0$ or $B = 0$,

(j) $AB = AC$ does not necessarily imply $B = C$.

See Problems 3-8.

PRODUCTS BY PARTITIONING. Let $A = [a_{ij}]$ be of order $m \times p$ and $B = [b_{ij}]$ be of order $p \times n$. In forming the product AB, the matrix A is in effect partitioned into m matrices of order $1 \times p$ and B into n matrices of order $p \times 1$. Other partitions may be used. For example, let A and B be partitioned into matrices of indicated orders by drawing in the dotted lines as

$$A = \begin{bmatrix} (m_1 \times p_1) & (m_1 \times p_2) & (m_1 \times p_3) \\ (m_2 \times p_1) & (m_2 \times p_2) & (m_2 \times p_3) \end{bmatrix}, \qquad B = \begin{bmatrix} (p_1 \times n_1) & (p_1 \times n_2) \\ (p_2 \times n_1) & (p_2 \times n_2) \\ (p_3 \times n_1) & (p_3 \times n_2) \end{bmatrix}$$

or

$$A = \begin{bmatrix} A_{11} & A_{12} & A_{13} \\ A_{21} & A_{22} & A_{23} \end{bmatrix}, \qquad B = \begin{bmatrix} B_{11} & B_{12} \\ B_{21} & B_{22} \\ B_{31} & B_{32} \end{bmatrix}$$

In any such partitioning, it is necessary that the columns of A and the rows of B be partitioned in exactly the same way; however m_1, m_2, n_1, n_2 may be any non-negative (including 0) integers such that $m_1 + m_2 = m$ and $n_1 + n_2 = n$. Then

$$AB = \begin{bmatrix} A_{11}B_{11} + A_{12}B_{21} + A_{13}B_{31} & A_{11}B_{12} + A_{12}B_{22} + A_{13}B_{32} \\ A_{21}B_{11} + A_{22}B_{21} + A_{23}B_{31} & A_{21}B_{12} + A_{22}B_{22} + A_{23}B_{32} \end{bmatrix} = \begin{bmatrix} C_{11} & C_{12} \\ C_{21} & C_{22} \end{bmatrix} = C$$

Example 5. Compute AB, given $A = \begin{bmatrix} 2 & 1 & 0 \\ 3 & 2 & 0 \\ 1 & 0 & 1 \end{bmatrix}$ and $B = \begin{bmatrix} 1 & 1 & 1 & 0 \\ 2 & 1 & 1 & 0 \\ 2 & 3 & 1 & 2 \end{bmatrix}$

Partitioning so that

$$A = \begin{bmatrix} A_{11} & A_{12} \\ A_{21} & A_{22} \end{bmatrix} = \left[\begin{array}{cc|c} 2 & 1 & 0 \\ 3 & 2 & 0 \\ \hline 1 & 0 & 1 \end{array}\right] \quad \text{and} \quad B = \begin{bmatrix} B_{11} & B_{12} \\ B_{21} & B_{22} \end{bmatrix} = \left[\begin{array}{ccc|c} 1 & 1 & 1 & 0 \\ 2 & 1 & 1 & 0 \\ \hline 2 & 3 & 1 & 2 \end{array}\right],$$

we have $AB = \begin{bmatrix} A_{11}B_{11} + A_{12}B_{21} & A_{11}B_{12} + A_{12}B_{22} \\ A_{21}B_{11} + A_{22}B_{21} & A_{21}B_{12} + A_{22}B_{22} \end{bmatrix}$

$$= \begin{bmatrix} \begin{bmatrix} 2 & 1 \\ 3 & 2 \end{bmatrix}\begin{bmatrix} 1 & 1 & 1 \\ 2 & 1 & 1 \end{bmatrix} + \begin{bmatrix} 0 \\ 0 \end{bmatrix}[2\ 3\ 1] & \begin{bmatrix} 2 & 1 \\ 3 & 2 \end{bmatrix}\begin{bmatrix} 0 \\ 0 \end{bmatrix} + \begin{bmatrix} 0 \\ 0 \end{bmatrix}[2] \\ [1\ 0]\begin{bmatrix} 1 & 1 & 1 \\ 2 & 1 & 1 \end{bmatrix} + [1][2\ 3\ 1] & [1\ 0]\begin{bmatrix} 0 \\ 0 \end{bmatrix} + [1][2] \end{bmatrix}$$

$$= \begin{bmatrix} \begin{bmatrix} 4 & 3 & 3 \\ 7 & 5 & 5 \end{bmatrix} + \begin{bmatrix} 0 & 0 & 0 \\ 0 & 0 & 0 \end{bmatrix} & \begin{bmatrix} 0 \\ 0 \end{bmatrix} + \begin{bmatrix} 0 \\ 0 \end{bmatrix} \\ [1\ 1\ 1] + [2\ 3\ 1] & [0] + [2] \end{bmatrix} = \begin{bmatrix} \begin{bmatrix} 4 & 3 & 3 \\ 7 & 5 & 5 \end{bmatrix} & \begin{bmatrix} 0 \\ 0 \end{bmatrix} \\ [3\ 4\ 2] & [2] \end{bmatrix} = \begin{bmatrix} 4 & 3 & 3 & 0 \\ 7 & 5 & 5 & 0 \\ 3 & 4 & 2 & 2 \end{bmatrix}$$

See also Problem 9.

Let $A, B, C, \ldots$ be n-square matrices. Let A be partitioned into matrices of the indicated orders

$$\begin{bmatrix} (p_1 \times p_1) & (p_1 \times p_2) & \ldots & (p_1 \times p_s) \\ (p_2 \times p_1) & (p_2 \times p_2) & \ldots & (p_2 \times p_s) \\ \ldots\ldots & \ldots\ldots & \ldots & \ldots\ldots \\ (p_s \times p_1) & (p_s \times p_2) & \ldots & (p_s \times p_s) \end{bmatrix} = \begin{bmatrix} A_{11} & A_{12} & \ldots & A_{1s} \\ A_{21} & A_{22} & \ldots & A_{2s} \\ \ldots\ldots & \ldots\ldots & \ldots & \ldots\ldots \\ A_{s1} & A_{s2} & \ldots & A_{ss} \end{bmatrix}$$

and let $B, C, \ldots$ be partitioned in exactly the same manner. Then sums, differences, and products may be formed using the matrices $A_{11}, A_{12}, \ldots;\ B_{11}, B_{12}, \ldots;\ C_{11}, C_{12}, \ldots$.

SOLVED PROBLEMS

1. (a) $\begin{bmatrix}1 & 2 & -1 & 0\\ 4 & 0 & 2 & 1\\ 2 & -5 & 1 & 2\end{bmatrix} + \begin{bmatrix}3 & -4 & 1 & 2\\ 1 & 5 & 0 & 3\\ 2 & -2 & 3 & -1\end{bmatrix} = \begin{bmatrix}1+3 & 2+(-4) & -1+1 & 0+2\\ 4+1 & 0+5 & 2+0 & 1+3\\ 2+2 & -5+(-2) & 1+3 & 2+(-1)\end{bmatrix} = \begin{bmatrix}4 & -2 & 0 & 2\\ 5 & 5 & 2 & 4\\ 4 & -7 & 4 & 1\end{bmatrix}$

(b) $\begin{bmatrix}1 & 2 & -1 & 0\\ 4 & 0 & 2 & 1\\ 2 & -5 & 1 & 2\end{bmatrix} - \begin{bmatrix}3 & -4 & 1 & 2\\ 1 & 5 & 0 & 3\\ 2 & -2 & 3 & -1\end{bmatrix} = \begin{bmatrix}1-3 & 2+4 & -1-1 & 0-2\\ 4-1 & 0-5 & 2-0 & 1-3\\ 2-2 & -5+2 & 1-3 & 2+1\end{bmatrix} = \begin{bmatrix}-2 & 6 & -2 & -2\\ 3 & -5 & 2 & -2\\ 0 & -3 & -2 & 3\end{bmatrix}$

(c) $3\begin{bmatrix}1 & 2 & -1 & 0\\ 4 & 0 & 2 & 1\\ 2 & -5 & 1 & 2\end{bmatrix} = \begin{bmatrix}3 & 6 & -3 & 0\\ 12 & 0 & 6 & 3\\ 6 & -15 & 3 & 6\end{bmatrix}$

(d) $-\begin{bmatrix}1 & 2 & -1 & 0\\ 4 & 0 & 2 & 1\\ 2 & -5 & 1 & 2\end{bmatrix} = \begin{bmatrix}-1 & -2 & 1 & 0\\ -4 & 0 & -2 & -1\\ -2 & 5 & -1 & -2\end{bmatrix}$

2. If $A = \begin{bmatrix}1 & 2\\ 3 & 4\\ 5 & 6\end{bmatrix}$ and $B = \begin{bmatrix}-3 & -2\\ 1 & -5\\ 4 & 3\end{bmatrix}$, find $D = \begin{bmatrix}p & q\\ r & s\\ t & u\end{bmatrix}$ such that $A + B - D = 0$.

If $A + B - D = \begin{bmatrix}1-3-p & 2-2-q\\ 3+1-r & 4-5-s\\ 5+4-t & 6+3-u\end{bmatrix} = \begin{bmatrix}-2-p & -q\\ 4-r & -1-s\\ 9-t & 9-u\end{bmatrix} = \begin{bmatrix}0 & 0\\ 0 & 0\\ 0 & 0\end{bmatrix}$, $-2-p = 0$ and $p = -2$, $4-r = 0$ and $r = 4$, ... Then $D = \begin{bmatrix}-2 & 0\\ 4 & -1\\ 9 & 9\end{bmatrix} = A + B$.

3. (a) $[4\ 5\ 6]\begin{bmatrix}2\\ 3\\ -1\end{bmatrix} = [4(2) + 5(3) + 6(-1)] = [17]$

(b) $\begin{bmatrix}2\\ 3\\ -1\end{bmatrix}[4\ 5\ 6] = \begin{bmatrix}2(4) & 2(5) & 2(6)\\ 3(4) & 3(5) & 3(6)\\ -1(4) & -1(5) & -1(6)\end{bmatrix} = \begin{bmatrix}8 & 10 & 12\\ 12 & 15 & 18\\ -4 & -5 & -6\end{bmatrix}$

(c) $[1\ 2\ 3]\begin{bmatrix}4 & -6 & 9 & 6\\ 0 & -7 & 10 & 7\\ 5 & 8 & -11 & -8\end{bmatrix}$

$= [1(4) + 2(0) + 3(5)\quad 1(-6) + 2(-7) + 3(8)\quad 1(9) + 2(10) + 3(-11)\quad 1(6) + 2(7) + 3(-8)]$
$= [19\ \ 4\ \ -4\ \ -4]$

(d) $\begin{bmatrix}2 & 3 & 4\\ 1 & 5 & 6\end{bmatrix}\begin{bmatrix}1\\ 2\\ 3\end{bmatrix} = \begin{bmatrix}2(1) + 3(2) + 4(3)\\ 1(1) + 5(2) + 6(3)\end{bmatrix} = \begin{bmatrix}20\\ 29\end{bmatrix}$

(e) $\begin{bmatrix}1 & 2 & 1\\ 4 & 0 & 2\end{bmatrix}\begin{bmatrix}3 & -4\\ 1 & 5\\ -2 & 2\end{bmatrix} = \begin{bmatrix}1(3) + 2(1) + 1(-2) & 1(-4) + 2(5) + 1(2)\\ 4(3) + 0(1) + 2(-2) & 4(-4) + 0(5) + 2(2)\end{bmatrix} = \begin{bmatrix}3 & 8\\ 8 & -12\end{bmatrix}$

4. Let $A = \begin{bmatrix}2 & -1 & 1\\ 0 & 1 & 2\\ 1 & 0 & 1\end{bmatrix}$. Then

$A^2 = \begin{bmatrix}2 & -1 & 1\\ 0 & 1 & 2\\ 1 & 0 & 1\end{bmatrix}\begin{bmatrix}2 & -1 & 1\\ 0 & 1 & 2\\ 1 & 0 & 1\end{bmatrix} = \begin{bmatrix}5 & -3 & 1\\ 2 & 1 & 4\\ 3 & -1 & 2\end{bmatrix}$ and $A^3 = A^2 \cdot A = \begin{bmatrix}5 & -3 & 1\\ 2 & 1 & 4\\ 3 & -1 & 2\end{bmatrix}\begin{bmatrix}2 & -1 & 1\\ 0 & 1 & 2\\ 1 & 0 & 1\end{bmatrix} = \begin{bmatrix}11 & -8 & 0\\ 8 & -1 & 8\\ 8 & -4 & 3\end{bmatrix}$

The reader will show that $A^3 = A \cdot A^2$ and $A^2 \cdot A^3 = A^3 \cdot A^2$.

5. Show that:

(a) $\sum_{k=1}^{2} a_{ik}(b_{kj}+c_{kj}) = \sum_{k=1}^{2} a_{ik}b_{kj} + \sum_{k=1}^{2} a_{ik}c_{kj}$,

(b) $\sum_{i=1}^{2}\sum_{j=1}^{3} a_{ij} = \sum_{j=1}^{3}\sum_{i=1}^{2} a_{ij}$,

(c) $\sum_{k=1}^{2} a_{ik}(\sum_{h=1}^{3} b_{kh}c_{hj}) = \sum_{h=1}^{3}(\sum_{k=1}^{2} a_{ik}b_{kh})c_{hj}$.

(a)
$$\sum_{k=1}^{2} a_{ik}(b_{kj}+c_{kj}) = a_{i1}(b_{1j}+c_{1j}) + a_{i2}(b_{2j}+c_{2j}) = (a_{i1}b_{1j}+a_{i2}b_{2j}) + (a_{i1}c_{1j}+a_{i2}c_{2j})$$
$$= \sum_{k=1}^{2} a_{ik}b_{kj} + \sum_{k=1}^{2} a_{ik}c_{kj}.$$

(b)
$$\sum_{i=1}^{2}\sum_{j=1}^{3} a_{ij} = \sum_{i=1}^{2}(a_{i1}+a_{i2}+a_{i3}) = (a_{11}+a_{12}+a_{13}) + (a_{21}+a_{22}+a_{23})$$
$$= (a_{11}+a_{21}) + (a_{12}+a_{22}) + (a_{13}+a_{23})$$
$$= \sum_{i=1}^{2} a_{i1} + \sum_{i=1}^{2} a_{i2} + \sum_{i=1}^{2} a_{i3} = \sum_{j=1}^{3}\sum_{i=1}^{2} a_{ij}.$$

This is simply the statement that in summing all of the elements of a matrix, one may sum first the elements of each row or the elements of each column.

(c)
$$\sum_{k=1}^{2} a_{ik}(\sum_{h=1}^{3} b_{kh}c_{hj}) = \sum_{k=1}^{2} a_{ik}(b_{k1}c_{1j} + b_{k2}c_{2j} + b_{k3}c_{3j})$$
$$= a_{i1}(b_{11}c_{1j} + b_{12}c_{2j} + b_{13}c_{3j}) + a_{i2}(b_{21}c_{1j} + b_{22}c_{2j} + b_{23}c_{3j})$$
$$= (a_{i1}b_{11} + a_{i2}b_{21})c_{1j} + (a_{i1}b_{12} + a_{i2}b_{22})c_{2j} + (a_{i1}b_{13} + a_{i2}b_{23})c_{3j}$$
$$= (\sum_{k=1}^{2} a_{ik}b_{k1})c_{1j} + (\sum_{k=1}^{2} a_{ik}b_{k2})c_{2j} + (\sum_{k=1}^{2} a_{ik}b_{k3})c_{3j}$$
$$= \sum_{h=1}^{3}(\sum_{k=1}^{2} a_{ik}b_{kh})c_{hj}.$$

6. Prove: If $A = [a_{ij}]$ is of order $m \times n$ and if $B = [b_{ij}]$ and $C = [c_{ij}]$ are of order $n \times p$, then $A(B+C) = AB + AC$.

The elements of the ith row of A are $a_{i1}, a_{i2}, \ldots, a_{in}$ and the elements of the jth column of $B+C$ are $b_{1j}+c_{1j},\ b_{2j}+c_{2j},\ \ldots,\ b_{nj}+c_{nj}$. Then the element standing in the ith row and jth column of $A(B+C)$ is $a_{i1}(b_{1j}+c_{1j}) + a_{i2}(b_{2j}+c_{2j}) + \ldots + a_{in}(b_{nj}+c_{nj}) = \sum_{k=1}^{n} a_{ik}(b_{kj}+c_{kj}) = \sum_{k=1}^{n} a_{ik}b_{kj} + \sum_{k=1}^{n} a_{ik}c_{kj}$, the sum of the elements standing in the ith row and jth column of AB and AC.

7. Prove: If $A = [a_{ij}]$ is of order $m \times n$, if $B = [b_{ij}]$ is of order $n \times p$, and if $C = [c_{ij}]$ is of order $p \times q$, then $A(BC) = (AB)C$.

The elements of the ith row of A are $a_{i1}, a_{i2}, \ldots, a_{in}$ and the elements of the jth column of BC are $\sum_{h=1}^{p} b_{1h}c_{hj}$, $\sum_{h=1}^{p} b_{2h}c_{hj}$, ..., $\sum_{h=1}^{p} b_{nh}c_{hj}$; hence the element standing in the ith row and jth column of $A(BC)$ is

$$a_{i1}\sum_{h=1}^{p} b_{1h}c_{hj} + a_{i2}\sum_{h=1}^{p} b_{2h}c_{hj} + \ldots + a_{in}\sum_{h=1}^{p} b_{nh}c_{hj} = \sum_{k=1}^{n} a_{ik}(\sum_{h=1}^{p} b_{kh}c_{hj})$$
$$= \sum_{h=1}^{p}(\sum_{k=1}^{n} a_{ik}b_{kh})c_{hj} = (\sum_{k=1}^{n} a_{ik}b_{k1})c_{1j} + (\sum_{k=1}^{n} a_{ik}b_{k2})c_{2j} + \ldots + (\sum_{k=1}^{n} a_{ik}b_{kp})c_{pj}$$

This is the element standing in the ith row and jth column of $(AB)C$; hence, $A(BC) = (AB)C$.

8. Assuming A, B, C, D conformable, show in two ways that $(A+B)(C+D) = AC + AD + BC + BD$.

Using (e) and then (f), $(A+B)(C+D) = (A+B)C + (A+B)D = AC + BC + AD + BD$.

Using (f) and then (e), $(A+B)(C+D) = A(C+D) + B(C+D) = AC + AD + BC + BD$
$= AC + BC + AD + BD$.

9. (a) $\left[\begin{array}{ccc|c} 1 & 0 & 0 & 1 \\ 0 & 1 & 0 & 2 \\ 0 & 0 & 1 & 3 \end{array}\right]\left[\begin{array}{ccc} 1 & 0 & 0 \\ 0 & 1 & 0 \\ 0 & 0 & 1 \\ \hline 3 & 1 & 2 \end{array}\right] = \begin{bmatrix} 1 & 0 & 0 \\ 0 & 1 & 0 \\ 0 & 0 & 1 \end{bmatrix}\begin{bmatrix} 1 & 0 & 0 \\ 0 & 1 & 0 \\ 0 & 0 & 1 \end{bmatrix} + \begin{bmatrix} 1 \\ 2 \\ 3 \end{bmatrix}[3\ 1\ 2] = \begin{bmatrix} 1 & 0 & 0 \\ 0 & 1 & 0 \\ 0 & 0 & 1 \end{bmatrix} + \begin{bmatrix} 3 & 1 & 2 \\ 6 & 2 & 4 \\ 9 & 3 & 6 \end{bmatrix} = \begin{bmatrix} 4 & 1 & 2 \\ 6 & 3 & 4 \\ 9 & 3 & 7 \end{bmatrix}$

(b) $$\left[\begin{array}{cc|cc|cc} 1 & 0 & 0 & 0 & 0 & 0 \\ 0 & 2 & 0 & 0 & 0 & 0 \\ \hline 0 & 0 & 3 & 0 & 0 & 0 \\ 0 & 0 & 0 & 4 & 0 & 0 \\ \hline 0 & 0 & 0 & 0 & 5 & 0 \\ 0 & 0 & 0 & 0 & 0 & 6 \end{array}\right]\left[\begin{array}{cc|cc|cc} 1 & 0 & 0 & 0 & 0 & 0 \\ 0 & 1 & 0 & 0 & 0 & 0 \\ \hline 0 & 0 & 1 & 0 & 0 & 0 \\ 0 & 0 & 0 & 3 & 0 & 0 \\ \hline 0 & 0 & 0 & 0 & 2 & 0 \\ 0 & 0 & 0 & 0 & 0 & 3 \end{array}\right] = \begin{bmatrix} \begin{bmatrix} 1 & 0 \\ 0 & 2 \end{bmatrix}\begin{bmatrix} 1 & 0 \\ 0 & 1 \end{bmatrix} & [0] & [0] \\ [0] & \begin{bmatrix} 3 & 0 \\ 0 & 4 \end{bmatrix}\begin{bmatrix} 1 & 0 \\ 0 & 3 \end{bmatrix} & [0] \\ [0] & [0] & \begin{bmatrix} 5 & 0 \\ 0 & 6 \end{bmatrix}\begin{bmatrix} 2 & 0 \\ 0 & 3 \end{bmatrix} \end{bmatrix}$$

$$= \begin{bmatrix} 1 & 0 & 0 & 0 & 0 & 0 \\ 0 & 2 & 0 & 0 & 0 & 0 \\ 0 & 0 & 3 & 0 & 0 & 0 \\ 0 & 0 & 0 & 12 & 0 & 0 \\ 0 & 0 & 0 & 0 & 10 & 0 \\ 0 & 0 & 0 & 0 & 0 & 18 \end{bmatrix}$$

(c) $$\left[\begin{array}{cc|ccc|c} 1 & 1 & 0 & 0 & 0 & 0 \\ 2 & 1 & 0 & 0 & 0 & 0 \\ \hline 0 & 0 & 3 & 1 & 2 & 0 \\ 0 & 0 & 1 & 2 & 1 & 0 \\ 0 & 0 & 0 & 1 & 1 & 0 \\ \hline 0 & 0 & 0 & 0 & 0 & 1 \end{array}\right]\left[\begin{array}{cc|ccc|c} 1 & 2 & 3 & 4 & 5 & 6 \\ 2 & 3 & 4 & 5 & 6 & 7 \\ \hline 3 & 4 & 5 & 6 & 7 & 8 \\ 4 & 5 & 6 & 7 & 8 & 9 \\ 9 & 8 & 7 & 6 & 5 & 4 \\ \hline 8 & 7 & 6 & 5 & 4 & 1 \end{array}\right] = \begin{bmatrix} \begin{bmatrix} 1 & 1 \\ 2 & 1 \end{bmatrix}\begin{bmatrix} 1 & 2 \\ 2 & 3 \end{bmatrix} & \begin{bmatrix} 1 & 1 \\ 2 & 1 \end{bmatrix}\begin{bmatrix} 3 & 4 & 5 \\ 4 & 5 & 6 \end{bmatrix} & \begin{bmatrix} 1 & 1 \\ 2 & 1 \end{bmatrix}\cdot\begin{bmatrix} 6 \\ 7 \end{bmatrix} \\ \begin{bmatrix} 3 & 1 & 2 \\ 1 & 2 & 1 \\ 0 & 1 & 1 \end{bmatrix}\begin{bmatrix} 3 & 4 \\ 4 & 5 \\ 9 & 8 \end{bmatrix} & \begin{bmatrix} 3 & 1 & 2 \\ 1 & 2 & 1 \\ 0 & 1 & 1 \end{bmatrix}\begin{bmatrix} 5 & 6 & 7 \\ 6 & 7 & 8 \\ 7 & 6 & 5 \end{bmatrix} & \begin{bmatrix} 3 & 1 & 2 \\ 1 & 2 & 1 \\ 0 & 1 & 1 \end{bmatrix}\cdot\begin{bmatrix} 8 \\ 9 \\ 4 \end{bmatrix} \\ [1]\cdot[8\ 7] & [1]\cdot[6\ 5\ 4] & [1]\cdot[1] \end{bmatrix}$$

$$= \begin{bmatrix} \begin{bmatrix} 3 & 5 \\ 4 & 7 \end{bmatrix} & \begin{bmatrix} 7 & 9 & 11 \\ 10 & 13 & 16 \end{bmatrix} & \begin{bmatrix} 13 \\ 19 \end{bmatrix} \\ \begin{bmatrix} 31 & 33 \\ 20 & 22 \\ 13 & 13 \end{bmatrix} & \begin{bmatrix} 35 & 37 & 39 \\ 24 & 26 & 28 \\ 13 & 13 & 13 \end{bmatrix} & \begin{bmatrix} 41 \\ 30 \\ 13 \end{bmatrix} \\ [8\ \ 7] & [6\ \ 5\ \ 4] & [1] \end{bmatrix} = \begin{bmatrix} 3 & 5 & 7 & 9 & 11 & 13 \\ 4 & 7 & 10 & 13 & 16 & 19 \\ 31 & 33 & 35 & 37 & 39 & 41 \\ 20 & 22 & 24 & 26 & 28 & 30 \\ 13 & 13 & 13 & 13 & 13 & 13 \\ 8 & 7 & 6 & 5 & 4 & 1 \end{bmatrix}$$

10. Let $\begin{cases} x_1 = a_{11}y_1 + a_{12}y_2 \\ x_2 = a_{21}y_1 + a_{22}y_2 \\ x_3 = a_{31}y_1 + a_{32}y_2 \end{cases}$ be three linear forms in y_1 and y_2 and let $\begin{cases} y_1 = b_{11}z_1 + b_{12}z_2 \\ y_2 = b_{21}z_1 + b_{22}z_2 \end{cases}$ be a linear transformation of the coordinates (y_1, y_2) into new coordinates (z_1, z_2). The result of applying the transformation to the given forms is the set of forms

$$\begin{cases} x_1 = (a_{11}b_{11} + a_{12}b_{21})z_1 + (a_{11}b_{12} + a_{12}b_{22})z_2 \\ x_2 = (a_{21}b_{11} + a_{22}b_{21})z_1 + (a_{21}b_{12} + a_{22}b_{22})z_2 \\ x_3 = (a_{31}b_{11} + a_{32}b_{21})z_1 + (a_{31}b_{12} + a_{32}b_{22})z_2 \end{cases}$$

Using matrix notation, we have the three forms $\begin{bmatrix} x_1 \\ x_2 \\ x_3 \end{bmatrix} = \begin{bmatrix} a_{11} & a_{12} \\ a_{21} & a_{22} \\ a_{31} & a_{32} \end{bmatrix}\begin{bmatrix} y_1 \\ y_2 \end{bmatrix}$ and the transformation $\begin{bmatrix} y_1 \\ y_2 \end{bmatrix} = \begin{bmatrix} b_{11} & b_{12} \\ b_{21} & b_{22} \end{bmatrix}\begin{bmatrix} z_1 \\ z_2 \end{bmatrix}$. The result of applying the transformation is the set of three forms

$$\begin{bmatrix} x_1 \\ x_2 \\ x_3 \end{bmatrix} = \begin{bmatrix} a_{11} & a_{12} \\ a_{21} & a_{22} \\ a_{31} & a_{32} \end{bmatrix}\begin{bmatrix} b_{11} & b_{12} \\ b_{21} & b_{22} \end{bmatrix}\begin{bmatrix} z_1 \\ z_2 \end{bmatrix}$$

Thus, when a set of m linear forms in n variables with matrix A is subjected to a linear transformation of the variables with matrix B, there results a set of m linear forms with matrix $C = AB$.

SUPPLEMENTARY PROBLEMS

11. Given $A = \begin{bmatrix} 1 & 2 & -3 \\ 5 & 0 & 2 \\ 1 & -1 & 1 \end{bmatrix}$, $B = \begin{bmatrix} 3 & -1 & 2 \\ 4 & 2 & 5 \\ 2 & 0 & 3 \end{bmatrix}$, and $C = \begin{bmatrix} 4 & 1 & 2 \\ 0 & 3 & 2 \\ 1 & -2 & 3 \end{bmatrix}$,

(*a*) Compute: $A+B = \begin{bmatrix} 4 & 1 & -1 \\ 9 & 2 & 7 \\ 3 & -1 & 4 \end{bmatrix}$, $A-C = \begin{bmatrix} -3 & 1 & -5 \\ 5 & -3 & 0 \\ 0 & 1 & -2 \end{bmatrix}$

(*b*) Compute: $-2A = \begin{bmatrix} -2 & -4 & 6 \\ -10 & 0 & -4 \\ -2 & 2 & -2 \end{bmatrix}$, $0 \cdot B = 0$

(*c*) Verify: $A + (B - C) = (A+B) - C$.

(*d*) Find the matrix D such that $A + D = B$. Verify that $D = B - A = -(A-B)$.

12. Given $A = \begin{bmatrix} 1 & -1 & 1 \\ -3 & 2 & -1 \\ -2 & 1 & 0 \end{bmatrix}$ and $B = \begin{bmatrix} 1 & 2 & 3 \\ 2 & 4 & 6 \\ 1 & 2 & 3 \end{bmatrix}$, compute $AB = 0$ and $BA = \begin{bmatrix} -11 & 6 & -1 \\ -22 & 12 & -2 \\ -11 & 6 & -1 \end{bmatrix}$. Hence, $AB \neq BA$ generally.

13. Given $A = \begin{bmatrix} 1 & -3 & 2 \\ 2 & 1 & -3 \\ 4 & -3 & -1 \end{bmatrix}$, $B = \begin{bmatrix} 1 & 4 & 1 & 0 \\ 2 & 1 & 1 & 1 \\ 1 & -2 & 1 & 2 \end{bmatrix}$, and $C = \begin{bmatrix} 2 & 1 & -1 & -2 \\ 3 & -2 & -1 & -1 \\ 2 & -5 & -1 & 0 \end{bmatrix}$, show that $AB = AC$. Thus, $AB = AC$ does not necessarily imply $B = C$.

14. Given $A = \begin{bmatrix} 1 & 1 & -1 \\ 2 & 0 & 3 \\ 3 & -1 & 2 \end{bmatrix}$, $B = \begin{bmatrix} 1 & 3 \\ 0 & 2 \\ -1 & 4 \end{bmatrix}$, and $C = \begin{bmatrix} 1 & 2 & 3 & -4 \\ 2 & 0 & -2 & 1 \end{bmatrix}$, show that $(AB)C = A(BC)$.

15. Using the matrices of Problem 11, show that $A(B+C) = AB + AC$ and $(A+B)C = AC + BC$.

16. Explain why, in general, $(A \pm B)^2 \neq A^2 \pm 2AB + B^2$ and $A^2 - B^2 \neq (A-B)(A+B)$.

17. Given $A = \begin{bmatrix} 2 & -3 & -5 \\ -1 & 4 & 5 \\ 1 & -3 & -4 \end{bmatrix}$, $B = \begin{bmatrix} -1 & 3 & 5 \\ 1 & -3 & -5 \\ -1 & 3 & 5 \end{bmatrix}$, and $C = \begin{bmatrix} 2 & -2 & -4 \\ -1 & 3 & 4 \\ 1 & -2 & -3 \end{bmatrix}$,

(*a*) show that $AB = BA = 0$, $AC = A$, $CA = C$.

(*b*) use the results of (*a*) to show that $ACB = CBA$, $A^2 - B^2 = (A-B)(A+B)$, $(A \pm B)^2 = A^2 + B^2$.

18. Given $A = \begin{bmatrix} i & 0 \\ 0 & i \end{bmatrix}$, where $i^2 = -1$, derive a formula for the positive integral powers of A.

Ans. $A^n = I, A, -I, -A$ according as $n = 4p, 4p+1, 4p+2, 4p+3$, where $I = \begin{bmatrix} 1 & 0 \\ 0 & 1 \end{bmatrix}$.

19. Show that the product of any two or more matrices of the set $\begin{bmatrix} 1 & 0 \\ 0 & 1 \end{bmatrix}$, $\begin{bmatrix} 0 & 1 \\ -1 & 0 \end{bmatrix}$, $\begin{bmatrix} 0 & -1 \\ 1 & 0 \end{bmatrix}$, $\begin{bmatrix} -1 & 0 \\ 0 & -1 \end{bmatrix}$, $\begin{bmatrix} i & 0 \\ 0 & -i \end{bmatrix}$, $\begin{bmatrix} -i & 0 \\ 0 & i \end{bmatrix}$, $\begin{bmatrix} 0 & -i \\ -i & 0 \end{bmatrix}$, $\begin{bmatrix} 0 & i \\ i & 0 \end{bmatrix}$ is a matrix of the set.

20. Given the matrices A of order $m \times n$, B of order $n \times p$, and C of order $r \times q$, under what conditions on p, q, and r would the matrices be conformable for finding the products and what is the order of each: (*a*) ABC, (*b*) ACB, (*c*) $A(B+C)$?

Ans. (*a*) $p = r$; $m \times q$ (*b*) $r = n = q$; $m \times p$ (*c*) $r = n$, $p = q$; $m \times q$

21. Compute AB, given:

(a) $A = \begin{bmatrix} 1 & 0 & 1 \\ 0 & 1 & 1 \\ 0 & 0 & 1 \end{bmatrix}$ and $B = \begin{bmatrix} 1 & 0 & 0 \\ 0 & 1 & 0 \\ 1 & 1 & 0 \end{bmatrix}$ *Ans.* $\begin{bmatrix} 2 & 1 & 0 \\ 1 & 2 & 0 \\ 1 & 1 & 0 \end{bmatrix}$

(b) $A = \begin{bmatrix} 1 & 0 & 3 \\ 0 & 1 & 2 \end{bmatrix}$ and $B = \begin{bmatrix} 1 & 0 \\ 0 & 1 \\ -1 & 2 \end{bmatrix}$ *Ans.* $\begin{bmatrix} -2 & 6 \\ -2 & 5 \end{bmatrix}$

(c) $A = \begin{bmatrix} 1 & 2 & 0 & 0 \\ 0 & 1 & 0 & 0 \\ 0 & 0 & 0 & 1 \\ 0 & 0 & 2 & 2 \end{bmatrix}$ and $B = \begin{bmatrix} 0 & 0 & 0 & 1 \\ 0 & 0 & 2 & 0 \\ 1 & 0 & 0 & 0 \\ 0 & 1 & 0 & 0 \end{bmatrix}$ *Ans.* $\begin{bmatrix} 0 & 0 & 4 & 1 \\ 0 & 0 & 2 & 0 \\ 0 & 1 & 0 & 0 \\ 2 & 2 & 0 & 0 \end{bmatrix}$

22. Prove: (a) trace $(A+B)$ = trace A + trace B, (b) trace $(kA) = k$ trace A.

23. If $\begin{cases} x_1 = y_1 - 2y_2 + y_3 \\ x_2 = 2y_1 + y_2 - 3y_3 \end{cases}$ and $\begin{cases} y_1 = z_1 + 2z_2 \\ y_2 = 2z_1 - z_2 \\ y_3 = 2z_1 + 3z_2 \end{cases}$, verify $\begin{bmatrix} x_1 \\ x_2 \end{bmatrix} = \begin{bmatrix} 1 & -2 & 1 \\ 2 & 1 & -3 \end{bmatrix} \begin{bmatrix} y_1 \\ y_2 \\ y_3 \end{bmatrix} = \begin{bmatrix} 1 & -2 & 1 \\ 2 & 1 & -3 \end{bmatrix} \begin{bmatrix} 1 & 2 \\ 2 & -1 \\ 2 & 3 \end{bmatrix} \begin{bmatrix} z_1 \\ z_2 \end{bmatrix}$

$= \begin{bmatrix} -z_1 + 7z_2 \\ -2z_1 - 6z_2 \end{bmatrix}$

24. If $A = [a_{ij}]$ and $B = [b_{ij}]$ are of order $m \times n$ and if $C = [c_{ij}]$ is of order $n \times p$, show that $(A+B)C = AC + BC$.

25. Let $A = [a_{ij}]$ and $B = [b_{jk}]$, where $(i = 1, 2, \dots, m;\ j = 1, 2, \dots, p;\ k = 1, 2, \dots, n)$. Denote by β_j the sum of the elements of the jth row of B, that is, let $\beta_j = \sum_{k=1}^{n} b_{jk}$. Show that the element in the ith row of $A \cdot \begin{bmatrix} \beta_1 \\ \beta_2 \\ \cdot \\ \cdot \\ \cdot \\ \beta_p \end{bmatrix}$

is the sum of the elements lying in the ith row of AB. Use this procedure to check the products formed in Problems 12 and 13.

26. A relation (such as parallelism, congruency) between mathematical entities possessing the following properties:

(i) Determination — Either a is in the relation to b or a is not in the relation to b.
(ii) Reflexivity — a is in the relation to a, for all a.
(iii) Symmetry — If a is in the relation to b then b is in the relation to a.
(iv) Transivity — If a is in the relation to b and b is in the relation to c then a is in the relation to c.

is called an **equivalence relation**.

Show that the parallelism of lines, similarity of triangles, and equality of matrices are equivalence relations. Show that perpendicularity of lines is not an equivalence relation.

27. Show that conformability for addition of matrices is an equivalence relation while conformability for multiplication is not.

28. Prove: If A, B, C are matrices such that $AC = CA$ and $BC = CB$, then $(AB \pm BA)C = C(AB \pm BA)$.

Chapter 2

Some Types of Matrices

THE IDENTITY MATRIX. A square matrix A whose elements $a_{ij} = 0$ for $i > j$ is called **upper triangular**; a square matrix A whose elements $a_{ij} = 0$ for $i < j$ is called **lower triangular**. Thus

$$\begin{bmatrix} a_{11} & a_{12} & a_{13} & \cdots & a_{1n} \\ 0 & a_{22} & a_{23} & \cdots & a_{2n} \\ 0 & 0 & a_{33} & \cdots & a_{3n} \\ \cdots & \cdots & \cdots & \cdots & \cdots \\ 0 & 0 & 0 & \cdots & a_{nn} \end{bmatrix} \quad \text{is upper triangular and}$$

$$\begin{bmatrix} a_{11} & 0 & 0 & \cdots & 0 \\ a_{21} & a_{22} & 0 & \cdots & 0 \\ a_{31} & a_{32} & a_{33} & \cdots & 0 \\ \cdots & \cdots & \cdots & \cdots & \cdots \\ a_{n1} & a_{n2} & a_{n3} & \cdots & a_{nn} \end{bmatrix} \quad \text{is lower triangular.}$$

The matrix $D = \begin{bmatrix} a_{11} & 0 & 0 & \cdots & 0 \\ 0 & a_{22} & 0 & \cdots & 0 \\ 0 & 0 & a_{33} & \cdots & 0 \\ \cdots & \cdots & \cdots & \cdots & \cdots \\ 0 & 0 & 0 & \cdots & a_{nn} \end{bmatrix}$, which is both upper and lower triangular, is called a **diagonal matrix**. It will frequently be written as

$$D = \text{diag}(a_{11}, a_{22}, a_{33}, \ldots, a_{nn})$$

See Problem 1.

If in the diagonal matrix D above, $a_{11} = a_{22} = \ldots = a_{nn} = k$, D is called a **scalar** matrix; if, in addition, $k = 1$, the matrix is called the **identity matrix** and is denoted by I_n. For example

$$I_2 = \begin{bmatrix} 1 & 0 \\ 0 & 1 \end{bmatrix} \quad \text{and} \quad I_3 = \begin{bmatrix} 1 & 0 & 0 \\ 0 & 1 & 0 \\ 0 & 0 & 1 \end{bmatrix}$$

When the order is evident or immaterial, an identity matrix will be denoted by I. Clearly, $I_n + I_n + \ldots$ to p terms $= p \cdot I_n = \text{diag}(p, p, p, \ldots, p)$ and $I^p = I \cdot I \ldots$ to p factors $= I$. Identity matrices have some of the properties of the integer 1. For example, if $A = \begin{bmatrix} 1 & 2 & 3 \\ 4 & 5 & 6 \end{bmatrix}$, then $I_2 \cdot A = A \cdot I_3 = I_2 A I_3 = A$, as the reader may readily show.

SPECIAL SQUARE MATRICES. If A and B are square matrices such that $AB = BA$, then A and B are called **commutative** or are said to **commute**. It is a simple matter to show that if A is any n-square matrix, it commutes with itself and also with I_n.

See Problem 2.

If A and B are such that $AB = -BA$, the matrices A and B are said to **anti-commute**.

A matrix A for which $A^{k+1} = A$, where k is a positive integer, is called **periodic**. If k is the least positive integer for which $A^{k+1} = A$, then A is said to be of **period** k.

If $k = 1$, so that $A^2 = A$, then A is called **idempotent**.

See Problems 3-4.

A matrix A for which $A^p = 0$, where p is a positive integer, is called **nilpotent**. If p is the least positive integer for which $A^p = 0$, then A is said to be nilpotent of **index** p.

See Problems 5-6.

THE INVERSE OF A MATRIX. If A and B are square matrices such that $AB = BA = I$, then B is called the **inverse** of A and we write $B = A^{-1}$ (B equals A inverse). The matrix B also has A as its inverse and we may write $A = B^{-1}$.

Example 1. Since $\begin{bmatrix} 1 & 2 & 3 \\ 1 & 3 & 3 \\ 1 & 2 & 4 \end{bmatrix}\begin{bmatrix} 6 & -2 & -3 \\ -1 & 1 & 0 \\ -1 & 0 & 1 \end{bmatrix} = \begin{bmatrix} 1 & 0 & 0 \\ 0 & 1 & 0 \\ 0 & 0 & 1 \end{bmatrix} = I$, each matrix in the product is the inverse of the other.

We shall find later (Chapter 7) that not every square matrix has an inverse. We can show here, however, that if A has an inverse then that inverse is unique.

See Problem 7.

If A and B are square matrices of the same order with inverses A^{-1} and B^{-1} respectively, then $(AB)^{-1} = B^{-1} \cdot A^{-1}$, that is,

I. The inverse of the product of two matrices, having inverses, is the product **in reverse order** of these inverses.

See Problem 8.

A matrix A such that $A^2 = I$ is called **involutory**. An identity matrix, for example, is involutory. An involutory matrix is its own inverse.

See Problem 9.

THE TRANSPOSE OF A MATRIX. The matrix of order $n \times m$ obtained by interchanging the rows and columns of an $m \times n$ matrix A is called the **transpose** of A and is denoted by A' (A transpose). For example, the transpose of $A = \begin{bmatrix} 1 & 2 & 3 \\ 4 & 5 & 6 \end{bmatrix}$ is $A' = \begin{bmatrix} 1 & 4 \\ 2 & 5 \\ 3 & 6 \end{bmatrix}$. Note that the element a_{ij} in the ith row and jth column of A stands in the jth row and ith column of A'.

If A' and B' are transposes respectively of A and B, and if k is a scalar, we have immediately

$$(a)\quad (A')' = A \qquad \text{and} \qquad (b)\quad (kA)' = kA'$$

In Problems 10 and 11, we prove:

II. The transpose of the sum of two matrices is the sum of their transposes, i.e.,

$$(A + B)' = A' + B'$$

and

III. The transpose of the product of two matrices is the product **in reverse order** of their transposes, i.e.,

$$(AB)' = B' \cdot A'$$

See Problems 10-12.

SYMMETRIC MATRICES. A square matrix A such that $A' = A$ is called **symmetric**. Thus, a square matrix $A = [a_{ij}]$ is symmetric provided $a_{ij} = a_{ji}$, for all values of i and j. For example,

$A = \begin{bmatrix} 1 & 2 & 3 \\ 2 & 4 & -5 \\ 3 & -5 & 6 \end{bmatrix}$ is symmetric and so also is kA for any scalar k.

In Problem 13, we prove

IV. If A is an n-square matrix, then $A + A'$ is symmetric.

A square matrix A such that $A' = -A$ is called **skew-symmetric**. Thus, a square matrix A is skew-symmetric provided $a_{ij} = -a_{ji}$ for all values of i and j. Clearly, the diagonal elements are zeros. For example, $A = \begin{bmatrix} 0 & -2 & 3 \\ 2 & 0 & 4 \\ -3 & -4 & 0 \end{bmatrix}$ is skew-symmetric and so also is kA for any scalar k.

With only minor changes in Problem 13, we can prove

V. If A is any n-square matrix, then $A - A'$ is skew-symmetric.

From Theorems **IV** and **V** follows

VI. Every square matrix A can be written as the sum of a symmetric matrix $B = \frac{1}{2}(A+A')$ and a skew-symmetric matrix $C = \frac{1}{2}(A-A')$.

See Problems 14-15.

THE CONJUGATE OF A MATRIX. Let a and b be real numbers and let $i = \sqrt{-1}$; then, $z = a+bi$ is called a **complex number**. The complex numbers $a+bi$ and $a-bi$ are called **conjugates**, each being the **conjugate** of the other. If $z = a+bi$, its conjugate is denoted by $\bar{z} = \overline{a+bi}$.

If $z_1 = a+bi$ and $z_2 = \bar{z}_1 = a-bi$, then $\bar{z}_2 = \bar{\bar{z}}_1 = \overline{a-bi} = a+bi$, that is, the conjugate of the conjugate of a complex number z is z itself.

If $z_1 = a+bi$ and $z_2 = c+di$, then

(i) $z_1 + z_2 = (a+c) + (b+d)i$ and $\overline{z_1 + z_2} = (a+c) - (b+d)i = (a-bi) + (c-di) = \bar{z}_1 + \bar{z}_2$,

that is, the conjugate of the sum of two complex numbers is the sum of their conjugates.

(ii) $z_1 \cdot z_2 = (ac-bd) + (ad+bc)i$ and $\overline{z_1 \cdot z_2} = (ac-bd) - (ad+bc)i = (a-bi)(c-di) = \bar{z}_1 \cdot \bar{z}_2$,

that is, the conjugate of the product of two complex numbers is the product of their conjugates.

When A is a matrix having complex numbers as elements, the matrix obtained from A by replacing each element by its conjugate is called the **conjugate** of A and is denoted by $\bar{A}$ (A conjugate).

Example 2. When $A = \begin{bmatrix} 1+2i & i \\ 3 & 2-3i \end{bmatrix}$ then $\bar{A} = \begin{bmatrix} 1-2i & -i \\ 3 & 2+3i \end{bmatrix}$.

If $\bar{A}$ and $\bar{B}$ are the conjugates of the matrices A and B and if k is any scalar, we have readily

$$(c)\quad (\bar{\bar{A}}) = A \qquad \text{and} \qquad (d)\quad (\overline{kA}) = \bar{k} \cdot \bar{A}$$

Using **(i)** and **(ii)** above, we may prove

VII. The conjugate of the sum of two matrices is the sum of their conjugates, i.e., $\overline{(A+B)} = \bar{A} + \bar{B}$.

VIII. The conjugate of the product of two matrices is the product, **in the same order**, of their conjugates, i.e., $\overline{(AB)} = \bar{A}\cdot\bar{B}$.

The transpose of $\bar{A}$ is denoted by $\bar{A}'$ (A conjugate transpose). It is sometimes written as A^*. We have

IX. The transpose of the conjugate of A is equal to the conjugate of the transpose of A, i.e., $(\bar{A})' = \overline{(A')}$.

Example 3. From Example 2

$$(\bar{A})' = \begin{bmatrix} 1-2i & 3 \\ -i & 2+3i \end{bmatrix} \quad \text{while} \quad A' = \begin{bmatrix} 1+2i & 3 \\ i & 2-3i \end{bmatrix} \quad \text{and} \quad \overline{(A')} = \begin{bmatrix} 1-2i & 3 \\ -i & 2+3i \end{bmatrix} = (\bar{A})'$$

HERMITIAN MATRICES. A square matrix $A = [a_{ij}]$ such that $\bar{A}' = A$ is called **Hermitian**. Thus, A is Hermitian provided $a_{ij} = \bar{a}_{ji}$ for all values of i and j. Clearly, the diagonal elements of an Hermitian matrix are real numbers.

Example 4. The matrix $A = \begin{bmatrix} 1 & 1-i & 2 \\ 1+i & 3 & i \\ 2 & -i & 0 \end{bmatrix}$ is Hermitian.

Is kA Hermitian if k is any real number? any complex number?

A square matrix $A = [a_{ij}]$ such that $\bar{A}' = -A$ is called **skew-Hermitian**. Thus, A is skew-Hermitian provided $a_{ij} = -\bar{a}_{ji}$ for all values of i and j. Clearly, the diagonal elements of a skew-Hermitian matrix are either zeros or pure imaginaries.

Example 5. The matrix $A = \begin{bmatrix} i & 1-i & 2 \\ -1-i & 3i & i \\ -2 & i & 0 \end{bmatrix}$ is skew-Hermitian. Is kA skew-Hermitian if k is any real number? any complex number? any pure imaginary?

By making minor changes in Problem 13, we may prove

X. If A is an n-square matrix then $A + \bar{A}'$ is Hermitian and $A - \bar{A}'$ is skew-Hermitian.

From Theorem **X** follows

XI. Every square matrix A with complex elements can be written as the sum of an Hermitian matrix $B = \frac{1}{2}(A + \bar{A}')$ and a skew-Hermitian matrix $C = \frac{1}{2}(A - \bar{A}')$.

DIRECT SUM. Let $A_1, A_2, \ldots, A_s$ be square matrices of respective orders $m_1, m_2, \ldots, m_s$. The generalization

$$A = \begin{bmatrix} A_1 & 0 & \ldots & 0 \\ 0 & A_2 & \ldots & 0 \\ \ldots & \ldots & \ldots & \ldots \\ 0 & 0 & \ldots & A_s \end{bmatrix} = \operatorname{diag}(A_1, A_2, \ldots, A_s)$$

of the diagonal matrix is called the **direct sum** of the A_i.

Example 6. Let $A_1 = [2]$, $A_2 = \begin{bmatrix} 1 & 2 \\ 3 & 4 \end{bmatrix}$, and $A_3 = \begin{bmatrix} 1 & 2 & -1 \\ 2 & 0 & 3 \\ 4 & 1 & -2 \end{bmatrix}$.

The direct sum of A_1, A_2, A_3 is $\text{diag}(A_1, A_2, A_3) = \begin{bmatrix} 2 & 0 & 0 & 0 & 0 & 0 \\ 0 & 1 & 2 & 0 & 0 & 0 \\ 0 & 3 & 4 & 0 & 0 & 0 \\ 0 & 0 & 0 & 1 & 2 & -1 \\ 0 & 0 & 0 & 2 & 0 & 3 \\ 0 & 0 & 0 & 4 & 1 & -2 \end{bmatrix}$

Problem 9(*b*), Chapter 1, illustrates

XII. If $A = \text{diag}(A_1, A_2, \ldots, A_s)$ and $B = \text{diag}(B_1, B_2, \ldots, B_s)$, where A_i and B_i have the same order for $(i = 1, 2, \ldots, s)$, then $AB = \text{diag}(A_1B_1, A_2B_2, \ldots, A_sB_s)$.

SOLVED PROBLEMS

1. Since $\begin{bmatrix} a_{11} & 0 & \ldots & 0 \\ 0 & a_{22} & \ldots & 0 \\ \ldots & \ldots & \ldots & \ldots \\ 0 & 0 & & a_{mm} \end{bmatrix} \begin{bmatrix} b_{11} & b_{12} & \ldots & b_{1n} \\ b_{21} & b_{22} & \ldots & b_{2n} \\ \ldots & \ldots & \ldots & \ldots \\ b_{m1} & b_{m2} & \ldots & b_{mn} \end{bmatrix} = \begin{bmatrix} a_{11}b_{11} & a_{11}b_{12} & \ldots & a_{11}b_{1n} \\ a_{22}b_{21} & a_{22}b_{22} & \ldots & a_{22}b_{2n} \\ \ldots & \ldots & \ldots & \ldots \\ a_{mm}b_{m1} & a_{mm}b_{m2} & \ldots & a_{mm}b_{mn} \end{bmatrix}$, the product AB of an m-square diagonal matrix $A = \text{diag}(a_{11}, a_{22}, \ldots, a_{mm})$ and any $m \times n$ matrix B is obtained by multiplying the first row of B by a_{11}, the second row of B by a_{22}, and so on.

2. Show that the matrices $\begin{bmatrix} a & b \\ b & a \end{bmatrix}$ and $\begin{bmatrix} c & d \\ d & c \end{bmatrix}$ commute for all values of a, b, c, d.

This follows from $\begin{bmatrix} a & b \\ b & a \end{bmatrix}\begin{bmatrix} c & d \\ d & c \end{bmatrix} = \begin{bmatrix} ac+bd & ad+bc \\ bc+ad & bd+ac \end{bmatrix} = \begin{bmatrix} c & d \\ d & c \end{bmatrix}\begin{bmatrix} a & b \\ b & a \end{bmatrix}$.

3. Show that $\begin{bmatrix} 2 & -2 & -4 \\ -1 & 3 & 4 \\ 1 & -2 & -3 \end{bmatrix}$ is idempotent.

$$A^2 = \begin{bmatrix} 2 & -2 & -4 \\ -1 & 3 & 4 \\ 1 & -2 & -3 \end{bmatrix}\begin{bmatrix} 2 & -2 & -4 \\ -1 & 3 & 4 \\ 1 & -2 & -3 \end{bmatrix} = \begin{bmatrix} 2 & -2 & -4 \\ -1 & 3 & 4 \\ 1 & -2 & -3 \end{bmatrix} = A$$

4. Show that if $AB = A$ and $BA = B$, then A and B are idempotent.

$ABA = (AB)A = A \cdot A = A^2$ and $ABA = A(BA) = AB = A$; then $A^2 = A$ and A is idempotent. Use BAB to show that B is idempotent.

5. Show that $A = \begin{bmatrix} 1 & 1 & 3 \\ 5 & 2 & 6 \\ -2 & -1 & -3 \end{bmatrix}$ is nilpotent of order 3.

$$A^2 = \begin{bmatrix} 1 & 1 & 3 \\ 5 & 2 & 6 \\ -2 & -1 & -3 \end{bmatrix}\begin{bmatrix} 1 & 1 & 3 \\ 5 & 2 & 6 \\ -2 & -1 & -3 \end{bmatrix} = \begin{bmatrix} 0 & 0 & 0 \\ 3 & 3 & 9 \\ -1 & -1 & -3 \end{bmatrix} \quad \text{and} \quad A^3 = A^2 \cdot A = \begin{bmatrix} 0 & 0 & 0 \\ 3 & 3 & 9 \\ -1 & -1 & -3 \end{bmatrix}\begin{bmatrix} 1 & 1 & 3 \\ 5 & 2 & 6 \\ -2 & -1 & -3 \end{bmatrix} = 0$$

6. If A is nilpotent of index 2, show that $A(I \pm A)^n = A$ for n any positive integer.

Since $A^2 = 0$, $A^3 = A^4 = \ldots = A^n = 0$. Then $A(I \pm A)^n = A(I \pm nA) = A \pm nA^2 = A$.

7. Let A, B, C be square matrices such that $AB = I$ and $CA = I$. Then $(CA)B = C(AB)$ so that $B = C$. Thus, $B = C = A^{-1}$ is the **unique** inverse of A. (What is B^{-1}?)

8. Prove: $(AB)^{-1} = B^{-1} \cdot A^{-1}$.

By definition $(AB)^{-1}(AB) = (AB)(AB)^{-1} = I$. Now

$$(B^{-1} \cdot A^{-1})AB = B^{-1}(A^{-1} \cdot A)B = B^{-1} \cdot I \cdot B = B^{-1} \cdot B = I$$

and

$$AB(B^{-1} \cdot A^{-1}) = A(B \cdot B^{-1})A^{-1} = A \cdot A^{-1} = I$$

By Problem 7, $(AB)^{-1}$ is unique; hence, $(AB)^{-1} = B^{-1} \cdot A^{-1}$

9. Prove: A matrix A is involutory if and only if $(I - A)(I + A) = 0$.

Suppose $(I-A)(I+A) = I - A^2 = 0$; then $A^2 = I$ and A is involutory.

Suppose A is involutory; then $A^2 = I$ and $(I-A)(I+A) = I - A^2 = I - I = 0$.

10. Prove: $(A + B)' = A' + B'$.

Let $A = [a_{ij}]$ and $B = [b_{ij}]$. We need only check that the element in the ith row and jth column of A', B', and $(A+B)'$ are respectively a_{ji}, b_{ji}, and $a_{ji} + b_{ji}$.

11. Prove: $(AB)' = B'A'$.

Let $A = [a_{ij}]$ be of order $m \times n$, $B = [b_{ij}]$ be of order $n \times p$; then $C = AB = [c_{ij}]$ is of order $m \times p$. The element standing in the ith row and jth column of AB is $c_{ij} = \sum_{k=1}^{n} a_{ik} \cdot b_{kj}$ and this is also the element standing in the jth row and ith column of $(AB)'$.

The elements of the jth row of B' are $b_{1j}, b_{2j}, \ldots, b_{nj}$ and the elements of the ith column of A' are a_{i1}, $a_{i2}, \ldots, a_{in}$. Then the element in the jth row and ith column of $B'A'$ is

$$\sum_{k=1}^{n} b_{kj} \cdot a_{ik} = \sum_{k=1}^{n} a_{ik} \cdot b_{kj} = c_{ij}$$

Thus, $(AB)' = B'A'$.

12. Prove: $(ABC)' = C'B'A'$.

Write $ABC = (AB)C$. Then, by Problem 11, $(ABC)' = \{(AB)C\}' = C'(AB)' = C'B'A'$.

13. Show that if $A = [a_{ij}]$ is n-square, then $B = [b_{ij}] = A + A'$ is symmetric.

First Proof.

The element in the ith row and jth column of A is a_{ij} and the corresponding element of A' is a_{ji}; hence $b_{ij} = a_{ij} + a_{ji}$. The element in the jth row and ith column of A is a_{ji} and the corresponding element of A' is a_{ij}; hence, $b_{ji} = a_{ji} + a_{ij}$. Thus, $b_{ij} = b_{ji}$ and B is symmetric.

Second Proof.

By Problem 10, $(A+A')' = A' + (A')' = A' + A = A + A'$ and $(A+A')$ is symmetric.

14. Prove: If A and B are n-square symmetric matrices then AB is symmetric if and only if A and B commute.

Suppose A and B commute so that $AB = BA$. Then $(AB)' = B'A' = BA = AB$ and AB is symmetric.

Suppose AB is symmetric so that $(AB)' = AB$. Now $(AB)' = B'A' = BA$; hence, $AB = BA$ and the matrices A and B commute.

15. Prove: If the m-square matrix A is symmetric (skew-symmetric) and if P is of order $m \times n$ then $B = P'AP$ is symmetric (skew-symmetric).

If A is symmetric then (see Problem 12) $B' = (P'AP)' = P'A'(P')' = P'A'P = P'AP$ and B is symmetric.

If A is skew-symmetric then $B' = (P'AP)' = -P'AP$ and B is skew-symmetric.

16. Prove: If A and B are n-square matrices then A and B commute if and only if $A - kI$ and $B - kI$ commute for every scalar k.

Suppose A and B commute; then $AB = BA$ and

$$\begin{aligned}(A-kI)(B-kI) &= AB - k(A+B) + k^2 I \\ &= BA - k(A+B) + k^2 I = (B-kI)(A-kI)\end{aligned}$$

Thus, $A - kI$ and $B - kI$ commute.

Suppose $A - kI$ and $B - kI$ commute; then

$$\begin{aligned}(A-kI)(B-kI) &= AB - k(A+B) + k^2 I \\ &= BA - k(A+B) + k^2 I = (B-kI)(A-kI)\end{aligned}$$

$AB = BA$, and A and B commute.

SUPPLEMENTARY PROBLEMS

17. Show that the product of two upper (lower) triangular matrices is upper (lower) triangular.

18. Derive a rule for forming the product BA of an $m\times n$ matrix B and $A = \text{diag}(a_{11}, a_{22}, \ldots, a_{nn})$.
Hint. See Problem 1.

19. Show that the scalar matrix with diagonal element k can be written as kI and that $kA = kIA = \text{diag}(k, k, \ldots, k)\,A$, where the order of I is the row order of A.

20. If A is n-square, show that $A^p \cdot A^q = A^q \cdot A^p$ where p and q are positive integers.

21. (*a*) Show that $A = \begin{bmatrix} 2 & -3 & -5 \\ -1 & 4 & 5 \\ 1 & -3 & -4 \end{bmatrix}$ and $B = \begin{bmatrix} -1 & 3 & 5 \\ 1 & -3 & -5 \\ -1 & 3 & 5 \end{bmatrix}$ are idempotent.

(*b*) Using A and B, show that the converse of Problem 4 does not hold.

22. If A is idempotent, show that $B = I - A$ is idempotent and that $AB = BA = 0$.

23. (*a*) If $A = \begin{bmatrix} 1 & 2 & 2 \\ 2 & 1 & 2 \\ 2 & 2 & 1 \end{bmatrix}$, show that $A^2 - 4A - 5I = 0$.

(*b*) If $A = \begin{bmatrix} 2 & 1 & 3 \\ 1 & -1 & 2 \\ 1 & 2 & 1 \end{bmatrix}$, show that $A^3 - 2A^2 - 9A = 0$. but $A^2 - 2A - 9I \neq 0$.

24. Show that $\begin{bmatrix} -1 & -1 & -1 \\ 0 & 1 & 0 \\ 0 & 0 & 1 \end{bmatrix}^2 = \begin{bmatrix} 0 & 1 & 0 \\ -1 & -1 & -1 \\ 0 & 0 & 1 \end{bmatrix}^3 = \begin{bmatrix} 0 & 1 & 0 \\ 0 & 0 & 1 \\ -1 & -1 & -1 \end{bmatrix}^4 = I$.

25. Show that $A = \begin{bmatrix} 1 & -2 & -6 \\ -3 & 2 & 9 \\ 2 & 0 & -3 \end{bmatrix}$ is periodic, of period 2.

26. Show that $\begin{bmatrix} 1 & -3 & -4 \\ -1 & 3 & 4 \\ 1 & -3 & -4 \end{bmatrix}$ is nilpotent.

27. Show that (*a*) $A = \begin{bmatrix} 1 & 2 & 3 \\ 3 & 2 & 0 \\ -1 & -1 & -1 \end{bmatrix}$ and $B = \begin{bmatrix} -2 & -1 & -6 \\ 3 & 2 & 9 \\ -1 & -1 & -4 \end{bmatrix}$ commute,

(*b*) $A = \begin{bmatrix} 1 & 1 & 2 \\ 2 & 3 & 1 \\ -1 & 2 & 4 \end{bmatrix}$ and $B = \begin{bmatrix} 2/3 & 0 & -1/3 \\ -3/5 & 2/5 & 1/5 \\ 7/15 & -1/5 & 1/15 \end{bmatrix}$ commute.

28. Show that $A = \begin{bmatrix} 1 & -1 \\ 2 & -1 \end{bmatrix}$ and $B = \begin{bmatrix} 1 & 1 \\ 4 & -1 \end{bmatrix}$ anti-commute and $(A+B)^2 = A^2 + B^2$.

29. Show that each of $\begin{bmatrix} 0 & 1 \\ 1 & 0 \end{bmatrix}, \begin{bmatrix} 0 & -i \\ i & 0 \end{bmatrix}, \begin{bmatrix} i & 0 \\ 0 & -i \end{bmatrix}$ anti-commutes with the others.

30. Prove: The only matrices which commute with every n-square matrix are the n-square scalar matrices.

31. (*a*) Find all matrices which commute with diag(1, 2, 3).
(*b*) Find all matrices which commute with $\text{diag}(a_{11}, a_{22}, \ldots, a_{nn})$.
Ans. (*a*) diag(a, b, c) where a, b, c are arbitrary.

32. Show that (a) $\begin{bmatrix} 1 & 2 & 3 \\ 2 & 5 & 7 \\ -2 & -4 & -5 \end{bmatrix}$ is the inverse of $\begin{bmatrix} 3 & -2 & -1 \\ -4 & 1 & -1 \\ 2 & 0 & 1 \end{bmatrix}$

(b) $\begin{bmatrix} 1 & 0 & 0 & 0 \\ 2 & 1 & 0 & 0 \\ 4 & 2 & 1 & 0 \\ -2 & 3 & 1 & 1 \end{bmatrix}$ is the inverse of $\begin{bmatrix} 1 & 0 & 0 & 0 \\ -2 & 1 & 0 & 0 \\ 0 & -2 & 1 & 0 \\ 8 & -1 & -1 & 1 \end{bmatrix}$.

33. Set $\begin{bmatrix} 1 & 2 \\ 3 & 4 \end{bmatrix}\begin{bmatrix} a & b \\ c & d \end{bmatrix} = \begin{bmatrix} 1 & 0 \\ 0 & 1 \end{bmatrix}$ to find the inverse of $\begin{bmatrix} 1 & 2 \\ 3 & 4 \end{bmatrix}$. *Ans.* $\begin{bmatrix} -2 & 1 \\ 3/2 & -1/2 \end{bmatrix}$

34. Show that the inverse of a diagonal matrix A, all of whose diagonal elements are different from zero, is a diagonal matrix whose diagonal elements are the inverses of those of A and in the same order. Thus, the inverse of I_n is I_n.

35. Show that $A = \begin{bmatrix} 0 & 1 & -1 \\ 4 & -3 & 4 \\ 3 & -3 & 4 \end{bmatrix}$ and $B = \begin{bmatrix} 4 & 3 & 3 \\ -1 & 0 & -1 \\ -4 & -4 & -3 \end{bmatrix}$ are involutory.

36. Let $A = \begin{bmatrix} 1 & 0 & 0 & 0 \\ 0 & 1 & 0 & 0 \\ a & b & -1 & 0 \\ c & d & 0 & -1 \end{bmatrix} = \begin{bmatrix} I_2 & 0 \\ A_{21} & -I_2 \end{bmatrix}$ by partitioning. Show that $A^2 = \begin{bmatrix} I_2 & 0 \\ 0 & I_2 \end{bmatrix} = I_4$.

37. Prove: (a) $(A')' = A$, (b) $(kA)' = kA'$, (c) $(A^p)' = (A')^p$ for p a positive integer.

38. Prove: $(ABC)^{-1} = C^{-1}B^{-1}A^{-1}$. *Hint.* Write $ABC = (AB)C$.

39. Prove: (a) $(A^{-1})^{-1} = A$, (b) $(kA)^{-1} = \frac{1}{k}A^{-1}$, (c) $(A^p)^{-1} = (A^{-1})^p$ for p a positive integer.

40. Show that every real symmetric matrix is Hermitian.

41. Prove: (a) $\overline{(\bar{A})} = A$, (b) $\overline{(A+B)} = \bar{A} + \bar{B}$, (c) $\overline{(kA)} = \bar{k}\,\bar{A}$, (d) $\overline{(AB)} = \bar{A}\,\bar{B}$.

42. Show: (a) $A = \begin{bmatrix} 1 & 1+i & 2+3i \\ 1-i & 2 & -i \\ 2-3i & i & 0 \end{bmatrix}$ is Hermitian,

(b) $B = \begin{bmatrix} i & 1+i & 2-3i \\ -1+i & 2i & 1 \\ -2-3i & -1 & 0 \end{bmatrix}$ is skew-Hermitian,

(c) iB is Hermitian,

(d) $\bar{A}$ is Hermitian and $\bar{B}$ is skew-Hermitian.

43. If A is n-square, show that (a) AA' and $A'A$ are symmetric, (b) $A+\bar{A}'$, $A\bar{A}'$, and $\bar{A}'A$ are Hermitian.

44. Prove: If H is Hermitian and A is any conformable matrix then $(\bar{A})'HA$ is Hermitian.

45. Prove: Every Hermitian matrix A can be written as $B+iC$ where B is real and symmetric and C is real and skew-symmetric.

46. Prove: (a) Every skew-Hermitian matrix A can be written as $A = B+iC$ where B is real and skew-symmetric and C is real and symmetric. (b) $A'A$ is real if and only if B and C anti-commute.

47. Prove: If A and B commute so also do A^{-1} and B^{-1}, A' and B', and $\bar{A}'$ and $\bar{B}'$.

48. Show that for m and n positive integers, A^m and B^n commute if A and B commute.

49. Show (a) $\begin{bmatrix} \lambda & 1 \\ 0 & \lambda \end{bmatrix}^n = \begin{bmatrix} \lambda^n & n\lambda^{n-1} \\ 0 & \lambda^n \end{bmatrix}$, (b) $\begin{bmatrix} \lambda & 1 & 0 \\ 0 & \lambda & 1 \\ 0 & 0 & \lambda \end{bmatrix}^n = \begin{bmatrix} \lambda^n & n\lambda^{n-1} & \frac{1}{2}n(n-1)\lambda^{n-2} \\ 0 & \lambda^n & n\lambda^{n-1} \\ 0 & 0 & \lambda^n \end{bmatrix}$

50. Prove: If A is symmetric or skew-symmetric then $AA' = A'A$ and A^2 are symmetric.

51. Prove: If A is symmetric so also is $aA^p + bA^{p-1} + \ldots + gI$ where $a, b, \ldots, g$ are scalars and p is a positive integer.

52. Prove: Every square matrix A can be written as $A = B + C$ where B is Hermitian and C is skew-Hermitian.

53. Prove: If A is real and skew-symmetric or if A is complex and skew-Hermitian then $\pm iA$ are Hermitian.

54. Show that the theorem of Problem 52 can be stated:
Every square matrix A can be written as $A = B + iC$ where B and C are Hermitian.

55. Prove: If A and B are such that $AB = A$ and $BA = B$ then (a) $B'A' = A'$ and $A'B' = B'$, (b) A' and B' are idempotent, (c) $A = B = I$ if A has an inverse.

56. If A is involutory, show that $\frac{1}{2}(I+A)$ and $\frac{1}{2}(I-A)$ are idempotent and $\frac{1}{2}(I+A) \cdot \frac{1}{2}(I-A) = 0$.

57. If the n-square matrix A has an inverse A^{-1}, show:
(a) $(A^{-1})' = (A')^{-1}$, (b) $(\bar{A})^{-1} = \overline{A^{-1}}$, (c) $(\bar{A}')^{-1} = \overline{(A^{-1})'}$
Hint. (a) From the transpose of $AA^{-1} = I$, obtain $(A^{-1})'$ as the inverse of A'.

58. Find all matrices which commute with (a) diag(1, 1, 2, 3), (b) diag(1, 1, 2, 2).
Ans. (a) diag(A, b, c), (b) diag(A, B) where A and B are 2-square matrices with arbitrary elements and b, c are scalars.

59. If $A_1, A_2, \ldots, A_s$ are scalar matrices of respective orders $m_1, m_2, \ldots, m_s$, find all matrices which commute with diag$(A_1, A_2, \ldots, A_s)$.
Ans. diag$(B_1, B_2, \ldots, B_s)$ where $B_1, B_2, \ldots, B_s$ are of respective orders $m_1, m_2, \ldots, m_s$ with arbitrary elements.

60. If $AB = 0$, where A and B are non-zero n-square matrices, then A and B are called **divisors of zero**. Show that the matrices A and B of Problem 21 are divisors of zero.

61. If $A = \text{diag}(A_1, A_2, \ldots, A_s)$ and $B = \text{diag}(B_1, B_2, \ldots, B_s)$ where A_i and B_i are of the same order, $(i = 1, 2, \ldots, s)$, show that
(a) $A + B = \text{diag}(A_1 + B_1, A_2 + B_2, \ldots, A_s + B_s)$
(b) $AB = \text{diag}(A_1B_1, A_2B_2, \ldots, A_sB_s)$
(c) trace AB = trace A_1B_1 + trace A_2B_2 + ... + trace A_sB_s.

62. Prove: If A and B are n-square skew-symmetric matrices then AB is symmetric if and only if A and B commute.

63. Prove: If A is n-square and $B = rA + sI$, where r and s are scalars, then A and B commute.

64. Let A and B be n-square matrices and let r_1, r_2, s_1, s_2 be scalars such that $r_1s_2 \neq r_2s_1$. Prove that $C_1 = r_1A + s_1B$, $C_2 = r_2A + s_2B$ commute if and only if A and B commute.

65. Show that the n-square matrix A will not have an inverse when (a) A has a row (column) of zero elements or (b) A has two identical rows (columns) or (c) A has a row (column) which is the sum of two other rows (columns).

66. If A and B are n-square matrices and A has an inverse, show that

$$(A+B)A^{-1}(A-B) = (A-B)A^{-1}(A+B)$$

Chapter 3

Determinant of a Square Matrix

PERMUTATIONS. Consider the $3! = 6$ permutations of the integers $1, 2, 3$ taken together

$$123 \quad 132 \quad 213 \quad 231 \quad 312 \quad 321 \tag{3.1}$$

and eight of the $4! = 24$ permutations of the integers $1, 2, 3, 4$ taken together

$$\begin{matrix} 1234 & 2134 & 3124 & 4123 \\ 1324 & 2314 & 3214 & 4213 \end{matrix} \tag{3.2}$$

If in a given permutation a larger integer precedes a smaller one, we say that there is an **inversion**. If in a given permutation the number of inversions is even (odd), the permutation is called **even (odd)**. For example, in **(3.1)** the permutation 123 is even since there is no inversion, the permutation 132 is odd since in it 3 precedes 2, the permutation 312 is even since in it 3 precedes 1 and 3 precedes 2. In **(3.2)** the permutation 4213 is even since in it 4 precedes 2, 4 precedes 1, 4 precedes 3, and 2 precedes 1.

THE DETERMINANT OF A SQUARE MATRIX. Consider the n-square matrix

$$A = \begin{bmatrix} a_{11} \; a_{12} \; a_{13} \; \cdots \; a_{1n} \\ a_{21} \; a_{22} \; a_{23} \; \cdots \; a_{2n} \\ \cdots\cdots\cdots\cdots\cdots\cdots \\ a_{n1} \; a_{n2} \; a_{n3} \; \cdots \; a_{nn} \end{bmatrix} \tag{3.3}$$

and a product

$$a_{1j_1} \, a_{2j_2} \, a_{3j_3} \cdots a_{nj_n} \tag{3.4}$$

of n of its elements, selected so that one and only one element comes from any row and one and only one element comes from any column. In **(3.4)**, as a matter of convenience, the factors have been arranged so that the sequence of first subscripts is the natural order $1, 2, \ldots, n$; the sequence $j_1, j_2, \ldots, j_n$ of second subscripts is then some one of the $n!$ permutations of the integers $1, 2, \ldots, n$. (Facility will be gained if the reader will parallel the work of this section beginning with a product arranged so that the sequence of second subscripts is in natural order.)

For a given permutation $j_1, j_2, \ldots, j_n$ of the second subscripts, define $\epsilon_{j_1 j_2 \cdots j_n} = +1$ or -1 according as the permutation is even or odd and form the signed product

$$\epsilon_{j_1 j_2 \cdots j_n} \, a_{1j_1} \, a_{2j_2} \cdots a_{nj_n} \tag{3.5}$$

By the **determinant** of A, denoted by $|A|$, is meant the sum of all the different signed products of the form **(3.5)**, called **terms** of $|A|$, which can be formed from the elements of A; thus,

$$|A| = \sum_{\rho} \epsilon_{j_1 j_2 \cdots j_n} \, a_{1j_1} \, a_{2j_2} \cdots a_{nj_n} \tag{3.6}$$

where the summation extends over $\rho = n!$ permutations $j_1 j_2 \cdots j_n$ of the integers $1, 2, \ldots, n$.

The determinant of a square matrix of order n is called a determinant of order n.

DETERMINANTS OF ORDER TWO AND THREE. From **(3.6)** we have for $n=2$ and $n=3$,

(3.7)
$$\begin{vmatrix} a_{11} & a_{12} \\ a_{21} & a_{22} \end{vmatrix} = \epsilon_{12}\, a_{11}a_{22} + \epsilon_{21}\, a_{12}a_{21} = a_{11}a_{22} - a_{12}a_{21}$$

and

(3.8)
$$\begin{aligned}\begin{vmatrix} a_{11} & a_{12} & a_{13} \\ a_{21} & a_{22} & a_{23} \\ a_{31} & a_{32} & a_{33} \end{vmatrix} &= \epsilon_{123}\, a_{11}a_{22}a_{33} + \epsilon_{132}\, a_{11}a_{23}a_{32} + \epsilon_{213}\, a_{12}a_{21}a_{33} \\ &\quad + \epsilon_{231}\, a_{12}a_{23}a_{31} + \epsilon_{312}\, a_{13}a_{21}a_{32} + \epsilon_{321}\, a_{13}a_{22}a_{31} \\ &= a_{11}a_{22}a_{33} - a_{11}a_{23}a_{32} - a_{12}a_{21}a_{33} \\ &\quad + a_{12}a_{23}a_{31} + a_{13}a_{21}a_{32} - a_{13}a_{22}a_{31} \\ &= a_{11}(a_{22}a_{33} - a_{23}a_{32}) - a_{12}(a_{21}a_{33} - a_{23}a_{31}) + a_{13}(a_{21}a_{32} - a_{22}a_{31}) \\ &= a_{11}\begin{vmatrix} a_{22} & a_{23} \\ a_{32} & a_{33} \end{vmatrix} - a_{12}\begin{vmatrix} a_{21} & a_{23} \\ a_{31} & a_{33} \end{vmatrix} + a_{13}\begin{vmatrix} a_{21} & a_{22} \\ a_{31} & a_{32} \end{vmatrix}\end{aligned}$$

Example 1.

(*a*) $\begin{vmatrix} 1 & 2 \\ 3 & 4 \end{vmatrix} = 1\cdot 4 - 2\cdot 3 = 4 - 6 = -2$

(*b*) $\begin{vmatrix} 2 & -1 \\ 3 & 0 \end{vmatrix} = 2\cdot 0 - (-1)3 = 0 + 3 = 3$

(*c*) $\begin{vmatrix} 2 & 3 & 5 \\ 1 & 0 & 1 \\ 2 & 1 & 0 \end{vmatrix} = 2\begin{vmatrix} 0 & 1 \\ 1 & 0 \end{vmatrix} - 3\begin{vmatrix} 1 & 1 \\ 2 & 0 \end{vmatrix} + 5\begin{vmatrix} 1 & 0 \\ 2 & 1 \end{vmatrix}$

$= 2(0\cdot 0 - 1\cdot 1) - 3(1\cdot 0 - 1\cdot 2) + 5(1\cdot 1 - 0\cdot 2) = 2(-1) - 3(-2) + 5(1) = 9$

(*d*) $\begin{vmatrix} 2 & -3 & -4 \\ 1 & 0 & -2 \\ 0 & -5 & -6 \end{vmatrix} = 2\{0(-6) - (-2)(-5)\} - (-3)\{1(-6) - (-2)0\} + (-4)\{1(-5) - 0\cdot 0\}$

$= -20 - 18 + 20 = -18$

See Problem 1.

PROPERTIES OF DETERMINANTS. Throughout this section, A is the square matrix whose determinant $|A|$ is given by **(3.6)**.

Suppose that every element of the ith row (every element of the jth column) is zero. Since every term of **(3.6)** contains one element from this row (column), every term in the sum is zero and we have

I. If every element of a row (column) of a square matrix A is zero, then $|A| = 0$.

Consider the transpose A' of A. It can be seen readily that every term of **(3.6)** can be obtained from A' by choosing properly the factors in order from the first, second, ... columns. Thus,

II. If A is a square matrix then $|A'| = |A|$; that is, for every theorem concerning the rows of a determinant there is a corresponding theorem concerning the columns and vice versa.

Denote by B the matrix obtained by multiplying each of the elements of the ith row of A by a scalar k. Since each term in the expansion of $|B|$ contains one and only one element from its ith row, that is, one and only one element having k as a factor,

$$|B| = k \sum_{\rho} \{\epsilon_{j_1 j_2 \ldots j_n} a_{1j_1} a_{2j_2} \ldots a_{nj_n}\} = k\,|A|$$

Thus,

III. If every element of a row (column) of a determinant $|A|$ is multiplied by a scalar k, the determinant is multiplied by k; if every element of a row (column) of a determinant $|A|$ has k as a factor then k may be factored from $|A|$. For example,

$$\begin{vmatrix} a_{11} & ka_{12} & a_{13} \\ a_{21} & ka_{22} & a_{23} \\ a_{31} & ka_{32} & a_{33} \end{vmatrix} = k \begin{vmatrix} a_{11} & a_{12} & a_{13} \\ a_{21} & a_{22} & a_{23} \\ a_{31} & a_{32} & a_{33} \end{vmatrix} = \begin{vmatrix} a_{11} & a_{12} & a_{13} \\ a_{21} & a_{22} & a_{23} \\ ka_{31} & ka_{32} & ka_{33} \end{vmatrix}$$

Let B denote the matrix obtained from A by interchanging its ith and $(i+1)$st rows. Each product in **(3.6)** of $|A|$ is a product of $|B|$, and vice versa; hence, except possibly for signs, **(3.6)** is the expansion of $|B|$. In counting the inversions in subscripts of any term of **(3.6)** as a term of $|B|$, i before $i+1$ in the row subscripts is an inversion; thus, each product of **(3.6)** with its sign changed is a term of $|B|$ and $|B| = -|A|$. Hence,

IV. If B is obtained from A by interchanging any two adjacent rows (columns), then $|B| = -|A|$.

As a consequence of Theorem **IV**, we have

V. If B is obtained from A by interchanging **any** two of its rows (columns), then $|B| = -|A|$.

VI. If B is obtained from A by carrying its ith row (column) over p rows (columns), then $|B| = (-1)^p\,|A|$.

VII. If two rows (columns) of A are identical, then $|A| = 0$.

Suppose that each element of the first row of A is expressed as a binomial $a_{1j} = b_{1j} + c_{1j}$, $(j = 1, 2, \ldots, n)$. Then

$$\begin{aligned} |A| &= \sum_{\rho} \epsilon_{j_1 j_2 \ldots j_n} (b_{1j_1} + c_{1j_1}) a_{2j_2} a_{3j_3} \cdots a_{nj_n} \\ &= \sum_{\rho} \epsilon_{j_1 j_2 \ldots j_n} b_{1j_1} a_{2j_2} a_{3j_3} \cdots a_{nj_n} + \sum_{\rho} \epsilon_{j_1 j_2 \ldots j_n} c_{1j_1} a_{2j_2} a_{3j_3} \cdots a_{nj_n} \\ &= \begin{vmatrix} b_{11} & b_{12} & b_{13} & \cdots & b_{1n} \\ a_{21} & a_{22} & a_{23} & \cdots & a_{2n} \\ \cdots & \cdots & \cdots & \cdots & \cdots \\ a_{n1} & a_{n2} & a_{n3} & \cdots & a_{nn} \end{vmatrix} + \begin{vmatrix} c_{11} & c_{12} & c_{13} & \cdots & c_{1n} \\ a_{21} & a_{22} & a_{23} & \cdots & a_{2n} \\ \cdots & \cdots & \cdots & \cdots & \cdots \\ a_{n1} & a_{n2} & a_{n3} & \cdots & a_{nn} \end{vmatrix} \end{aligned}$$

In general,

VIII. If every element of the ith row (column) of A is the sum of p terms, then $|A|$ can be expressed as the sum of p determinants. The elements in the ith rows (columns) of these p determinants are respectively the first, second, ..., pth terms of the sums and all other rows (columns) are those of A.

The most useful theorem is

IX. If B is obtained from A by adding to the elements of its ith row (column), a scalar multiple of the corresponding elements of another row (column), then $|B| = |A|$. For example,

$$\begin{vmatrix} a_{11} & a_{12} & a_{13} \\ a_{21} & a_{22} & a_{23} \\ a_{31} & a_{32} & a_{33} \end{vmatrix} = \begin{vmatrix} a_{11}+ka_{13} & a_{12} & a_{13} \\ a_{21}+ka_{23} & a_{22} & a_{23} \\ a_{31}+ka_{33} & a_{32} & a_{33} \end{vmatrix} = \begin{vmatrix} a_{11} & a_{12} & a_{13} \\ a_{21} & a_{22} & a_{23} \\ a_{31}+ka_{21} & a_{32}+ka_{22} & a_{33}+ka_{23} \end{vmatrix}$$

See Problems 2-7.

FIRST MINORS AND COFACTORS. Let A be the n-square matrix **(3.3)** whose determinant $|A|$ is given by **(3.6)**. When from A the elements of its ith row and jth column are removed, the determinant of the remaining $(n-1)$-square matrix is called a **first minor** of A or of $|A|$ and denoted by $|M_{ij}|$.

More frequently, it is called **the minor** of a_{ij}. The signed minor, $(-1)^{i+j}|M_{ij}|$ is called the **cofactor** of a_{ij} and is denoted by α_{ij}.

Example 2. If $A = \begin{vmatrix} a_{11} & a_{12} & a_{13} \\ a_{21} & a_{22} & a_{23} \\ a_{31} & a_{32} & a_{33} \end{vmatrix}$,

$$|M_{11}| = \begin{vmatrix} a_{22} & a_{23} \\ a_{32} & a_{33} \end{vmatrix}, \quad |M_{12}| = \begin{vmatrix} a_{21} & a_{23} \\ a_{31} & a_{33} \end{vmatrix}, \quad |M_{13}| = \begin{vmatrix} a_{21} & a_{22} \\ a_{31} & a_{32} \end{vmatrix}$$

and

$$\alpha_{11} = (-1)^{1+1}|M_{11}| = |M_{11}|, \quad \alpha_{12} = (-1)^{1+2}|M_{12}| = -|M_{12}|,$$

$$\alpha_{13} = (-1)^{1+3}|M_{13}| = |M_{13}|$$

Then **(3.8)** is

$$\begin{aligned} |A| &= a_{11}|M_{11}| - a_{12}|M_{12}| + a_{13}|M_{13}| \\ &= a_{11}\alpha_{11} + a_{12}\alpha_{12} + a_{13}\alpha_{13} \end{aligned}$$

In Problem 9, we prove

X. The value of the determinant $|A|$, where A is the matrix of **(3.3)**, is the sum of the products obtained by multiplying each element of a row (column) of $|A|$ by its cofactor, i.e.,

$$\textbf{(3.9)} \qquad |A| = a_{i1}\alpha_{i1} + a_{i2}\alpha_{i2} + \cdots + a_{in}\alpha_{in} = \sum_{k=1}^{n} a_{ik}\alpha_{ik}$$

$$\textbf{(3.10)} \qquad |A| = a_{1j}\alpha_{1j} + a_{2j}\alpha_{2j} + \cdots + a_{nj}\alpha_{nj} = \sum_{k=1}^{n} a_{kj}\alpha_{kj} \qquad (i, j, = 1, 2, \ldots, n)$$

Using Theorem **VII**, we can prove

XI. The sum of the products formed by multiplying the elements of a row (column) of an n-square matrix A by the corresponding cofactors of **another** row (column) of A is zero.

Example 3. If A is the matrix of Example 2, we have

$$a_{31}\alpha_{31} + a_{32}\alpha_{32} + a_{33}\alpha_{33} = |A|$$

and

$$a_{12}\alpha_{12} + a_{22}\alpha_{22} + a_{32}\alpha_{32} = |A|$$

while

$$a_{31}\alpha_{21} + a_{32}\alpha_{22} + a_{33}\alpha_{23} = 0$$

and

$$a_{12}\alpha_{13} + a_{22}\alpha_{23} + a_{32}\alpha_{33} = 0$$

See Problems 10-11.

MINORS AND ALGEBRAIC COMPLEMENTS. Consider the matrix **(3.3)**. Let $i_1, i_2, \ldots, i_m$, arranged in order of magnitude, be m, $(1 \leq m < n)$, of the row indices $1, 2, \ldots, n$ and let $j_1, j_2, \ldots, j_m$ arranged in order of magnitude, be m of the column indices. Let the remaining row and column indices, arranged in order of magnitude, be respectively $i_{m+1}, i_{m+2}, \ldots, i_n$ and $j_{m+1}, j_{m+2}, \ldots, j_n$. Such a separation of the row and column indices determines uniquely two matrices

$$\textbf{(3.11)} \qquad A^{j_1, j_2, \ldots, j_m}_{i_1, i_2, \ldots, i_m} = \begin{bmatrix} a_{i_1, j_1} & a_{i_1, j_2} & \cdots & a_{i_1, j_m} \\ a_{i_2, j_1} & a_{i_2, j_2} & \cdots & a_{i_2, j_m} \\ \cdots & \cdots & \cdots & \cdots \\ a_{i_m, j_1} & a_{i_m, j_2} & \cdots & a_{i_m, j_m} \end{bmatrix}$$

and

$$(3.12) \qquad A^{j_{m+1},j_{m+2},\ldots,j_n}_{i_{m+1},i_{m+2},\ldots,i_n} = \begin{bmatrix} a_{i_{m+1},j_{m+1}} & a_{i_{m+1},j_{m+2}} & \cdots & a_{i_{m+1},j_n} \\ a_{i_{m+2},j_{m+1}} & a_{i_{m+2},j_{m+2}} & \cdots & a_{i_{m+2},j_n} \\ \cdots & \cdots & \cdots & \cdots \\ a_{i_n,j_{m+1}} & a_{i_n,j_{m+2}} & \cdots & a_{i_n,j_n} \end{bmatrix}$$

called sub-matrices of A.

The determinant of each of these sub-matrices is called a **minor** of A and the pair of minors

$$\left| A^{j_1,j_2,\ldots,j_m}_{i_1,i_2,\ldots,i_m} \right| \quad \text{and} \quad \left| A^{j_{m+1},j_{m+2},\ldots,j_n}_{i_{m+1},i_{m+2},\ldots,i_n} \right|$$

are called **complementary minors** of A, each being the complement of the other.

Example 3. For the 5-square matrix $A = \left[a_{ij} \right]$,

$$\left| A^{1,3}_{2,5} \right| = \begin{vmatrix} a_{21} & a_{23} \\ a_{51} & a_{53} \end{vmatrix} \quad \text{and} \quad \left| A^{2,4,5}_{1,3,4} \right| = \begin{vmatrix} a_{12} & a_{14} & a_{15} \\ a_{32} & a_{34} & a_{35} \\ a_{42} & a_{44} & a_{45} \end{vmatrix}$$

are a pair of complementary minors.

Let

$$(3.13) \qquad p = i_1 + i_2 + \cdots + i_m + j_1 + j_2 + \cdots + j_m$$

and

$$(3.14) \qquad q = i_{m+1} + i_{m+2} + \cdots + i_n + j_{m+1} + j_{m+2} + \cdots + j_n$$

The signed minor $(-1)^p \left| A^{j_1,j_2,\ldots,j_m}_{i_1,i_2,\ldots,i_m} \right|$ is called the **algebraic complement** of $\left| A^{j_{m+1},j_{m+2},\ldots,j_n}_{i_{m+1},i_{m+2},\ldots,i_n} \right|$

and $(-1)^q \left| A^{j_{m+1},j_{m+2},\ldots,j_n}_{i_{m+1},i_{m+2},\ldots,i_n} \right|$ is called the **algebraic complement** of $\left| A^{j_1,j_2,\ldots,j_m}_{i_1,i_2,\ldots,i_m} \right|$

Example 4. For the minors of Example 3, $(-1)^{2+5+1+3} \left| A^{1,3}_{2,5} \right| = -\left| A^{1,3}_{2,5} \right|$ is the algebraic complement of $\left| A^{2,4,5}_{1,3,4} \right|$ and $(-1)^{1+3+4+2+4+5} \left| A^{2,4,5}_{1,3,4} \right| = -\left| A^{2,4,5}_{1,3,4} \right|$ is the algebraic complement of $\left| A^{1,3}_{2,5} \right|$. Note that the sign given to the two complementary minors is the same. Is this always true?

When $m=1$, **(3.11)** becomes $A^{j_1}_{i_1} = \left[a_{i_1 j_1} \right]$ and $\left| A^{j_1}_{i_1} \right| = a_{i_1 j_1}$, an element of A. The complementary minor $\left| A^{j_2,j_3,\ldots,j_n}_{i_2,i_3,\ldots,i_n} \right|$ is $\left| M_{i_1,j_1} \right|$ in the notation of the section above, and the algebraic complement is the cofactor $\alpha_{i_1 j_1}$.

A minor of A, whose diagonal elements are also diagonal elements of A, is called a **principal minor** of A. The complement of a principal minor of A is also a principal minor of A; the algebraic complement of a principal minor is its complement.

Example 5. For the 5-square matrix $A = [a_{ij}]$,

$$|A^{1,3}_{1,3}| = \begin{vmatrix} a_{11} & a_{13} \\ a_{31} & a_{33} \end{vmatrix} \quad \text{and} \quad |A^{2,4,5}_{2,4,5}| = \begin{vmatrix} a_{22} & a_{24} & a_{25} \\ a_{42} & a_{44} & a_{45} \\ a_{52} & a_{54} & a_{55} \end{vmatrix}$$

are a pair of complementary principal minors of A. What is the algebraic complement of each?

The terms minor, complementary minor, algebraic complement, and principal minor as defined above for a square matrix A will also be used without change in connection with $|A|$.

See Problems 12-13.

SOLVED PROBLEMS

1. (a) $\begin{vmatrix} 2 & 3 \\ -1 & 4 \end{vmatrix} = 2\cdot4 - 3(-1) = 11$

(b) $\begin{vmatrix} 1 & 0 & 2 \\ 3 & 4 & 5 \\ 5 & 6 & 7 \end{vmatrix} = (1)\begin{vmatrix} 4 & 5 \\ 6 & 7 \end{vmatrix} - 0\begin{vmatrix} 3 & 5 \\ 5 & 7 \end{vmatrix} + 2\begin{vmatrix} 3 & 4 \\ 5 & 6 \end{vmatrix} = (1)(4\cdot7 - 5\cdot6) - 0 + 2(3\cdot6 - 4\cdot5)$
$= -2 - 4 = -6$

(c) $\begin{vmatrix} 1 & 0 & 6 \\ 3 & 4 & 15 \\ 5 & 6 & 21 \end{vmatrix} = 1(4\cdot21 - 15\cdot6) + 6(3\cdot6 - 4\cdot5) = -18$

(d) $\begin{vmatrix} 1 & 0 & 0 \\ 2 & 3 & 5 \\ 4 & 1 & 3 \end{vmatrix} = 1(3\cdot3 - 5\cdot1) = 4$

2. Adding to the elements of the first column the corresponding elements of the other columns,

$$\begin{vmatrix} -4 & 1 & 1 & 1 & 1 \\ 1 & -4 & 1 & 1 & 1 \\ 1 & 1 & -4 & 1 & 1 \\ 1 & 1 & 1 & -4 & 1 \\ 1 & 1 & 1 & 1 & -4 \end{vmatrix} = \begin{vmatrix} 0 & 1 & 1 & 1 & 1 \\ 0 & -4 & 1 & 1 & 1 \\ 0 & 1 & -4 & 1 & 1 \\ 0 & 1 & 1 & -4 & 1 \\ 0 & 1 & 1 & 1 & -4 \end{vmatrix} = 0$$

by Theorem **I**.

3. Adding the second column to the third, removing the common factor from this third column, and using Theorem **VII**

$$\begin{vmatrix} 1 & a & b+c \\ 1 & b & c+a \\ 1 & c & a+b \end{vmatrix} = \begin{vmatrix} 1 & a & a+b+c \\ 1 & b & a+b+c \\ 1 & c & a+b+c \end{vmatrix} = (a+b+c)\begin{vmatrix} 1 & a & 1 \\ 1 & b & 1 \\ 1 & c & 1 \end{vmatrix} = 0$$

4. Adding to the third row the first and second rows, then removing the common factor 2; subtracting the second row from the third; subtracting the third row from the first; subtracting the first row from the second; finally carrying the third row over the other rows

$$\begin{vmatrix} a_1+b_1 & a_2+b_2 & a_3+b_3 \\ b_1+c_1 & b_2+c_2 & b_3+c_3 \\ c_1+a_1 & c_2+a_2 & c_3+a_3 \end{vmatrix} = 2\begin{vmatrix} a_1+b_1 & a_2+b_2 & a_3+b_3 \\ b_1+c_1 & b_2+c_2 & b_3+c_3 \\ a_1+b_1+c_1 & a_2+b_2+c_2 & a_3+b_3+c_3 \end{vmatrix} = 2\begin{vmatrix} a_1+b_1 & a_2+b_2 & a_3+b_3 \\ b_1+c_1 & b_2+c_2 & b_3+c_3 \\ a_1 & a_2 & a_3 \end{vmatrix}$$

$$= 2\begin{vmatrix} b_1 & b_2 & b_3 \\ b_1+c_1 & b_2+c_2 & b_3+c_3 \\ a_1 & a_2 & a_3 \end{vmatrix} = 2\begin{vmatrix} b_1 & b_2 & b_3 \\ c_1 & c_2 & c_3 \\ a_1 & a_2 & a_3 \end{vmatrix} = 2\begin{vmatrix} a_1 & a_2 & a_3 \\ b_1 & b_2 & b_3 \\ c_1 & c_2 & c_3 \end{vmatrix}$$

5. Without expanding, show that $|A| = \begin{vmatrix} a_1^2 & a_1 & 1 \\ a_2^2 & a_2 & 1 \\ a_3^2 & a_3 & 1 \end{vmatrix} = -(a_1-a_2)(a_2-a_3)(a_3-a_1)$.

Subtract the second row from the first; then

$$|A| = \begin{vmatrix} a_1^2-a_2^2 & a_1-a_2 & 0 \\ a_2^2 & a_2 & 1 \\ a_3^2 & a_3 & 1 \end{vmatrix} = (a_1-a_2)\begin{vmatrix} a_1+a_2 & 1 & 0 \\ a_2^2 & a_2 & 1 \\ a_3^2 & a_3 & 1 \end{vmatrix} \quad \text{by Theorem III}$$

and a_1-a_2 is a factor of $|A|$. Similarly, a_2-a_3 and a_3-a_1 are factors. Now $|A|$ is of order three in the letters; hence,

(*i*) $$|A| = k(a_1-a_2)(a_2-a_3)(a_3-a_1)$$

The product of the diagonal elements, $a_1^2a_2$, is a term of $|A|$ and, from (i), the term is $-ka_1^2a_2$. Thus, $k=-1$ and $|A| = -(a_1-a_2)(a_2-a_3)(a_3-a_1)$. Note that $|A|$ vanishes if and only if two of the a_1, a_2, a_3 are equal.

6. Prove: If A is skew-symmetric and of odd order $2p-1$, then $|A| = 0$.

Since A is skew-symmetric, $A' = -A$; then $|A'| = |-A| = (-1)^{2p-1}|A| = -|A|$. But, by Theorem II, $|A'| = |A|$; hence, $|A| = -|A|$ and $|A| = 0$.

7. Prove: If A is Hermitian, then $|A|$ is a real number.

Since A is Hermitian, $\bar{A} = A'$, and $|\bar{A}| = |A'| = |A|$ by Theorem II. But if

$$|A| = \sum_{\rho} \epsilon_{j_1 j_2 \dots j_n} a_{1j_1} a_{2j_2} \dots a_{nj_n} = a + bi$$

then

$$|\bar{A}| = \sum_{\rho} \epsilon_{j_1 j_2 \dots j_n} \bar{a}_{1j_1} \bar{a}_{2j_2} \dots \bar{a}_{nj_n} = a - bi$$

Now $|\bar{A}| = |A|$ requires $b = 0$; hence, $|A|$ is a real number.

8. For the matrix $A = \begin{bmatrix} 1 & 2 & 3 \\ 2 & 3 & 2 \\ 1 & 2 & 2 \end{bmatrix}$,

$$\alpha_{11} = (-1)^{1+1}\begin{vmatrix} 3 & 2 \\ 2 & 2 \end{vmatrix} = 2, \quad \alpha_{12} = (-1)^{1+2}\begin{vmatrix} 2 & 2 \\ 1 & 2 \end{vmatrix} = -2, \quad \alpha_{13} = (-1)^{1+3}\begin{vmatrix} 2 & 3 \\ 1 & 2 \end{vmatrix} = 1$$

$$\alpha_{21} = (-1)^{2+1}\begin{vmatrix} 2 & 3 \\ 2 & 2 \end{vmatrix} = 2, \quad \alpha_{22} = (-1)^{2+2}\begin{vmatrix} 1 & 3 \\ 1 & 2 \end{vmatrix} = -1, \quad \alpha_{23} = (-1)^{2+3}\begin{vmatrix} 1 & 2 \\ 1 & 2 \end{vmatrix} = 0$$

$$\alpha_{31} = (-1)^{3+1}\begin{vmatrix} 2 & 3 \\ 3 & 2 \end{vmatrix} = -5, \quad \alpha_{32} = (-1)^{3+2}\begin{vmatrix} 1 & 3 \\ 2 & 2 \end{vmatrix} = 4, \quad \alpha_{33} = (-1)^{3+3}\begin{vmatrix} 1 & 2 \\ 2 & 3 \end{vmatrix} = -1$$

Note that the signs given to the minors of the elements in forming the cofactors follow the pattern

$$\begin{matrix} + & - & + \\ - & + & - \\ + & - & + \end{matrix}$$

where each sign occupies the same position in the display as the element, whose cofactor is required, occupies in A. Write the display of signs for a 5-square matrix.

. Prove: The value of the determinant $|A|$ of an n-square matrix A is the sum of the products obtained by multiplying each element of a row (column) of A by its cofactor.

We shall prove this for a row. The terms of (3.6) having a_{11} as a factor are

$$(a) \qquad a_{11} \sum \epsilon_{1, j_2 j_3 \ldots j_n} a_{2j_2} a_{3j_3} \ldots a_{nj_n}$$

Now $\epsilon_{1, j_2 j_3 \ldots j_n} = \epsilon_{j_2 j_3 \ldots j_n}$ since in a permutation $1, j_1, j_2, \ldots j_n$, the 1 is in natural order. Then (a) may be written as

$$(b) \qquad a_{11} \sum_{\sigma} \epsilon_{j_2 j_3 \ldots j_n} a_{2j_2} a_{3j_3} \ldots a_{nj_n}$$

where the summation extends over the $\sigma = (n-1)!$ permutations of the integers $2, 3, \ldots, n$, and hence, as

$$(c) \qquad a_{11} \begin{vmatrix} a_{22} & a_{23} & \cdots & a_{2n} \\ a_{32} & a_{33} & \cdots & a_{3n} \\ \cdots & \cdots & \cdots & \cdots \\ a_{n2} & a_{n3} & \cdots & a_{nn} \end{vmatrix} = a_{11}|M_{11}|$$

Consider the matrix B obtained from A by moving its sth column over the first $s-1$ columns. By Theorem VI, $|B| = (-1)^{s-1}|A|$. Moreover, the element standing in the first row and first column of B is a_{1s} and the minor of a_{1s} in B is precisely the minor $|M_{1s}|$ of a_{1s} in A. By the argument leading to (c), the terms of $a_{1s}|M_{1s}|$ are all the terms of $|B|$ having a_{1s} as a factor and, thus, all the terms of $(-1)^{s-1}|A|$ having a_{1s} as a factor. Then the terms of $a_{1s}\{(-1)^{s-1}|M_{1s}|\}$ are all the terms of $|A|$ having a_{1s} as a factor. Thus,

$$(3.15) \qquad \begin{aligned} |A| &= a_{11}\{(-1)^{1+1}|M_{11}|\} + a_{12}\{(-1)^{1+2}|M_{12}|\} \\ &\quad + \cdots + a_{1s}\{(-1)^{1+s}|M_{1s}|\} + \cdots + a_{1n}\{(-1)^{1+n}|M_{1n}|\} \\ &= a_{11}\alpha_{11} + a_{12}\alpha_{12} + \cdots + a_{1n}\alpha_{1n} \end{aligned}$$

since $(-1)^{s-1} = (-1)^{s+1}$. We have (3.9) with $i = 1$. We shall call (3.15) the expansion of $|A|$ along its first row.

The expansion of $|A|$ along its rth row (that is, (3.9) for $i = r$) is obtained by repeating the above arguments. Let B be the matrix obtained from A by moving its rth row over the first $r-1$ rows and then its sth column over the first $s-1$ columns. Then

$$|B| = (-1)^{r-1} \cdot (-1)^{s-1}|A| = (-1)^{r+s}|A|$$

The element standing in the first row and the first column of B is a_{rs} and the minor of a_{rs} in B is precisely the minor of a_{rs} in A. Thus, the terms of

$$a_{rs}\{(-1)^{r+s}|M_{rs}|\}$$

are all the terms of $|A|$ having a_{rs} as a factor. Then

$$|A| = \sum_{k=1}^{n} a_{rk}\{(-1)^{r+k}|M_{rk}|\} = \sum_{k=1}^{n} a_{rk}\alpha_{rk}$$

and we have (3.9) for $i = r$.

10. When α_{ij} is the cofactor of a_{ij} in the n-square matrix $A = [a_{ij}]$, show that

$$(i) \qquad k_1\alpha_{1j} + k_2\alpha_{2j} + \cdots + k_n\alpha_{nj} = \begin{vmatrix} a_{11} & a_{12} & \cdots & a_{1,j-1} & k_1 & a_{1,j+1} & \cdots & a_{1n} \\ a_{21} & a_{22} & \cdots & a_{2,j-1} & k_2 & a_{2,j+1} & \cdots & a_{2n} \\ \cdots & \cdots & \cdots & \cdots & \cdots & \cdots & \cdots & \cdots \\ a_{n1} & a_{n2} & \cdots & a_{n,j-1} & k_n & a_{n,j+1} & \cdots & a_{nn} \end{vmatrix}$$

This relation follows from **(3.10)** by replacing a_{1j} with k_1, a_{2j} with k_2, ..., a_{nj} with k_n. In making these replacements none of the cofactors $\alpha_{1j}, \alpha_{2j}, \ldots, \alpha_{nj}$ appearing is affected since none contains an element from the jth column of A.

By Theorem **VII**, the determinant in (i) is 0 when $k_r = a_{rs}$, $(r = 1, 2, \ldots, n$ and $s \neq j)$. By Theorems **VIII**, and **VII**, the determinant in (i) is $|A|$ when $k_r = a_{rj} + k a_{rs}$, $(r = 1, 2, \ldots, n$ and $s \neq j)$.

Write the equality similar to (i) obtained from **(3.9)** when the elements of the ith row of A are replaced by $k_1, k_2, \ldots, k_n$.

11. Evaluate: $(a)\ |A| = \begin{vmatrix} 1 & 0 & 2 \\ 3 & 0 & 4 \\ 2 & -5 & 1 \end{vmatrix}$ $\quad (c)\ |A| = \begin{vmatrix} 3 & 4 & 5 \\ 1 & 2 & 3 \\ -2 & 5 & -4 \end{vmatrix}$ $\quad (e)\ |A| = \begin{vmatrix} 28 & 25 & 38 \\ 42 & 38 & 65 \\ 56 & 47 & 83 \end{vmatrix}$

$(b)\ |A| = \begin{vmatrix} 1 & 4 & 8 \\ -2 & 1 & 5 \\ -3 & 2 & 4 \end{vmatrix}$ $\quad (d)\ |A| = \begin{vmatrix} 2 & 3 & -4 \\ 5 & -6 & 3 \\ 4 & 2 & -3 \end{vmatrix}$

(a) Expanding along the second column (see Theorem **X**)

$$|A| = \begin{vmatrix} 1 & 0 & 2 \\ 3 & 0 & 4 \\ 2 & -5 & 1 \end{vmatrix} = a_{12}\alpha_{12} + a_{22}\alpha_{22} + a_{32}\alpha_{32} = 0\cdot\alpha_{12} + 0\cdot\alpha_{22} + (-5)\alpha_{32}$$

$$= -5(-1)^{3+2}\begin{vmatrix} 1 & 2 \\ 3 & 4 \end{vmatrix} = 5(4-6) = -10$$

(b) Subtracting twice the second column from the third (see Theorem **IX**)

$$|A| = \begin{vmatrix} 1 & 4 & 8 \\ -2 & 1 & 5 \\ -3 & 2 & 4 \end{vmatrix} = \begin{vmatrix} 1 & 4 & 8-2\cdot4 \\ -2 & 1 & 5-2\cdot1 \\ -3 & 2 & 4-2\cdot2 \end{vmatrix} = \begin{vmatrix} 1 & 4 & 0 \\ -2 & 1 & 3 \\ -3 & 2 & 0 \end{vmatrix} = 3(-1)^{2+3}\begin{vmatrix} 1 & 4 \\ -3 & 2 \end{vmatrix}$$

$$= -3(14) = -42$$

(c) Subtracting three times the second row from the first and adding twice the second row to the third

$$|A| = \begin{vmatrix} 3 & 4 & 5 \\ 1 & 2 & 3 \\ -2 & 5 & -4 \end{vmatrix} = \begin{vmatrix} 3-3(1) & 4-3(2) & 5-3(3) \\ 1 & 2 & 3 \\ -2+2(1) & 5+2(2) & -4+2(3) \end{vmatrix} = \begin{vmatrix} 0 & -2 & -4 \\ 1 & 2 & 3 \\ 0 & 9 & 2 \end{vmatrix} = -\begin{vmatrix} -2 & -4 \\ 9 & 2 \end{vmatrix}$$

$$= -(-4+36) = -32$$

(d) Subtracting the first column from the second and then proceeding as in (c)

$$|A| = \begin{vmatrix} 2 & 3 & -4 \\ 5 & -6 & 3 \\ 4 & 2 & -3 \end{vmatrix} = \begin{vmatrix} 2 & 1 & -4 \\ 5 & -11 & 3 \\ 4 & -2 & -3 \end{vmatrix} = \begin{vmatrix} 2-2(1) & 1 & -4+4(1) \\ 5-2(-11) & -11 & 3+4(-11) \\ 4-2(-2) & -2 & -3+4(-2) \end{vmatrix}$$

$$= \begin{vmatrix} 0 & 1 & 0 \\ 27 & -11 & -41 \\ 8 & -2 & -11 \end{vmatrix} = -\begin{vmatrix} 27 & -41 \\ 8 & -11 \end{vmatrix} = -31$$

(*e*) Factoring 14 from the first column, then using Theorem **IX** to reduce the elements in the remaining columns

$$|A| = \begin{vmatrix} 28 & 25 & 38 \\ 42 & 38 & 65 \\ 56 & 47 & 83 \end{vmatrix} = 14\begin{vmatrix} 2 & 25 & 38 \\ 3 & 38 & 65 \\ 4 & 47 & 83 \end{vmatrix} = 14\begin{vmatrix} 2 & 25-12(2) & 38-20(2) \\ 3 & 38-12(3) & 65-20(3) \\ 4 & 47-12(4) & 83-20(4) \end{vmatrix}$$

$$= 14\begin{vmatrix} 2 & 1 & -2 \\ 3 & 2 & 5 \\ 4 & -1 & 3 \end{vmatrix} = 14\begin{vmatrix} 0 & 1 & 0 \\ -1 & 2 & 9 \\ 6 & -1 & 1 \end{vmatrix} = -14\begin{vmatrix} -1 & 9 \\ 6 & 1 \end{vmatrix} = -14(-1-54) = 770$$

12. Show that p and q, given by **(3.13)** and **(3.14)**, are either both even or both odd.

Since each row (column) index is found in either p or q but never in both,

$$p+q = (1+2+\cdots+n) + (1+2+\cdots+n) = 2\cdot\tfrac{1}{2}n(n+1) = n(n+1)$$

Now $p+q$ is even (either n or $n+1$ is even); hence, p and q are either both even or both odd. Thus, $(-1)^p = (-1)^q$ and only one need be computed.

13. For the matrix $A = [a_{ij}] = \begin{bmatrix} 1 & 2 & 3 & 4 & 5 \\ 6 & 7 & 8 & 9 & 10 \\ 11 & 12 & 13 & 14 & 15 \\ 16 & 17 & 18 & 19 & 20 \\ 21 & 22 & 23 & 24 & 25 \end{bmatrix}$, the algebraic complement of $|A_{2,3}^{2,4}|$ is

$$(-1)^{2+3+2+4}|A_{1,4,5}^{1,3,5}| = -\begin{vmatrix} 1 & 3 & 5 \\ 16 & 18 & 20 \\ 21 & 23 & 25 \end{vmatrix} \qquad \text{(see Problem 12)}$$

and the algebraic complement of $|A_{1,4,5}^{1,3,5}|$ is $-|A_{2,3}^{2,4}| = -\begin{vmatrix} 7 & 9 \\ 12 & 14 \end{vmatrix}$.

SUPPLEMENTARY PROBLEMS

14. Show that the permutation 12534 of the integers 1, 2, 3, 4, 5 is even, 24135 is odd, 41532 is even, 53142 is odd, and 52314 is even.

15. List the complete set of permutations of 1, 2, 3, 4, taken together; show that half are even and half are odd.

16. Let the elements of the diagonal of a 5-square matrix A be a, b, c, d, e. Show, using **(3.6)**, that when A is diagonal, upper triangular, or lower triangular then $|A| = abcde$.

17. Given $A = \begin{bmatrix} 1 & 2 \\ 3 & 4 \end{bmatrix}$ and $B = \begin{bmatrix} 5 & 6 \\ 7 & 8 \end{bmatrix}$, show that $AB \neq BA \neq A'B \neq AB' \neq A'B' \neq B'A'$ but that the determinant of each product is 4.

18. Evaluate, as in Problem 1,

(*a*) $\begin{vmatrix} 2 & -1 & 1 \\ 3 & 2 & 4 \\ -1 & 0 & 3 \end{vmatrix} = 27$ (*b*) $\begin{vmatrix} 2 & 2 & -2 \\ 1 & 2 & 3 \\ 2 & 3 & 4 \end{vmatrix} = 4$ (*c*) $\begin{vmatrix} 0 & 2 & 3 \\ -2 & 0 & 4 \\ -3 & -4 & 0 \end{vmatrix} = 0$

19. (*a*) Evaluate $|A| = \begin{vmatrix} 1 & 2 & 10 \\ 2 & 3 & 9 \\ 4 & 5 & 11 \end{vmatrix} = -4$.

(*b*) Denote by $|B|$ the determinant obtained from $|A|$ by multiplying the elements of its second column by 5. Evaluate $|B|$ to verify Theorem **III**.

(*c*) Denote by $|C|$ the determinant obtained from $|A|$ by interchanging its first and third columns. Evaluate $|C|$ to verify Theorem **V**.

(*d*) Show that $|A| = \begin{vmatrix} 1 & 2 & 7 \\ 2 & 3 & 5 \\ 4 & 5 & 8 \end{vmatrix} + \begin{vmatrix} 1 & 2 & 3 \\ 2 & 3 & 4 \\ 4 & 5 & 3 \end{vmatrix}$, thus verifying Theorem **VIII**.

(*e*) Obtain from $|A|$ the determinant $|D| = \begin{vmatrix} 1 & 2 & 7 \\ 2 & 3 & 3 \\ 4 & 5 & -1 \end{vmatrix}$ by subtracting three times the elements of the first column from the corresponding elements of the third column. Evaluate $|D|$ to verify Theorem **IX**.

(*f*) In $|A|$ subtract twice the first row from the second and four times the first row from the third. Evaluate the resulting determinant.

(*g*) In $|A|$ multiply the first column by three and from it subtract the third column. Evaluate to show that $|A|$ has been tripled. Compare with (*e*). Do not confuse (*e*) and (*g*).

20. If A is an n-square matrix and k is a scalar, use (**3.6**) to show that $|kA| = k^n|A|$.

21. Prove: (*a*) If $|A| = k$, then $|\bar{A}| = \bar{k} = |\bar{A}'|$.

(*b*) If A is skew-Hermitian, then $|A|$ is either real or is a pure imaginary number.

22. (*a*) Count the number of interchanges of adjacent rows (columns) necessary to obtain B from A in Theorem **V** and thus prove the theorem.

(*b*) Same, for Theorem **VI**.

23. Prove Theorem **VII**. Hint: Interchange the identical rows and use Theorem **V**.

24. Prove: If any two rows (columns) of a square matrix A are proportional, then $|A| = 0$.

25. Use Theorems **VIII**, **III**, and **VII** to prove Theorem **IX**.

26. Evaluate the determinants of Problem 18 as in Problem 11.

27. Use (**3.6**) to evaluate $|A| = \begin{vmatrix} a & b & 0 & 0 \\ c & d & 0 & 0 \\ 0 & 0 & e & f \\ 0 & 0 & g & h \end{vmatrix}$; then check that $|A| = \begin{vmatrix} a & b \\ c & d \end{vmatrix} \begin{vmatrix} e & f \\ g & h \end{vmatrix}$. Thus, if $A = \mathrm{diag}(A_1, A_2)$, where A_1, A_2 are 2-square matrices, $|A| = |A_1| \cdot |A_2|$.

28. Show that the cofactor of each element of $\begin{bmatrix} -1/3 & -2/3 & -2/3 \\ 2/3 & 1/3 & -2/3 \\ 2/3 & -2/3 & 1/3 \end{bmatrix}$ is that element.

29. Show that the cofactor of an element of any row of $\begin{bmatrix} -4 & -3 & -3 \\ 1 & 0 & 1 \\ 4 & 4 & 3 \end{bmatrix}$ is the corresponding element of the same numbered column.

30. Prove: (*a*) If A is symmetric then $\alpha_{ij} = \alpha_{ji}$ when $i \neq j$.

(*b*) If A is n-square and skew-symmetric then $\alpha_{ij} = (-1)^{n-1}\alpha_{ji}$ when $i \neq j$.

31. For the matrix A of Problem 8;
(*a*) show that $|A| = 1$

(*b*) form the matrix $C = \begin{bmatrix} \alpha_{11} & \alpha_{21} & \alpha_{31} \\ \alpha_{12} & \alpha_{22} & \alpha_{32} \\ \alpha_{13} & \alpha_{23} & \alpha_{33} \end{bmatrix}$ and show that $AC = I$.

(*c*) explain why the result in (*b*) is known as soon as (*a*) is known.

32. Multiply the columns of $|A| = \begin{vmatrix} bc & a^2 & a^2 \\ b^2 & ca & b^2 \\ c^2 & c^2 & ab \end{vmatrix}$ respectively by a, b, c; remove the common factor from each of the rows to show that $|A| = \begin{vmatrix} bc & ab & ca \\ ab & ca & bc \\ ca & bc & ab \end{vmatrix}$.

33. Without evaluating show that $\begin{vmatrix} a^2 & a & 1 & bcd \\ b^2 & b & 1 & acd \\ c^2 & c & 1 & abd \\ d^2 & d & 1 & abc \end{vmatrix} = \begin{vmatrix} a^3 & a^2 & a & 1 \\ b^3 & b^2 & b & 1 \\ c^3 & c^2 & c & 1 \\ d^3 & d^2 & d & 1 \end{vmatrix} = (a-b)(a-c)(a-d)(b-c)(b-d)(c-d)$.

34. Show that the n-square determinant $|A| = \begin{vmatrix} 0 & 1 & 1 & \ldots & 1 \\ 1 & 0 & 1 & \ldots & 1 \\ 1 & 1 & 0 & \ldots & 1 \\ \ldots & & & & \\ \ldots & & & & \\ 1 & 1 & 1 & \ldots & 0 \end{vmatrix} = (n-1) \begin{vmatrix} 0 & 1 & 1 & \ldots & 1 & 1 \\ 1 & 0 & 1 & \ldots & 1 & 1 \\ 1 & 1 & 0 & \ldots & 1 & 1 \\ \ldots & & & & & \\ 1 & 1 & 1 & \ldots & 0 & 1 \\ 1 & 1 & 1 & \ldots & 1 & 1 \end{vmatrix} = (-1)^{n-1}(n-1)$.

35. Prove: $\begin{vmatrix} a_1^{n-1} & a_1^{n-2} & \ldots & a_1 & 1 \\ a_2^{n-1} & a_2^{n-2} & \ldots & a_2 & 1 \\ \ldots & & & & \\ a_n^{n-1} & a_n^{n-2} & \ldots & a_n & 1 \end{vmatrix} = \{(a_1-a_2)(a_1-a_3)\ldots(a_1-a_n)\}\{(a_2-a_3)(a_2-a_4)\ldots(a_2-a_n)\} \cdots \{a_{n-1}-a_n\}$.

36. Without expanding, show that $\begin{vmatrix} na_1+b_1 & na_2+b_2 & na_3+b_3 \\ nb_1+c_1 & nb_2+c_2 & nb_3+c_3 \\ nc_1+a_1 & nc_2+a_2 & nc_3+a_3 \end{vmatrix} = (n+1)(n^2-n+1) \begin{vmatrix} a_1 & a_2 & a_3 \\ b_1 & b_2 & b_3 \\ c_1 & c_2 & c_3 \end{vmatrix}$.

37. Without expanding, show that the equation $\begin{vmatrix} 0 & x-a & x-b \\ x+a & 0 & x-c \\ x+b & x+c & 0 \end{vmatrix} = 0$ has 0 as a root.

38. Prove $\begin{vmatrix} a+b & a & a & \cdots & a \\ a & a+b & a & \cdots & a \\ \ldots & & & & \\ \ldots & & & & \\ a & a & a & \cdots & a+b \end{vmatrix} = b^{n-1}(na+b)$.

Chapter 4

Evaluation of Determinants

PROCEDURES FOR EVALUATING determinants of orders two and three are found in Chapter 3. In Problem 11 of that chapter, two uses of Theorem **IX** were illustrated: (*a*) to obtain an element 1 or −1 if the given determinant contains no such element, (*b*) to replace an element of a given determinant with 0.

For determinants of higher orders, the general procedure is to replace, by repeated use of Theorem **IX**, Chapter 3, the given determinant $|A|$ by another $|B| = |b_{ij}|$ having the property that all elements, except one, in some row (column) are zero. If b_{pq} is this non-zero element and β_{pq} is its cofactor,

$$|A| \quad = \quad |B| \quad = \quad b_{pq} \cdot \beta_{pq} \quad = \quad (-1)^{p+q} b_{pq} \cdot \text{minor of } b_{pq}$$

Then the minor of b_{pq} is treated in similar fashion and the process is continued until a determinant of order two or three is obtained.

Example 1.

$$\begin{vmatrix} 2 & 3 & -2 & 4 \\ 3 & -2 & 1 & 2 \\ 3 & 2 & 3 & 4 \\ -2 & 4 & 0 & 5 \end{vmatrix} = \begin{vmatrix} 2+2(3) & 3+2(-2) & -2+2(1) & 4+2(2) \\ 3 & -2 & 1 & 2 \\ 3-3(3) & 2-3(-2) & 3-3(1) & 4-3(2) \\ -2 & 4 & 0 & 5 \end{vmatrix} = \begin{vmatrix} 8 & -1 & 0 & 8 \\ 3 & -2 & 1 & 2 \\ -6 & 8 & 0 & -2 \\ -2 & 4 & 0 & 5 \end{vmatrix}$$

$$= (-1)^{2+3}\begin{vmatrix} 8 & -1 & 8 \\ -6 & 8 & -2 \\ -2 & 4 & 5 \end{vmatrix} = -\begin{vmatrix} 8+8(-1) & -1 & 8+8(-1) \\ -6+8(8) & 8 & -2+8(8) \\ -2+8(4) & 4 & 5+8(4) \end{vmatrix} = -\begin{vmatrix} 0 & -1 & 0 \\ 58 & 8 & 62 \\ 30 & 4 & 37 \end{vmatrix}$$

$$= -(-1)^{1+2}(-1)\begin{vmatrix} 58 & 62 \\ 30 & 37 \end{vmatrix} = -286$$

See Problems 1-3

For determinants having elements of the type in Example 2 below, the following variation may be used: divide the first row by one of its non-zero elements and proceed to obtain zero elements in a row or column.

Example 2.

$$\begin{vmatrix} 0.921 & 0.185 & 0.476 & 0.614 \\ 0.782 & 0.157 & 0.527 & 0.138 \\ 0.872 & 0.484 & 0.637 & 0.799 \\ 0.312 & 0.555 & 0.841 & 0.448 \end{vmatrix} = 0.921\begin{vmatrix} 1 & 0.201 & 0.517 & 0.667 \\ 0.782 & 0.157 & 0.527 & 0.138 \\ 0.872 & 0.484 & 0.637 & 0.799 \\ 0.312 & 0.555 & 0.841 & 0.448 \end{vmatrix} = 0.921\begin{vmatrix} 1 & 0.201 & 0.517 & 0.667 \\ 0 & 0 & 0.123 & -0.384 \\ 0 & 0.309 & 0.196 & 0.217 \\ 0 & 0.492 & 0.680 & 0.240 \end{vmatrix}$$

$$= 0.921\begin{vmatrix} 0 & 0.123 & -0.384 \\ 0.309 & 0.196 & 0.217 \\ 0.492 & 0.680 & 0.240 \end{vmatrix} = 0.921(-0.384)\begin{vmatrix} 0 & -0.320 & 1 \\ 0.309 & 0.196 & 0.217 \\ 0.492 & 0.680 & 0.240 \end{vmatrix}$$

$$= 0.921(-0.384)\begin{vmatrix} 0 & 0 & 1 \\ 0.309 & 0.265 & 0.217 \\ 0.492 & 0.757 & 0.240 \end{vmatrix} = 0.921(-0.384)\begin{vmatrix} 0.309 & 0.265 \\ 0.492 & 0.757 \end{vmatrix}$$

$$= 0.921(-0.384)(0.104) = -0.037$$

THE LAPLACE EXPANSION. The expansion of a determinant $|A|$ of order n along a row (column) is a special case of the Laplace expansion. Instead of selecting one row of $|A|$, let m rows numbered $i_1, i_2, \ldots, i_m$, when arranged in order of magnitude, be selected. From these m rows

$$\rho = \frac{n(n-1)\ldots(n-m+1)}{1\cdot 2 \ldots\ m} \text{ minors} \left| A_{i_1, i_2, \ldots, i_m}^{j_1, j_2, \ldots, j_m} \right|$$

can be formed by making all possible selections of m columns from the n columns. Using these minors and their algebraic complements, we have the Laplace expansion

$$(4.1) \qquad |A| = \sum_{\rho} (-1)^s \left| A_{i_1, i_2, \ldots, i_m}^{j_1, j_2, \ldots, j_m} \right| \cdot \left| A_{i_{m+1}, i_{m+2}, \ldots, i_n}^{j_{m+1}, j_{m+2}, \ldots, j_n} \right|$$

where $s = i_1 + i_2 + \cdots + i_m + j_1 + j_2 + \cdots + j_m$ and the summation extends over the ρ selections of the column indices taken m at a time.

Example 3.

Evaluate $|A| = \begin{vmatrix} 2 & 3 & -2 & 4 \\ 3 & -2 & 1 & 2 \\ 3 & 2 & 3 & 4 \\ -2 & 4 & 0 & 5 \end{vmatrix}$, using minors of the first two rows.

From (4.1),

$$\begin{aligned} |A| &= (-1)^{1+2+1+2} \left| A_{1,2}^{1,2} \right| \cdot \left| A_{3,4}^{3,4} \right| + (-1)^{1+2+1+3} \left| A_{1,2}^{1,3} \right| \cdot \left| A_{3,4}^{2,4} \right| \\ &\quad + (-1)^{1+2+1+4} \left| A_{1,2}^{1,4} \right| \cdot \left| A_{3,4}^{2,3} \right| + (-1)^{1+2+2+3} \left| A_{1,2}^{2,3} \right| \cdot \left| A_{3,4}^{1,4} \right| \\ &\quad + (-1)^{1+2+2+4} \left| A_{1,2}^{2,4} \right| \cdot \left| A_{3,4}^{1,3} \right| + (-1)^{1+2+3+4} \left| A_{1,2}^{3,4} \right| \cdot \left| A_{3,4}^{1,2} \right| \\ &= \begin{vmatrix} 2 & 3 \\ 3 & -2 \end{vmatrix} \cdot \begin{vmatrix} 3 & 4 \\ 0 & 5 \end{vmatrix} - \begin{vmatrix} 2 & -2 \\ 3 & 1 \end{vmatrix} \cdot \begin{vmatrix} 2 & 4 \\ 4 & 5 \end{vmatrix} + \begin{vmatrix} 2 & 4 \\ 3 & 2 \end{vmatrix} \cdot \begin{vmatrix} 2 & 3 \\ 4 & 0 \end{vmatrix} \\ &\quad + \begin{vmatrix} 3 & -2 \\ -2 & 1 \end{vmatrix} \cdot \begin{vmatrix} 3 & 4 \\ -2 & 5 \end{vmatrix} - \begin{vmatrix} 3 & 4 \\ -2 & 2 \end{vmatrix} \cdot \begin{vmatrix} 3 & 3 \\ -2 & 0 \end{vmatrix} + \begin{vmatrix} -2 & 4 \\ 1 & 2 \end{vmatrix} \cdot \begin{vmatrix} 3 & 2 \\ -2 & 4 \end{vmatrix} \\ &= (-13)(15) - (8)(-6) + (-8)(-12) + (-1)(23) - (14)(6) + (-8)(16) \\ &= -286 \end{aligned}$$

See Problems 4-6

DETERMINANT OF A PRODUCT. If A and B are n-square matrices, then

$$(4.2) \qquad |AB| = |A| \cdot |B|$$

See Problem 7

EXPANSION ALONG THE FIRST ROW AND COLUMN. If $A = [a_{ij}]$ is n-square, then

$$(4.3) \qquad |A| = a_{11}\alpha_{11} - \sum_{i=2}^{n} \sum_{j=2}^{n} a_{i1} a_{1j} \alpha_{1j}^{i1}$$

where α_{11} is the cofactor of a_{11} and α_{1j}^{i1} is the algebraic complement of the minor $\begin{vmatrix} a_{11} & a_{1j} \\ a_{i1} & a_{ij} \end{vmatrix}$ of A.

DERIVATIVE OF A DETERMINANT. Let the n-square matrix $A = [a_{ij}]$ have as elements differentiable functions of a variable x. Then

I. The derivative, $\frac{d}{dx}|A|$, of $|A|$ with respect to x is the sum of n determinants obtained by replacing in all possible ways the elements of one row (column) of $|A|$ by their derivatives with respect to x.

Example 4.

$$\frac{d}{dx}\begin{vmatrix} x^2 & x+1 & 3 \\ 1 & 2x-1 & x^3 \\ 0 & x & -2 \end{vmatrix} = \begin{vmatrix} 2x & 1 & 0 \\ 1 & 2x-1 & x^3 \\ 0 & x & -2 \end{vmatrix} + \begin{vmatrix} x^2 & x+1 & 3 \\ 0 & 2 & 3x^2 \\ 0 & x & -2 \end{vmatrix} + \begin{vmatrix} x^2 & x+1 & 3 \\ 1 & 2x-1 & x^3 \\ 0 & 1 & 0 \end{vmatrix}$$

$$= 5 + 4x - 12x^2 - 6x^5$$

See Problem 8

SOLVED PROBLEMS

1. $$\begin{vmatrix} 2 & 3 & -2 & 4 \\ 7 & 4 & -3 & 10 \\ 3 & 2 & 3 & 4 \\ -2 & 4 & 0 & 5 \end{vmatrix} = \begin{vmatrix} 2 & 3 & -2 & 4 \\ 7-2(2) & 4-2(3) & -3-2(-2) & 10-2(4) \\ 3 & 2 & 3 & 4 \\ -2 & 4 & 0 & 5 \end{vmatrix} = \begin{vmatrix} 2 & 3 & -2 & 4 \\ 3 & -2 & 1 & 2 \\ 3 & 2 & 3 & 4 \\ -2 & 4 & 0 & 5 \end{vmatrix} = -286 \quad \text{(See Example 1)}$$

There are, of course, many other ways of obtaining an element +1 or −1; for example, subtract the first column from the second, the fourth column from the second, the first row from the second, etc.

2. $$\begin{vmatrix} 1 & 0 & -1 & 2 \\ 2 & 3 & 2 & -2 \\ 2 & 4 & 2 & 1 \\ 3 & 1 & 5 & -3 \end{vmatrix} = \begin{vmatrix} 1 & 0 & -1+1 & 2-2(1) \\ 2 & 3 & 2+2 & -2-2(2) \\ 2 & 4 & 2+2 & 1-2(2) \\ 3 & 1 & 5+3 & -3-2(3) \end{vmatrix} = \begin{vmatrix} 1 & 0 & 0 & 0 \\ 2 & 3 & 4 & -6 \\ 2 & 4 & 4 & -3 \\ 3 & 1 & 8 & -9 \end{vmatrix}$$

$$= \begin{vmatrix} 3 & 4 & -6 \\ 4 & 4 & -3 \\ 1 & 8 & -9 \end{vmatrix} = \begin{vmatrix} 3-2(4) & 4-2(4) & -6-2(-3) \\ 4 & 4 & -3 \\ 1-3(4) & 8-3(4) & -9-3(-3) \end{vmatrix} = \begin{vmatrix} -5 & -4 & 0 \\ 4 & 4 & -3 \\ -11 & -4 & 0 \end{vmatrix}$$

$$= 3\begin{vmatrix} -5 & -4 \\ -11 & -4 \end{vmatrix} = -72$$

3. Evaluate $|A| = \begin{vmatrix} 0 & 1+i & 1+2i \\ 1-i & 0 & 2-3i \\ 1-2i & 2+3i & 0 \end{vmatrix}$.

Multiply the second row by $1+i$ and the third row by $1+2i$; then

$$(1+i)(1+2i)|A| = (-1+3i)|A| = \begin{vmatrix} 0 & 1+i & 1+2i \\ 2 & 0 & 5-i \\ 5 & -4+7i & 0 \end{vmatrix} = \begin{vmatrix} 0 & 1+i & 1+2i \\ 2 & 0 & 5-i \\ 1 & -4+7i & -10+2i \end{vmatrix} = \begin{vmatrix} 0 & 1+i & 1+2i \\ 0 & 8-14i & 25-5i \\ 1 & -4+7i & -10+2i \end{vmatrix}$$

$$= \begin{vmatrix} 1+i & 1+2i \\ 8-14i & 25-5i \end{vmatrix} = -6 + 18i$$

and $|A| = 6$.

4. Derive the Laplace expansion of $|A| = |a_{ij}|$ of order n, using minors of order $m<n$.

Consider the m-square minor $\left|A_{i_1,i_2,\ldots,i_m}^{j_1,j_2,\ldots,j_m}\right|$ of $|A|$ in which the row and column indices are arranged in order of magnitude. Now by i_1-1 interchanges of adjacent rows of $|A|$, the row numbered i_1 can be brought into the first row, by i_2-2 interchanges of adjacent rows the row numbered i_2 can be brought into the second row, ..., by i_m-m interchanges of adjacent rows the row numbered i_m can be brought into the mth row. Thus, after $(i_1-1)+(i_2-2)+\cdots+(i_m-m) = i_1+i_2+\cdots+i_m-\frac{1}{2}m(m+1)$ interchanges of adjacent rows the rows numbered $i_1, i_2, \ldots, i_m$ occupy the position of the first m rows. Similarly, after $j_1+j_2+\cdots+j_m-\frac{1}{2}m(m+1)$ interchanges of adjacent columns, the columns numbered $j_1, j_2, \ldots, j_m$ occupy the position of the first m columns. As a result of the interchanges of adjacent rows and adjacent columns, the minor selected above occupies the upper left corner and its complement occupies the lower right corner of the determinant; moreover, $|A|$ has changed sign $\sigma = i_1+i_2+\cdots+i_m+j_1+j_2+\cdots+j_m-m(m+1)$ times which is equivalent to $s = i_1+i_2+\cdots+i_m+j_1+j_2+\cdots+j_m$ changes. Thus

$$\left|A_{i_1,i_2,\ldots,i_m}^{j_1,j_2,\ldots,j_m}\right|\cdot\left|A_{i_{m+1},i_{m+2},\ldots,i_n}^{j_{m+1},j_{m+2},\ldots,j_n}\right| \quad \text{yields} \quad m!(n-m)! \text{ terms of } (-1)^s|A| \quad \text{or}$$

$$(a) \qquad (-1)^s\left|A_{i_1,i_2,\ldots,i_m}^{j_1,j_2,\ldots,j_m}\right|\cdot\left|A_{i_{m+1},i_{m+2},\ldots,i_n}^{j_{m+1},j_{m+2},\ldots,j_n}\right| \quad \text{yields} \quad m!(n-m)! \text{ terms of } |A|.$$

Let $i_1, i_2, \ldots, i_m$ be held fixed. From these rows $\rho = \dfrac{n(n-1)\ldots(n-m+1)}{1\cdot 2\ldots m} = \dfrac{n!}{m!(n-m)!}$ different m-square minors may be selected. Each of these minors when multiplied by its algebraic complement yields $m!(n-m)!$ terms of $|A|$. Since, by their formation, there are no duplicate terms of $|A|$ among these products,

$$|A| = \sum_{\rho}(-1)^s\left|A_{i_1,i_2,\ldots,i_m}^{j_1,j_2,\ldots,j_m}\right|\cdot\left|A_{i_{m+1},i_{m+2},\ldots,i_n}^{j_{m+1},j_{m+2},\ldots,j_n}\right|$$

where $s = i_1+i_2+\cdots+i_m+j_1+j_2+\cdots+j_m$ and the summation extends over the ρ different selections $j_1, j_2, \ldots, j_m$ of the column indices.

5. Evaluate $|A| = \begin{vmatrix} 1 & 2 & 3 & 4 \\ 2 & 1 & 2 & 1 \\ 0 & 0 & 1 & 1 \\ 3 & 4 & 1 & 2 \end{vmatrix}$, using minors of the first two columns.

$$\begin{aligned} |A| &= (-1)^{1+2+1+2}\begin{vmatrix} 1 & 2 \\ 2 & 1 \end{vmatrix}\cdot\begin{vmatrix} 1 & 1 \\ 1 & 2 \end{vmatrix} + (-1)^{1+4+1+2}\begin{vmatrix} 1 & 2 \\ 3 & 4 \end{vmatrix}\cdot\begin{vmatrix} 2 & 1 \\ 1 & 1 \end{vmatrix} + (-1)^{2+4+1+2}\begin{vmatrix} 2 & 1 \\ 3 & 4 \end{vmatrix}\cdot\begin{vmatrix} 3 & 4 \\ 1 & 1 \end{vmatrix} \\ &= (-3)(1) + (-2)(1) - (5)(-1) \\ &= 0 \end{aligned}$$

6. If A, B, and C are n-square matrices, prove

$$|P| = \begin{vmatrix} A & O \\ C & B \end{vmatrix} = |A|\cdot|B|$$

From the first n rows of $|P|$ only one non-zero n-square minor, $|A|$, can be formed. Its algebraic complement is $|B|$. Hence, by the Laplace expansion, $|P| = |A|\cdot|B|$.

7. Prove $|AB| = |A|\cdot|B|$.

Suppose $A = [a_{ij}]$ and $B = [b_{ij}]$ are n-square. Let $C = [c_{ij}] = AB$ so that $c_{ij} = \sum_{1}^{n} a_{ik}b_{kj}$. From Problem 6,

$$|P| = \begin{vmatrix} a_{11} & a_{12} & \cdots & a_{1n} & 0 & 0 & \cdots & 0 \\ a_{21} & a_{22} & \cdots & a_{2n} & 0 & 0 & \cdots & 0 \\ \cdots & \cdots & \cdots & \cdots & \cdots & \cdots & \cdots & \cdots \\ a_{n1} & a_{n2} & \cdots & a_{nn} & 0 & 0 & \cdots & 0 \\ -1 & 0 & \cdots & 0 & b_{11} & b_{12} & \cdots & b_{1n} \\ 0 & -1 & \cdots & 0 & b_{21} & b_{22} & \cdots & b_{2n} \\ \cdots & \cdots & \cdots & \cdots & \cdots & \cdots & \cdots & \cdots \\ 0 & 0 & \cdots & -1 & b_{n1} & b_{n2} & \cdots & b_{nn} \end{vmatrix} = |A| \cdot |B|$$

To the $(n+1)$st column of $|P|$ add b_{11} times the first column, b_{21} times the second column, ..., b_{n1} times the nth column; we have

$$|P| = \begin{vmatrix} a_{11} & a_{12} & \cdots & a_{1n} & c_{11} & 0 & \cdots & 0 \\ a_{21} & a_{22} & \cdots & a_{2n} & c_{21} & 0 & \cdots & 0 \\ \cdots & \cdots & \cdots & \cdots & \cdots & \cdots & \cdots & \cdots \\ a_{n1} & a_{n2} & \cdots & a_{nn} & c_{n1} & 0 & \cdots & 0 \\ -1 & 0 & \cdots & 0 & 0 & b_{12} & \cdots & b_{1n} \\ 0 & -1 & \cdots & 0 & 0 & b_{22} & \cdots & b_{2n} \\ \cdots & \cdots & \cdots & \cdots & \cdots & \cdots & \cdots & \cdots \\ 0 & 0 & \cdots & -1 & 0 & b_{n2} & \cdots & b_{nn} \end{vmatrix}$$

Next, to the $(n+2)$nd column of $|P|$ add b_{12} times the first column, b_{22} times the second column, ..., b_{n2} times the nth column. We have

$$|P| = \begin{vmatrix} a_{11} & a_{12} & \cdots & a_{1n} & c_{11} & c_{12} & 0 & \cdots & 0 \\ a_{21} & a_{22} & \cdots & a_{2n} & c_{21} & c_{22} & 0 & \cdots & 0 \\ \cdots & \cdots & \cdots & \cdots & \cdots & \cdots & \cdots & \cdots & \cdots \\ a_{n1} & a_{n2} & \cdots & a_{nn} & c_{n1} & c_{n2} & 0 & \cdots & 0 \\ -1 & 0 & \cdots & 0 & 0 & 0 & b_{13} & \cdots & b_{1n} \\ 0 & -1 & \cdots & 0 & 0 & 0 & b_{23} & \cdots & b_{2n} \\ \cdots & \cdots & \cdots & \cdots & \cdots & \cdots & \cdots & \cdots & \cdots \\ 0 & 0 & \cdots & -1 & 0 & 0 & b_{n3} & \cdots & b_{nn} \end{vmatrix}$$

Continuing this process, we obtain finally $|P| = \begin{vmatrix} A & C \\ -I_n & 0 \end{vmatrix}$. From the last n rows of $|P|$ only one non-zero n-square minor, $|-I_n| = (-1)^n$ can be formed. Its algebraic complement is $(-1)^{1+2+\cdots+n+(n+1)+\cdots+2n}|C| = (-1)^{n(2n+1)}|C|$. Hence, $|P| = (-1)^n(-1)^{n(2n+1)}|C| = |C|$ and $|C| = |AB| = |A| \cdot |B|$.

8. Let $A = \begin{bmatrix} a_{11} & a_{12} & a_{13} \\ a_{21} & a_{22} & a_{23} \\ a_{31} & a_{32} & a_{33} \end{bmatrix}$ where $a_{ij} = a_{ij}(x)$, $(i, j = 1, 2, 3)$, are differentiable functions of x. Then

$$|A| = a_{11}a_{22}a_{33} + a_{12}a_{23}a_{31} + a_{13}a_{32}a_{21} - a_{11}a_{23}a_{32} - a_{12}a_{21}a_{33} - a_{13}a_{22}a_{31}$$

and, denoting $\frac{d}{dx}a_{ij}$ by a'_{ij},

$$\frac{d}{dx}|A| = a'_{11}a_{22}a_{33} + a'_{22}a_{11}a_{33} + a'_{33}a_{11}a_{22} + a'_{12}a_{23}a_{31} + a'_{23}a_{12}a_{31} + a'_{31}a_{12}a_{23}$$
$$+ a'_{13}a_{32}a_{21} + a'_{32}a_{13}a_{21} + a'_{21}a_{13}a_{32} - a'_{11}a_{23}a_{32} - a'_{23}a_{11}a_{32} - a'_{32}a_{11}a_{23}$$
$$- a'_{12}a_{21}a_{33} - a'_{21}a_{12}a_{33} - a'_{33}a_{12}a_{21} - a'_{13}a_{22}a_{31} - a'_{22}a_{13}a_{31} - a'_{31}a_{13}a_{22}$$

$$= a'_{11}\alpha_{11} + a'_{12}\alpha_{12} + a'_{13}\alpha_{13} + a'_{21}\alpha_{21} + a'_{22}\alpha_{22} + a'_{23}\alpha_{23} + a'_{31}\alpha_{31} + a'_{32}\alpha_{32} + a'_{33}\alpha_{33}$$

$$= \begin{vmatrix} a'_{11} & a'_{12} & a'_{13} \\ a_{21} & a_{22} & a_{23} \\ a_{31} & a_{32} & a_{33} \end{vmatrix} + \begin{vmatrix} a_{11} & a_{12} & a_{13} \\ a'_{21} & a'_{22} & a'_{23} \\ a_{31} & a_{32} & a_{33} \end{vmatrix} + \begin{vmatrix} a_{11} & a_{12} & a_{13} \\ a_{21} & a_{22} & a_{23} \\ a'_{31} & a'_{32} & a'_{33} \end{vmatrix}$$

by Problem 10, Chapter 3.

SUPPLEMENTARY PROBLEMS

9. Evaluate:

(a) $\begin{vmatrix} 3 & 5 & 7 & 2 \\ 2 & 4 & 1 & 1 \\ -2 & 0 & 0 & 0 \\ 1 & 1 & 3 & 4 \end{vmatrix} = 156$ (c) $\begin{vmatrix} 1 & -2 & 3 & -4 \\ 2 & -1 & 4 & -3 \\ 2 & 3 & -4 & -5 \\ 3 & -4 & 5 & 6 \end{vmatrix} = -304$

(b) $\begin{vmatrix} 1 & 1 & 1 & 6 \\ 2 & 4 & 1 & 6 \\ 4 & 1 & 2 & 9 \\ 2 & 4 & 2 & 7 \end{vmatrix} = 41$ (d) $\begin{vmatrix} 1 & -2 & 3 & -2 & -2 \\ 2 & -1 & 1 & 3 & 2 \\ 1 & 1 & 2 & 1 & 1 \\ 1 & -4 & -3 & -2 & -5 \\ 3 & -2 & 2 & 2 & -2 \end{vmatrix} = 118$

10. If A is n-square, show that $|\bar{A}'A|$ is real and non-negative.

11. Evaluate the determinant of Problem 9(a) using minors from the first two rows; also using minors from the first two columns.

12. (a) Let $A = \begin{bmatrix} a_1 & a_2 \\ -a_2 & a_1 \end{bmatrix}$ and $B = \begin{bmatrix} b_1 & b_2 \\ -b_2 & b_1 \end{bmatrix}$.

Use $|AB| = |A|\cdot|B|$ to show that $(a_1^2+a_2^2)(b_1^2+b_2^2) = (a_1b_1 - a_2b_2)^2 + (a_2b_1 + a_1b_2)^2$.

(b) Let $A = \begin{bmatrix} a_1+ia_3 & a_2+ia_4 \\ -a_2+ia_4 & a_1-ia_3 \end{bmatrix}$ and $B = \begin{bmatrix} b_1+ib_3 & b_2+ib_4 \\ -b_2+ib_4 & b_1-ib_3 \end{bmatrix}$.

Use $|AB| = |A|\cdot|B|$ to express $(a_1^2+a_2^2+a_3^2+a_4^2)(b_1^2+b_2^2+b_3^2+b_4^2)$ as a sum of four squares.

13. Evaluate $\begin{vmatrix} 0 & 0 & 0 & 0 & 0 & 1 \\ 0 & 0 & 0 & 0 & 2 & 1 \\ 0 & 0 & 0 & 3 & 2 & 1 \\ 0 & 0 & 4 & 3 & 2 & 1 \\ 0 & 5 & 4 & 3 & 2 & 1 \\ 6 & 5 & 4 & 3 & 2 & 1 \end{vmatrix}$ using minors from the first three rows. *Ans.* -720

14. Evaluate $\begin{vmatrix} 1 & 2 & 1 & 2 & 1 \\ 0 & 0 & 1 & 1 & 1 \\ 1 & 1 & 0 & 0 & 0 \\ 0 & 0 & 1 & 1 & 2 \\ 1 & 2 & 2 & 1 & 1 \end{vmatrix}$ using minors from the first two columns. *Ans.* 2

15. If $A_1, A_2, \ldots, A_s$ are square matrices, use the Laplace expansion to prove

$$|\,\text{diag}(A_1, A_2, \ldots, A_s)\,| \;=\; |A_1|\cdot|A_2| \cdots |A_s|$$

16. Expand $\begin{vmatrix} a_1 & a_2 & a_3 & a_4 \\ b_1 & b_2 & b_3 & b_4 \\ a_1 & a_2 & a_3 & a_4 \\ b_1 & b_2 & b_3 & b_4 \end{vmatrix}$ using minors of the first two rows and show that

$$\begin{vmatrix} a_1 & a_2 \\ b_1 & b_2 \end{vmatrix}\cdot\begin{vmatrix} a_3 & a_4 \\ b_3 & b_4 \end{vmatrix} - \begin{vmatrix} a_1 & a_3 \\ b_1 & b_3 \end{vmatrix}\cdot\begin{vmatrix} a_2 & a_4 \\ b_2 & b_4 \end{vmatrix} + \begin{vmatrix} a_1 & a_4 \\ b_1 & b_4 \end{vmatrix}\cdot\begin{vmatrix} a_2 & a_3 \\ b_2 & b_3 \end{vmatrix} = 0$$

17. Use the Laplace expansion to show that the n-square determinant $\begin{vmatrix} 0 & A \\ B & C \end{vmatrix}$, where 0 is k-square, is zero when $k > \frac{1}{2}n$.

18. In $|A| = a_{11}\alpha_{11} + a_{12}\alpha_{12} + a_{13}\alpha_{13} + a_{14}\alpha_{14}$ expand each of the cofactors $\alpha_{12}, \alpha_{13}, \alpha_{14}$ along its first column to show

$$|A| \;=\; a_{11}\alpha_{11} \;-\; \sum_{i=2}^{4}\sum_{j=2}^{4} a_{i1}\,a_{1j}\,\alpha_{1j}^{i1}$$

where α_{1j}^{i1} is the algebraic complement of the minor $\begin{vmatrix} a_{11} & a_{1j} \\ a_{i1} & a_{ij} \end{vmatrix}$ of $|A|$.

19. If α_{ij} denotes the cofactor of a_{ij} in the n-square matrix $A = [a_{ij}]$, show that the bordered determinant

$$\begin{vmatrix} a_{11} & a_{12} & \cdots & a_{1n} & p_1 \\ a_{21} & a_{22} & \cdots & a_{2n} & p_2 \\ \cdots & \cdots & \cdots & \cdots & \cdots \\ a_{n1} & a_{n2} & \cdots & a_{nn} & p_n \\ q_1 & q_2 & \cdots & q_n & 0 \end{vmatrix} = \begin{vmatrix} 0 & q_1 & q_2 & \cdots & q_n \\ p_1 & a_{11} & a_{12} & \cdots & a_{1n} \\ \cdots & \cdots & \cdots & \cdots & \cdots \\ p_n & a_{n1} & a_{n2} & \cdots & a_{nn} \end{vmatrix} = -\sum_{i=1}^{n}\sum_{j=1}^{n} p_i q_j \alpha_{ij}$$

Hint. Use **(4.3)**.

20. For each of the determinants $|A|$, find the derivative.

(*a*) $\begin{vmatrix} x^2 & x^3 \\ 2x & 3x+1 \end{vmatrix}$ (*b*) $\begin{vmatrix} x & 1 & 2 \\ x^2 & 2x+1 & x^3 \\ 0 & 3x-2 & x^2+1 \end{vmatrix}$ (*c*) $\begin{vmatrix} x^2-1 & x-1 & 1 \\ x^4 & x^3 & 2x+5 \\ x+1 & x^2 & x \end{vmatrix}$

Ans. (*a*) $2x + 9x^2 - 8x^3$, (*b*) $1 - 6x + 21x^2 + 12x^3 - 15x^4$, (*c*) $6x^5 - 5x^4 - 28x^3 + 9x^2 + 20x - 2$

21. Prove: If A and B are real n-square matrices with A non-singular and if $H = A + iB$ is Hermitian, then

$$|H|^2 \;=\; |A|^2 \cdot |I + (A^{-1}B)^2|$$

Chapter 5

Equivalence

THE RANK OF A MATRIX. A non-zero matrix A is said to have **rank** r if at least one of its r-square minors is different from zero while every $(r+1)$-square minor, if any, is zero. A zero matrix is said to have rank 0.

Example 1. The rank of $A = \begin{bmatrix} 1 & 2 & 3 \\ 2 & 3 & 4 \\ 3 & 5 & 7 \end{bmatrix}$ is $r = 2$ since $\begin{vmatrix} 1 & 2 \\ 2 & 3 \end{vmatrix} = -1 \neq 0$ while $|A| = 0$.

See Problem 1.

An n-square matrix A is called **non-singular** if its rank $r = n$, that is, if $|A| \neq 0$. Otherwise, A is called **singular**. The matrix of Example 1 is singular.

From $|AB| = |A| \cdot |B|$ follows

I. The product of two or more non-singular n-square matrices is non-singular; the product of two or more n-square matrices is singular if at least one of the matrices is singular.

ELEMENTARY TRANSFORMATIONS. The following operations, called **elementary transformations**, on a matrix do not change either its order or its rank:

(1) The interchange of the ith and jth rows, denoted by H_{ij};
The interchange of the ith and jth columns, denoted by K_{ij}.

(2) The multiplication of every element of the ith row by a non-zero scalar k, denoted by $H_i(k)$;
The multiplication of every element of the ith column by a non-zero scalar k, denoted by $K_i(k)$.

(3) The addition to the elements of the ith row of k, a scalar, times the corresponding elements of the jth row, denoted by $H_{ij}(k)$;
The addition to the elements of the ith column of k, a scalar, times the corresponding elements of the jth column, denoted by $K_{ij}(k)$.

The transformations H are called **elementary row transformations**; the transformations K are called **elementary column transformations**.

The elementary transformations, being precisely those performed on the rows (columns) of a determinant, need no elaboration. It is clear that an elementary transformation cannot alter the order of a matrix. In Problem 2, it is shown that an elementary transformation does not alter its rank.

THE INVERSE OF AN ELEMENTARY TRANSFORMATION. The **inverse** of an elementary transformation is an operation which undoes the effect of the elementary transformation; that is, after A has been subjected to one of the elementary transformations and then the resulting matrix has been subjected to the inverse of that elementary transformation, the final result is the matrix A.

Example 2. Let $A = \begin{bmatrix} 1 & 2 & 3 \\ 4 & 5 & 6 \\ 7 & 8 & 9 \end{bmatrix}$

The effect of the elementary row transformation $H_{21}(-2)$ is to produce $B = \begin{bmatrix} 1 & 2 & 3 \\ 2 & 1 & 0 \\ 7 & 8 & 9 \end{bmatrix}$.

The effect of the elementary row transformation $H_{21}(+2)$ on B is to produce A again. Thus, $H_{21}(-2)$ and $H_{21}(+2)$ are inverse elementary row transformations.

The inverse elementary transformations are:

(1') $\quad H_{ij}^{-1} = H_{ij} \qquad K_{ij}^{-1} = K_{ij}$

(2') $\quad H_i^{-1}(k) = H_i(1/k) \qquad K_i^{-1}(k) = K_i(1/k)$

(3') $\quad H_{ij}^{-1}(k) = H_{ij}(-k) \qquad K_{ij}^{-1}(k) = K_{ij}(-k)$

We have

II. The inverse of an elementary transformation is an elementary transformation of the same type.

EQUIVALENT MATRICES. Two matrices A and B are called **equivalent**, $A \sim B$, if one can be obtained from the other by a sequence of elementary transformations.

Equivalent matrices have the same order and the same rank.

Example 3. Applying in turn the elementary transformations $H_{21}(-2)$, $H_{31}(1)$, $H_{32}(-1)$,

$$A = \begin{bmatrix} 1 & 2 & -1 & 4 \\ 2 & 4 & 3 & 5 \\ -1 & -2 & 6 & -7 \end{bmatrix} \sim \begin{bmatrix} 1 & 2 & -1 & 4 \\ 0 & 0 & 5 & -3 \\ -1 & -2 & 6 & -7 \end{bmatrix} \sim \begin{bmatrix} 1 & 2 & -1 & 4 \\ 0 & 0 & 5 & -3 \\ 0 & 0 & 5 & -3 \end{bmatrix} \sim \begin{bmatrix} 1 & 2 & -1 & 4 \\ 0 & 0 & 5 & -3 \\ 0 & 0 & 0 & 0 \end{bmatrix} = B$$

Since all 3-square minors of B are zero while $\begin{vmatrix} -1 & 4 \\ 5 & -3 \end{vmatrix} \neq 0$, the rank of B is 2; hence, the rank of A is 2. This procedure of obtaining from A an equivalent matrix B from which the rank is evident by inspection is to be compared with that of computing the various minors of A.

See Problem 3.

ROW EQUIVALENCE. If a matrix A is reduced to B by the use of elementary row transformations alone, B is said to be **row equivalent** to A and conversely. The matrices A and B of Example 3 are row equivalent.

Any non-zero matrix A of rank r is row equivalent to a **canonical matrix** C in which

(a) one or more elements of each of the first r rows are non-zero while all other rows have only zero elements.

(b) in the ith row, $(i = 1, 2, \ldots, r)$, the first non-zero element is 1; let the column in which this element stands be numbered j_i.

(c) $j_1 < j_2 < \cdots < j_r$.

(d) the only non-zero element in the column numbered j_i, $(i = 1, 2, \ldots, r)$, is the element 1 of the ith row.

To reduce A to C, suppose j_1 is the number of the first non-zero column of A.

(i₁) If $a_{1j_1} \neq 0$, use $H_1(1/a_{1j_1})$ to reduce it to 1, when necessary.

(i₂) If $a_{ij_1} = 0$ but $a_{pj_1} \neq 0$, use H_{1p} and proceed as in **(i₁)**.

(ii) Use row transformations of type **(3)** with appropriate multiples of the first row to obtain zeros elsewhere in the j_1st column.

If non-zero elements of the resulting matrix B occur only in the first row, $B = C$. Otherwise, suppose j_2 is the number of the first column in which this does not occur. If $b_{2j_2} \neq 0$, use $H_2(1/b_{2j_2})$ as in **(i₁)**; if $b_{2j_2} = 0$ but $b_{qj_2} \neq 0$, use H_{2q} and proceed as in **(i₁)**. Then, as in **(ii)**, clear the j_2nd column of all other non-zero elements.

If non-zero elements of the resulting matrix occur only in the first two rows, we have C. Otherwise, the procedure is repeated until C is reached.

Example 4. The sequence of row transformations $H_{21}(-2), H_{31}(1)$; $H_2(1/5)$; $H_{12}(1), H_{32}(-5)$ applied to A of Example 3 yields

$$A = \begin{bmatrix} 1 & 2 & -1 & 4 \\ 2 & 4 & 3 & 5 \\ -1 & -2 & 6 & -7 \end{bmatrix} \sim \begin{bmatrix} 1 & 2 & -1 & 4 \\ 0 & 0 & 5 & -3 \\ 0 & 0 & 5 & -3 \end{bmatrix} \sim \begin{bmatrix} 1 & 2 & -1 & 4 \\ 0 & 0 & 1 & -3/5 \\ 0 & 0 & 5 & -3 \end{bmatrix} \sim \begin{bmatrix} 1 & 2 & 0 & 17/5 \\ 0 & 0 & 1 & -3/5 \\ 0 & 0 & 0 & 0 \end{bmatrix} = C$$

having the properties (a)-(d).

See Problem 4.

THE NORMAL FORM OF A MATRIX. By means of elementary transformations any matrix A of rank $r > 0$ can be reduced to one of the forms

$$\textbf{(5.1)} \qquad I_r, \quad \begin{bmatrix} I_r & 0 \\ 0 & 0 \end{bmatrix}, \quad [I_r \;\; 0], \quad \begin{bmatrix} I_r \\ 0 \end{bmatrix}$$

called its **normal form**. A zero matrix is its own normal form.

Since both row and column transformations may be used here, the element 1 of the first row obtained in the section above can be moved into the first column. Then both the first row **and** first column can be cleared of other non-zero elements. Similarly, the element 1 of the second row can be brought into the second column, and so on.

For example, the sequence $H_{21}(-2)$, $H_{31}(1)$, $K_{21}(-2)$, $K_{31}(1)$, $K_{41}(-4)$, K_{23}, $K_2(1/5)$, $H_{32}(-1)$, $K_{42}(3)$ applied to A of Example 3 yields $\begin{bmatrix} I_2 & 0 \\ 0 & 0 \end{bmatrix}$, the normal form.

See Problem 5.

ELEMENTARY MATRICES. The matrix which results when an elementary row (column) transformation is applied to the identity matrix I_n is called an **elementary row (column) matrix**. Here, an elementary matrix will be denoted by the symbol introduced to denote the elementary transformation which produces the matrix.

Example 5. Examples of elementary matrices obtained from $I_3 = \begin{bmatrix} 1 & 0 & 0 \\ 0 & 1 & 0 \\ 0 & 0 & 1 \end{bmatrix}$ are:

$$H_{12} = \begin{bmatrix} 0 & 1 & 0 \\ 1 & 0 & 0 \\ 0 & 0 & 1 \end{bmatrix} = K_{12}, \qquad H_3(k) = \begin{bmatrix} 1 & 0 & 0 \\ 0 & 1 & 0 \\ 0 & 0 & k \end{bmatrix} = K_3(k), \qquad H_{23}(k) = \begin{bmatrix} 1 & 0 & 0 \\ 0 & 1 & k \\ 0 & 0 & 1 \end{bmatrix} = K_{32}(k)$$

Every elementary matrix is non-singular. (Why?)

The effect of applying an elementary transformation to an $m \times n$ matrix A can be produced by multiplying A by an elementary matrix.

To effect a given elementary row transformation on A of order $m \times n$, apply the transformation to I_m to form the corresponding elementary matrix H and multiply A **on the left** by H.

To effect a given elementary column transformation on A, apply the transformation to I_n to form the corresponding elementary matrix K and multiply A **on the right** by K.

Example 6. When $A = \begin{bmatrix} 1 & 2 & 3 \\ 4 & 5 & 6 \\ 7 & 8 & 9 \end{bmatrix}$, $H_{13} \cdot A = \begin{bmatrix} 0 & 0 & 1 \\ 0 & 1 & 0 \\ 1 & 0 & 0 \end{bmatrix} \begin{bmatrix} 1 & 2 & 3 \\ 4 & 5 & 6 \\ 7 & 8 & 9 \end{bmatrix} = \begin{bmatrix} 7 & 8 & 9 \\ 4 & 5 & 6 \\ 1 & 2 & 3 \end{bmatrix}$ interchanges the first and third rows of A; $AK_{13}(2) = \begin{bmatrix} 1 & 2 & 3 \\ 4 & 5 & 6 \\ 7 & 8 & 9 \end{bmatrix} \cdot \begin{bmatrix} 1 & 0 & 0 \\ 0 & 1 & 0 \\ 2 & 0 & 1 \end{bmatrix} = \begin{bmatrix} 7 & 2 & 3 \\ 16 & 5 & 6 \\ 25 & 8 & 9 \end{bmatrix}$ adds to the first column of A two times the third column.

LET A AND B BE EQUIVALENT MATRICES. Let the elementary row and column matrices corresponding to the elementary row and column transformations which reduce A to B be designated as $H_1, H_2, \ldots, H_s$; $K_1, K_2, \ldots, K_t$ where H_1 is the first row transformation, H_2 is the second, ...; K_1 is the first column transformation, K_2 is the second, ... Then

(5.2)
$$H_s \ldots H_2 \cdot H_1 \cdot A \cdot K_1 \cdot K_2 \ldots K_t \;=\; PAQ \;=\; B$$

where

(5.3)
$$P \;=\; H_s \ldots H_2 \cdot H_1 \quad \text{and} \quad Q \;=\; K_1 \cdot K_2 \ldots K_t$$

We have

III. Two matrices A and B are equivalent if and only if there exist non-singular matrices P and Q defined in **(5.3)** such that $PAQ = B$.

Example 7. When $A = \begin{bmatrix} 1 & 2 & -1 & 2 \\ 2 & 5 & -2 & 3 \\ 1 & 2 & 1 & 2 \end{bmatrix}$, $H_{31}(-1) \cdot H_{21}(-2) \cdot A \cdot K_{21}(-2) \cdot K_{31}(1) \cdot K_{41}(-2) \cdot K_{42}(1) \cdot K_3(\tfrac{1}{2})$

$$= \begin{bmatrix} 1 & 0 & 0 \\ 0 & 1 & 0 \\ -1 & 0 & 1 \end{bmatrix} \cdot \begin{bmatrix} 1 & 0 & 0 \\ -2 & 1 & 0 \\ 0 & 0 & 1 \end{bmatrix} \cdot A \cdot \begin{bmatrix} 1 & -2 & 0 & 0 \\ 0 & 1 & 0 & 0 \\ 0 & 0 & 1 & 0 \\ 0 & 0 & 0 & 1 \end{bmatrix} \cdot \begin{bmatrix} 1 & 0 & 1 & 0 \\ 0 & 1 & 0 & 0 \\ 0 & 0 & 1 & 0 \\ 0 & 0 & 0 & 1 \end{bmatrix} \cdot \begin{bmatrix} 1 & 0 & 0 & -2 \\ 0 & 1 & 0 & 0 \\ 0 & 0 & 1 & 0 \\ 0 & 0 & 0 & 1 \end{bmatrix} \cdot \begin{bmatrix} 1 & 0 & 0 & 0 \\ 0 & 1 & 0 & 1 \\ 0 & 0 & 1 & 0 \\ 0 & 0 & 0 & 1 \end{bmatrix} \cdot \begin{bmatrix} 1 & 0 & 0 & 0 \\ 0 & 1 & 0 & 0 \\ 0 & 0 & \frac{1}{2} & 0 \\ 0 & 0 & 0 & 1 \end{bmatrix}$$

$$= \begin{bmatrix} 1 & 0 & 0 \\ -2 & 1 & 0 \\ -1 & 0 & 1 \end{bmatrix} \cdot A \cdot \begin{bmatrix} 1 & -2 & \frac{1}{2} & -4 \\ 0 & 1 & 0 & 1 \\ 0 & 0 & \frac{1}{2} & 0 \\ 0 & 0 & 0 & 1 \end{bmatrix} = PAQ = \begin{bmatrix} 1 & 0 & 0 & 0 \\ 0 & 1 & 0 & 0 \\ 0 & 0 & 1 & 0 \end{bmatrix} = B$$

Since any matrix is equivalent to its normal form, we have

IV. If A is an n-square non-singular matrix, there exist non-singular matrices P and Q as defined in **(5.3)** such that $PAQ = I_n$.

See Problem 6.

INVERSE OF A PRODUCT OF ELEMENTARY MATRICES. Let

$$P \;=\; H_s \ldots H_2 \cdot H_1 \qquad \text{and} \qquad Q \;=\; K_1 \cdot K_2 \ldots K_t$$

as in (**5.3**). Since each H and K has an inverse and since the inverse of a product is the product in reverse order of the inverses of the factors

(**5.4**) $$P^{-1} \;=\; H_1^{-1} \cdot H_2^{-1} \ldots H_s^{-1} \qquad \text{and} \qquad Q^{-1} \;=\; K_t^{-1} \ldots K_2^{-1} \cdot K_1^{-1}.$$

Let A be an n-square non-singular matrix and let P and Q defined above be such that $PAQ = I_n$. Then

(**5.5**) $$A \;=\; P^{-1}(PAQ)Q^{-1} \;=\; P^{-1} \cdot I_n \cdot Q^{-1} \;=\; P^{-1} \cdot Q^{-1}$$

We have proved

V. Every non-singular matrix can be expressed as a product of elementary matrices.

See Problem 7.

From this follow

VI. If A is non-singular, the rank of AB (also of BA) is that of B.

VII. If P and Q are non-singular, the rank of PAQ is that of A.

CANONICAL SETS UNDER EQUIVALENCE. In Problem 8, we prove

VIII. Two $m \times n$ matrices A and B are equivalent if and only if they have the same rank.

A set of $m \times n$ matrices is called a **canonical set** under equivalence if every $m \times n$ matrix is equivalent to one and only one matrix of the set. Such a canonical set is given by (**5.1**) as r ranges over the values $1, 2, \ldots, m$ or $1, 2, \ldots, n$ whichever is the smaller.

See Problem 9.

RANK OF A PRODUCT. Let A be an $m \times p$ matrix of rank r. By Theorem **III** there exist non-singular matrices P and Q such that

$$PAQ \;=\; N \;=\; \begin{bmatrix} I_r & 0 \\ 0 & 0 \end{bmatrix}$$

Then $A = P^{-1}NQ^{-1}$. Let B be a $p \times n$ matrix and consider the rank of

(**5.6**) $$AB \;=\; P^{-1}NQ^{-1}B$$

By Theorem **VI**, the rank of AB is that of $NQ^{-1}B$. Now the rows of $NQ^{-1}B$ consist of the first r rows of $Q^{-1}B$ and $m-r$ rows of zeros. Hence, the rank of AB cannot exceed r, the rank of A. Similarly, the rank of AB cannot exceed that of B. We have proved

IX. The rank of the product of two matrices cannot exceed the rank of either factor.

Suppose $AB = 0$; then from (**5.6**), $NQ^{-1}B = 0$. This requires that the first r rows of $Q^{-1}B$ be zeros while the remaining rows may be arbitrary. Thus, the rank of $Q^{-1}B$ and, hence, the rank of B cannot exceed $p-r$. We have proved

X. If the $m \times p$ matrix A is of rank r and if the $p \times n$ matrix B is such that $AB = 0$, the rank of B cannot exceed $p-r$.

SOLVED PROBLEMS

1. (a) The rank of $A = \begin{bmatrix} 1 & 2 & 3 \\ -4 & 0 & 5 \end{bmatrix}$ is 2 since $\begin{vmatrix} 1 & 2 \\ -4 & 0 \end{vmatrix} \neq 0$ and there are no minors of order three.

(b) The rank of $A = \begin{bmatrix} 1 & 2 & 3 \\ 1 & 2 & 5 \\ 2 & 4 & 8 \end{bmatrix}$ is 2 since $|A| = 0$ and $\begin{vmatrix} 2 & 3 \\ 2 & 5 \end{vmatrix} \neq 0$.

(c) The rank of $A = \begin{bmatrix} 0 & 2 & 3 \\ 0 & 4 & 6 \\ 0 & 6 & 9 \end{bmatrix}$ is 1 since $|A| = 0$, each of the nine 2-square minors is 0, but not every element is 0.

2. Show that the elementary transformations do not alter the rank of a matrix.

We shall consider only row transformations here and leave consideration of the column transformations as an exercise. Let the rank of the $m \times n$ matrix A be r so that every $(r+1)$-square minor of A, if any, is zero. Let B be the matrix obtained from A by a row transformation. Denote by $|R|$ any $(r+1)$-square minor of A and by $|S|$ the $(r+1)$-square minor of B having the same position as $|R|$.

Let the row transformation be H_{ij}. Its effect on $|R|$ is either **(i)** to leave it unchanged, **(ii)** to interchange two of its rows, or **(iii)** to interchange one of its rows with a row not of $|R|$. In the case **(i)**, $|S| = |R| = 0$; in the case **(ii)**, $|S| = -|R| = 0$; in the case **(iii)**, $|S|$ is, except possibly for sign, another $(r+1)$-square minor of $|A|$ and, hence, is 0.

Let the row transformation be $H_i(k)$. Its effect on $|R|$ is either **(i)** to leave it unchanged or **(ii)** to multiply one of its rows by k. Then, respectively, $|S| = |R| = 0$ or $|S| = k|R| = 0$.

Let the row transformation be $H_{ij}(k)$. Its effect on $|R|$ is either **(i)** to leave it unchanged, **(ii)** to increase one of its rows by k times another of its rows, or **(iii)** to increase one of its rows by k times a row not of $|R|$. In the cases **(i)** and **(ii)**, $|S| = |R| = 0$; in the case **(iii)**, $|S| = |R| \pm k$ (another $(r+1)$-square minor of A) $= 0 \pm k \cdot 0 = 0$.

Thus, an elementary row transformation cannot raise the rank of a matrix. On the other hand, it cannot lower the rank for, if it did, the inverse transformation would have to raise it. Hence, an elementary row transformation does not alter the rank of a matrix.

3. For each of the matrices A obtain an equivalent matrix B and from it, by inspection, determine the rank of A.

(a) $A = \begin{bmatrix} 1 & 2 & 3 \\ 2 & 1 & 3 \\ 3 & 2 & 1 \end{bmatrix} \sim \begin{bmatrix} 1 & 2 & 3 \\ 0 & -3 & -3 \\ 0 & -4 & -8 \end{bmatrix} \sim \begin{bmatrix} 1 & 2 & 3 \\ 0 & 1 & 1 \\ 0 & 1 & 2 \end{bmatrix} \sim \begin{bmatrix} 1 & 2 & 3 \\ 0 & 1 & 1 \\ 0 & 0 & 1 \end{bmatrix} = B$

The transformations used were $H_{21}(-2)$, $H_{31}(-3)$; $H_2(-1/3)$, $H_3(-1/4)$; $H_{32}(-1)$. The rank is 3.

(b) $A = \begin{bmatrix} 1 & 2 & 3 & 0 \\ 2 & 4 & 3 & 2 \\ 3 & 2 & 1 & 3 \\ 6 & 8 & 7 & 5 \end{bmatrix} \sim \begin{bmatrix} 1 & 2 & 3 & 0 \\ 0 & 0 & -3 & 2 \\ 0 & -4 & -8 & 3 \\ 0 & -4 & -11 & 5 \end{bmatrix} \sim \begin{bmatrix} 1 & 2 & 3 & 0 \\ 0 & -4 & -8 & 3 \\ 0 & 0 & -3 & 2 \\ 0 & -4 & -11 & 5 \end{bmatrix} \sim \begin{bmatrix} 1 & 2 & 3 & 0 \\ 0 & -4 & -8 & 3 \\ 0 & 0 & -3 & 2 \\ 0 & 0 & -3 & 2 \end{bmatrix} \sim \begin{bmatrix} 1 & 2 & 3 & 0 \\ 0 & -4 & -8 & 3 \\ 0 & 0 & -3 & 2 \\ 0 & 0 & 0 & 0 \end{bmatrix} = B$. The rank is 3.

(c) $A = \begin{bmatrix} 1 & 1+i & -i \\ 0 & i & 1+2i \\ 1 & 1+2i & 1+i \end{bmatrix} \sim \begin{bmatrix} 1 & 0 & 0 \\ 0 & i & 1+2i \\ 1 & i & 1+2i \end{bmatrix} \sim \begin{bmatrix} 1 & 0 & 0 \\ 0 & i & 1+2i \\ 0 & 0 & 0 \end{bmatrix} = B$. The rank is 2.

Note. The equivalent matrices B obtained here are not unique. In particular, since in (a) and (b) only row transformations were used, the reader may obtain others by using only column transformations. When the elements are rational numbers, there generally is no gain in mixing row and column transformations.

4. Obtain the canonical matrix C row equivalent to each of the given matrices A.

$$(a)\quad A = \begin{bmatrix} 0&0&1&3&-2\\ 0&1&2&6&0\\ 0&2&3&9&2\\ 0&1&1&3&2 \end{bmatrix} \sim \begin{bmatrix} 0&1&1&3&2\\ 0&1&2&6&0\\ 0&2&3&9&2\\ 0&0&1&3&-2 \end{bmatrix} \sim \begin{bmatrix} 0&1&1&3&2\\ 0&0&1&3&-2\\ 0&0&1&3&-2\\ 0&0&1&3&-2 \end{bmatrix} \sim \begin{bmatrix} 0&1&0&0&4\\ 0&0&1&3&-2\\ 0&0&0&0&0\\ 0&0&0&0&0 \end{bmatrix} = C$$

$$(b)\quad A = \begin{bmatrix} 1&2&-2&3&1\\ 1&3&-2&3&0\\ 2&4&-3&6&4\\ 1&1&-1&4&6 \end{bmatrix} \sim \begin{bmatrix} 1&2&-2&3&1\\ 0&1&0&0&-1\\ 0&0&1&0&2\\ 0&-1&1&1&5 \end{bmatrix} \sim \begin{bmatrix} 1&0&-2&3&3\\ 0&1&0&0&-1\\ 0&0&1&0&2\\ 0&0&1&1&4 \end{bmatrix} \sim \begin{bmatrix} 1&0&0&3&7\\ 0&1&0&0&-1\\ 0&0&1&0&2\\ 0&0&0&1&2 \end{bmatrix} \sim \begin{bmatrix} 1&0&0&0&1\\ 0&1&0&0&-1\\ 0&0&1&0&2\\ 0&0&0&1&2 \end{bmatrix} = C$$

5. Reduce each of the following to normal form.

$$(a)\quad A = \begin{bmatrix} 1&2&0&-1\\ 3&4&1&2\\ -2&3&2&5 \end{bmatrix} \sim \begin{bmatrix} 1&2&0&-1\\ 0&-2&1&5\\ 0&7&2&3 \end{bmatrix} \sim \begin{bmatrix} 1&0&0&0\\ 0&-2&1&5\\ 0&7&2&3 \end{bmatrix} \sim \begin{bmatrix} 1&0&0&0\\ 0&1&-2&5\\ 0&2&7&3 \end{bmatrix} \sim \begin{bmatrix} 1&0&0&0\\ 0&1&-2&5\\ 0&0&11&-7 \end{bmatrix} \sim \begin{bmatrix} 1&0&0&0\\ 0&1&0&0\\ 0&0&11&-7 \end{bmatrix} \sim \begin{bmatrix} 1&0&0&0\\ 0&1&0&0\\ 0&0&1&0 \end{bmatrix}$$

$$= [I_3\ 0]$$

The elementary transformations are:

$H_{21}(-3),\ H_{31}(2);\ K_{21}(-2),\ K_{41}(1);\ K_{23};\ H_{32}(-2);\ K_{32}(2),\ K_{42}(-5);\ K_3(1/11),\ K_{43}(7)$

$$(b)\quad A = \begin{bmatrix} 0&2&3&4\\ 2&3&5&4\\ 4&8&13&12 \end{bmatrix} \sim \begin{bmatrix} 2&3&5&4\\ 0&2&3&4\\ 4&8&13&12 \end{bmatrix} \sim \begin{bmatrix} 1&3&5&4\\ 0&2&3&4\\ 2&8&13&12 \end{bmatrix} \sim \begin{bmatrix} 1&3&5&4\\ 0&2&3&4\\ 0&2&3&4 \end{bmatrix} \sim \begin{bmatrix} 1&0&0&0\\ 0&2&3&4\\ 0&2&3&4 \end{bmatrix} \sim \begin{bmatrix} 1&0&0&0\\ 0&1&3&4\\ 0&1&3&4 \end{bmatrix} \sim \begin{bmatrix} 1&0&0&0\\ 0&1&0&0\\ 0&1&0&0 \end{bmatrix} \sim \begin{bmatrix} 1&0&0&0\\ 0&1&0&0\\ 0&0&0&0 \end{bmatrix}$$

$$= \begin{bmatrix} I_2 & 0\\ 0 & 0 \end{bmatrix}$$

The elementary transformations are:

$H_{12};\ K_1(\tfrac{1}{2});\ H_{31}(-2);\ K_{21}(-3),\ K_{31}(-5),\ K_{41}(-4);\ K_2(\tfrac{1}{2});\ K_{32}(-3),\ K_{42}(-4);\ H_{32}(-1)$

6. Reduce $A = \begin{bmatrix} 1&2&3&-2\\ 2&-2&1&3\\ 3&0&4&1 \end{bmatrix}$ to normal form N and compute the matrices P_1 and Q_1 such that $P_1 A Q_1 = N$.

Since A is 3×4, we shall work with the array $\begin{array}{cc} I_4 & \\ A & I_3 \end{array}$. Each row transformation is performed on a row of seven elements and each column transformation is performed on a column of seven elements.

$$\begin{array}{ccccccc} 1&0&0&0&&&\\ 0&1&0&0&&&\\ 0&0&1&0&&&\\ 0&0&0&1&&&\\ 1&2&3&-2&1&0&0\\ 2&-2&1&3&0&1&0\\ 3&0&4&1&0&0&1 \end{array} \rightarrow \begin{array}{ccccccc} 1&0&0&0&&&\\ 0&1&0&0&&&\\ 0&0&1&0&&&\\ 0&0&0&1&&&\\ 1&2&3&-2&1&0&0\\ 0&-6&-5&7&-2&1&0\\ 0&-6&-5&7&-3&0&1 \end{array} \rightarrow \begin{array}{ccccccc} 1&-2&-3&2&&&\\ 0&1&0&0&&&\\ 0&0&1&0&&&\\ 0&0&0&1&&&\\ 1&0&0&0&1&0&0\\ 0&-6&-5&7&-2&1&0\\ 0&-6&-5&7&-3&0&1 \end{array} \rightarrow \begin{array}{ccccccc} 1&-2&-3&2&&&\\ 0&1&0&0&&&\\ 0&0&1&0&&&\\ 0&0&0&1&&&\\ 1&0&0&0&1&0&0\\ 0&-6&-5&7&-2&1&0\\ 0&0&0&0&-1&-1&1 \end{array}$$

$$\rightarrow \begin{array}{ccccccc} 1&1/3&-3&2&&&\\ 0&-1/6&0&0&&&\\ 0&0&1&0&&&\\ 0&0&0&1&&&\\ 1&0&0&0&1&0&0\\ 0&1&-5&7&-2&1&0\\ 0&0&0&0&-1&-1&1 \end{array} \rightarrow \begin{array}{ccccccc} 1&1/3&-4/3&-1/3&&&\\ 0&-1/6&-5/6&7/6&&&\\ 0&0&1&0&&&\\ 0&0&0&1&&&\\ 1&0&0&0&1&0&0\\ 0&1&0&0&-2&1&0\\ 0&0&0&0&-1&-1&1 \end{array} \quad \text{or} \quad \begin{array}{cc} Q_1 & \\ N & P_1 \end{array}$$

Thus, $P_1 = \begin{bmatrix} 1 & 0 & 0 \\ -2 & 1 & 0 \\ -1 & -1 & 1 \end{bmatrix}$, $Q_1 = \begin{bmatrix} 1 & 1/3 & -4/3 & -1/3 \\ 0 & -1/6 & -5/6 & 7/6 \\ 0 & 0 & 1 & 0 \\ 0 & 0 & 0 & 1 \end{bmatrix}$ and $P_1AQ_1 = \begin{bmatrix} 1 & 0 & 0 & 0 \\ 0 & 1 & 0 & 0 \\ 0 & 0 & 0 & 0 \end{bmatrix} = N$.

7. Express $A = \begin{bmatrix} 1 & 3 & 3 \\ 1 & 4 & 3 \\ 1 & 3 & 4 \end{bmatrix}$ as a product of elementary matrices.

The elementary transformations $H_{21}(-1)$, $H_{31}(-1)$; $K_{21}(-3)$, $K_{31}(-3)$ reduce A to I_3, that is, [see **(5.2)**]

$$I = H_2 \cdot H_1 \cdot A \cdot K_1 \cdot K_2 = H_{31}(-1) \cdot H_{21}(-1) \cdot A \cdot K_{21}(-3) \cdot K_{31}(-3)$$

From **(5.5)**, $\quad A = H_1^{-1} \cdot H_2^{-1} \cdot K_2^{-1} \cdot K_1^{-1} = \begin{bmatrix} 1 & 0 & 0 \\ 1 & 1 & 0 \\ 0 & 0 & 1 \end{bmatrix} \begin{bmatrix} 1 & 0 & 0 \\ 0 & 1 & 0 \\ 1 & 0 & 1 \end{bmatrix} \begin{bmatrix} 1 & 0 & 3 \\ 0 & 1 & 0 \\ 0 & 0 & 1 \end{bmatrix} \begin{bmatrix} 1 & 3 & 0 \\ 0 & 1 & 0 \\ 0 & 0 & 1 \end{bmatrix}$

8. Prove: Two $m \times n$ matrices A and B are equivalent if and only if they have the same rank.

If A and B have the same rank, both are equivalent to the same matrix **(5.1)** and are equivalent to each other. Conversely, if A and B are equivalent, there exist non-singular matrices P and Q such that $B = PAQ$. By Theorem **VII**, A and B have the same rank.

9. A canonical set for non-zero matrices of order 3 is

$$I_3, \quad \begin{bmatrix} I_2 & 0 \\ 0 & 0 \end{bmatrix} = \begin{bmatrix} 1 & 0 & 0 \\ 0 & 1 & 0 \\ 0 & 0 & 0 \end{bmatrix}, \quad \begin{bmatrix} I_1 & 0 \\ 0 & 0 \end{bmatrix} = \begin{bmatrix} 1 & 0 & 0 \\ 0 & 0 & 0 \\ 0 & 0 & 0 \end{bmatrix}$$

A canonical set for non-zero 3×4 matrices is

$$[I_3 \ 0] = \begin{bmatrix} 1 & 0 & 0 & 0 \\ 0 & 1 & 0 & 0 \\ 0 & 0 & 1 & 0 \end{bmatrix}, \quad \begin{bmatrix} I_2 & 0 \\ 0 & 0 \end{bmatrix} = \begin{bmatrix} 1 & 0 & 0 & 0 \\ 0 & 1 & 0 & 0 \\ 0 & 0 & 0 & 0 \end{bmatrix}, \quad \begin{bmatrix} I_1 & 0 \\ 0 & 0 \end{bmatrix} = \begin{bmatrix} 1 & 0 & 0 & 0 \\ 0 & 0 & 0 & 0 \\ 0 & 0 & 0 & 0 \end{bmatrix}$$

10. If from a square matrix A of order n and rank r_A, a submatrix B consisting of s rows (columns) of A is selected, the rank r_B of B is equal to or greater than $r_A + s - n$.

The normal form of A has $n - r_A$ rows whose elements are zeros and the normal form of B has $s - r_B$ rows whose elements are zeros. Clearly

$$n - r_A \geq s - r_B$$

from which follows $r_B \geq r_A + s - n$ as required.

SUPPLEMENTARY PROBLEMS

11. Find the rank of (*a*) $\begin{bmatrix} 1 & 2 & 3 & 2 \\ 2 & 3 & 5 & 1 \\ 1 & 3 & 4 & 5 \end{bmatrix}$, (*b*) $\begin{bmatrix} 1 & 2 & 1 & 2 \\ 1 & 3 & 2 & 2 \\ 2 & 4 & 3 & 4 \\ 3 & 7 & 4 & 6 \end{bmatrix}$, (*c*) $\begin{bmatrix} 1 & 2 & -2 & 3 \\ 2 & 5 & -4 & 6 \\ -1 & -3 & 2 & -2 \\ 2 & 4 & -1 & 6 \end{bmatrix}$, (*d*) $\begin{bmatrix} 3 & 4 & 5 & 6 & 7 \\ 4 & 5 & 6 & 7 & 8 \\ 5 & 6 & 7 & 8 & 9 \\ 10 & 11 & 12 & 13 & 14 \\ 15 & 16 & 17 & 18 & 19 \end{bmatrix}$.

Ans. (*a*) 2, (*b*) 3, (*c*) 4, (*d*) 2

12. Show by considering minors that $A, A', \bar{A}$, and $\bar{A}'$ have the same rank.

13. Show that the canonical matrix C, row equivalent to a given matrix A, is uniquely determined by A.

14. Find the canonical matrix row equivalent to each of the following:

(*a*) $\begin{bmatrix} 1 & 2 & -3 \\ 2 & 5 & -4 \end{bmatrix} \sim \begin{bmatrix} 1 & 0 & -7 \\ 0 & 1 & 2 \end{bmatrix}$ (*b*) $\begin{bmatrix} 1 & 2 & 3 & 4 \\ 3 & 4 & 1 & 2 \\ 4 & 3 & 1 & 2 \end{bmatrix} \sim \begin{bmatrix} 1 & 0 & 0 & 1/9 \\ 0 & 1 & 0 & 1/9 \\ 0 & 0 & 1 & 11/9 \end{bmatrix}$ (*c*) $\begin{bmatrix} 1 & 1 & 1 & 2 \\ 2 & 1 & -3 & -6 \\ 3 & -3 & 1 & 2 \end{bmatrix} \sim \begin{bmatrix} 1 & 0 & 0 & 0 \\ 0 & 1 & 0 & 0 \\ 0 & 0 & 1 & 2 \end{bmatrix}$

(*d*) $\begin{bmatrix} 1 & 2 & 1 & 0 \\ 3 & 2 & 1 & 2 \\ 2 & -1 & 2 & 5 \\ 5 & 6 & 3 & 2 \\ 1 & 3 & -1 & -3 \end{bmatrix} \sim \begin{bmatrix} 1 & 0 & 0 & 1 \\ 0 & 1 & 0 & -1 \\ 0 & 0 & 1 & 1 \\ 0 & 0 & 0 & 0 \\ 0 & 0 & 0 & 0 \end{bmatrix}$ (*e*) $\begin{bmatrix} 1 & -1 & 1 & 1 & 1 \\ 1 & -1 & 2 & 3 & 1 \\ 2 & -2 & 1 & 0 & 2 \\ 1 & 1 & -1 & -3 & 3 \end{bmatrix} \sim \begin{bmatrix} 1 & 0 & 0 & -1 & 2 \\ 0 & 1 & 0 & 0 & 1 \\ 0 & 0 & 1 & 2 & 0 \\ 0 & 0 & 0 & 0 & 0 \end{bmatrix}$

15. Write the normal form of each of the matrices of Problem 14.

Ans. (*a*) $[I_2 \ 0]$, (*b*),(*c*) $[I_3 \ 0]$ (*d*) $\begin{bmatrix} I_3 & 0 \\ 0 & 0 \end{bmatrix}$ (*e*) $\begin{bmatrix} I_3 & 0 \\ 0 & 0 \end{bmatrix}$

16. Let $A = \begin{bmatrix} 1 & 2 & 3 & 4 \\ 2 & 3 & 4 & 1 \\ 3 & 4 & 1 & 2 \end{bmatrix}$

(*a*) From I_3 form $H_{12}, H_2(3), H_{13}(-4)$ and check that each HA effects the corresponding row transformation.
(*b*) From I_4 form $K_{24}, K_3(-1), K_{42}(3)$ and show that each AK effects the corresponding column transformation.
(*c*) Write the inverses $H_{12}^{-1}, H_2^{-1}(3), H_{13}^{-1}(-4)$ of the elementary matrices of (*a*). Check that for each H, $H \cdot H^{-1} = I$.
(*d*) Write the inverses $K_{24}^{-1}, K_3^{-1}(-1), K_{42}^{-1}(3)$ of the elementary matrices of (*b*). Check that for each K, $KK^{-1} = I$.
(*e*) Compute $B = H_{12} \cdot H_2(3) \cdot H_{13}(-4) = \begin{bmatrix} 0 & 3 & 0 \\ 1 & 0 & -4 \\ 0 & 0 & 1 \end{bmatrix}$ and $C = H_{13}^{-1}(-4) \cdot H_2^{-1}(3) \cdot H_{12}^{-1} = \begin{bmatrix} 0 & 1 & 4 \\ 1/3 & 0 & 0 \\ 0 & 0 & 1 \end{bmatrix}$.
(*f*) Show that $BC = CB = I$.

17. (*a*) Show that $K'_{ij} = H_{ij}$, $K'_i(k) = H_i(k)$, and $K'_{ij}(k) = H_{ij}(k)$.
(*b*) Show that if R is a product of elementary column matrices, R' is the product in reverse order of the same elementary row matrices.

18. Prove: (*a*) AB and BA are non-singular if A and B are non-singular n-square matrices.
(*b*) AB and BA are singular if at least one of the n-square matrices A and B is singular.

19. If P and Q are non-singular, show that A, PA, AQ, and PAQ have the same rank.
Hint. Express P and Q as products of elementary matrices.

20. Reduce $B = \begin{bmatrix} 1 & 3 & 6 & -1 \\ 1 & 4 & 5 & 1 \\ 1 & 5 & 4 & 3 \end{bmatrix}$ to normal form N and compute the matrices P_2 and Q_2 such that $P_2BQ_2 = N$.

21. (*a*) Show that the number of matrices in a canonical set of n-square matrices under equivalence is $n+1$.
(*b*) Show that the number of matrices in a canonical set of $m\times n$ matrices under equivalence is the smaller of $m+1$ and $n+1$.

22. Given $A = \begin{bmatrix} 1 & 2 & 4 & 4 \\ 1 & 3 & 2 & 6 \\ 2 & 5 & 6 & 10 \end{bmatrix}$ of rank 2. Find a 4-square matrix $B \neq 0$ such that $AB = 0$.

Hint. Follow the proof of Theorem **X** and take

$$Q^{-1}B = \begin{bmatrix} 0 & 0 & 0 & 0 \\ 0 & 0 & 0 & 0 \\ a & b & c & d \\ e & f & g & h \end{bmatrix}$$

where $a, b, \ldots, h$ are arbitrary.

23. The matrix A of Problem 6 and the matrix B of Problem 20 are equivalent. Find P and Q such that $B = PAQ$.

24. If the $m\times n$ matrices A and B are of rank r_A and r_B respectively, show that the rank of $A+B$ cannot exceed $r_A + r_B$.

25. Let A be an arbitrary n-square matrix and B be an n-square elementary matrix. By considering each of the six different types of matrix B, show that $|AB| = |A|\cdot|B|$.

26. Let A and B be n-square matrices. (*a*) If at least one is singular show that $|AB| = |A|\cdot|B|$; (*b*) If both are non-singular, use (**5.5**) and Problem 25 to show that $|AB| = |A|\cdot|B|$.

27. Show that equivalence of matrices is an equivalence relation.

28. Prove: The row equivalent canonical form of a non-singular matrix A is I and conversely.

29. Prove: Not every matrix A can be reduced to normal form by row transformations alone.
Hint. Exhibit a matrix which cannot be so reduced.

30. Show how to effect on any matrix A the transformation H_{ij} by using a succession of row transformations of types (**2**) and (**3**).

31. Prove: If A is an $m\times n$ matrix, $(m \leqq n)$, of rank m then AA' is a non-singular symmetric matrix. State the theorem when the rank of A is $< m$.

Chapter 6

The Adjoint of a Square Matrix

THE ADJOINT. Let $A = [a_{ij}]$ be an n-square matrix and α_{ij} be the cofactor of a_{ij}; then by definition,

(6.1)
$$\text{adjoint } A \quad = \quad \text{adj } A \quad = \quad \begin{bmatrix} \alpha_{11} & \alpha_{21} & \cdots & \alpha_{n1} \\ \alpha_{12} & \alpha_{22} & \cdots & \alpha_{n2} \\ \cdots & \cdots & \cdots & \cdots \\ \alpha_{1n} & \alpha_{2n} & & \alpha_{nn} \end{bmatrix}$$

Note carefully that the cofactors of the elements of the ith row (column) of A are the elements of the ith column (row) of adj A.

Example 1. For the matrix $A = \begin{bmatrix} 1 & 2 & 3 \\ 2 & 3 & 2 \\ 3 & 3 & 4 \end{bmatrix}$,

$$\alpha_{11} = 6, \quad \alpha_{12} = -2, \quad \alpha_{13} = -3, \quad \alpha_{21} = 1, \quad \alpha_{22} = -5, \quad \alpha_{23} = 3, \quad \alpha_{31} = -5, \quad \alpha_{32} = 4, \quad \alpha_{33} = -1$$

and
$$\text{adj } A \quad = \quad \begin{bmatrix} 6 & 1 & -5 \\ -2 & -5 & 4 \\ -3 & 3 & -1 \end{bmatrix}$$

See Problems 1-2.

Using Theorems **X** and **XI** of Chapter 3, we find

(6.2)
$$A(\text{adj } A) \quad = \quad \begin{bmatrix} a_{11} & a_{12} & \cdots & a_{1n} \\ a_{21} & a_{22} & \cdots & a_{2n} \\ \cdots & \cdots & \cdots & \cdots \\ a_{n1} & a_{n2} & \cdots & a_{nn} \end{bmatrix} \begin{bmatrix} \alpha_{11} & \alpha_{21} & \cdots & \alpha_{n1} \\ \alpha_{12} & \alpha_{22} & \cdots & \alpha_{n2} \\ \cdots & \cdots & \cdots & \cdots \\ \alpha_{1n} & \alpha_{2n} & \cdots & \alpha_{nn} \end{bmatrix}$$

$$= \text{diag}(|A|, |A|, \ldots, |A|) \quad = \quad |A| \cdot I_n \quad = \quad (\text{adj } A)\,A$$

Example 2. For the matrix A of Example 1, $|A| = -7$ and

$$A(\text{adj } A) \quad = \quad \begin{bmatrix} 1 & 2 & 3 \\ 2 & 3 & 2 \\ 3 & 3 & 4 \end{bmatrix} \begin{bmatrix} 6 & 1 & -5 \\ -2 & -5 & 4 \\ -3 & 3 & -1 \end{bmatrix} \quad = \quad \begin{bmatrix} -7 & 0 & 0 \\ 0 & -7 & 0 \\ 0 & 0 & -7 \end{bmatrix} \quad = \quad -7I$$

By taking determinants in **(6.2)**, we have

(6.3)
$$|A| \cdot |\text{adj } A| \quad = \quad |A|^n \quad = \quad |\text{adj } A| \cdot |A|$$

There follow

I. If A is n-square and non-singular, then

(6.4)
$$|\text{adj } A| \quad = \quad |A|^{n-1}$$

II. If A is n-square and singular, then

$$A\,(\text{adj}\,A) \quad = \quad (\text{adj}\,A)\,A \quad = \quad 0$$

If A is of rank $< n-1$, then $\text{adj}\,A = 0$. If A is of rank $n-1$, then $\text{adj}\,A$ is of rank 1.

See Problem 3.

THE ADJOINT OF A PRODUCT. In Problem 4, we prove

III. If A and B are n-square matrices,

(6.5)
$$\text{adj}\,AB \quad = \quad \text{adj}\,B \cdot \text{adj}\,A$$

MINOR OF AN ADJOINT. In Problem 6, we prove

IV. Let $\left|A^{j_1, j_2, \ldots, j_m}_{i_1, i_2, \ldots, i_m}\right|$ be an m-square minor of the n-square matrix $A = [a_{ij}]$,

let $\left|A^{j_{m+1}, j_{m+2}, \ldots, j_n}_{i_{m+1}, i_{m+2}, \ldots, i_n}\right|$ be its complement in A, and

let $\left|M^{j_1, j_2, \ldots, j_m}_{i_1, i_2, \ldots, i_m}\right|$ denote the m-square minor of $\text{adj}\,A$ whose elements occupy **the same position** in $\text{adj}\,A$ as those of $\left|A^{j_1, j_2, \ldots, j_m}_{i_1, i_2, \ldots, i_m}\right|$ occupy in A.

Then

(6.6)
$$|A| \cdot \left|M^{j_1, j_2, \ldots, j_m}_{i_1, i_2, \ldots, i_m}\right| \quad = \quad (-1)^s |A|^m \cdot \left|A^{j_{m+1}, j_{m+2}, \ldots, j_n}_{i_{m+1}, i_{m+2}, \ldots, i_n}\right|$$

where $s = i_1 + i_2 + \cdots + i_m + j_1 + j_2 + \cdots + j_m$.

If in **(6.6)**, A is non-singular, then

(6.7)
$$\left|M^{j_1, j_2, \ldots, j_m}_{i_1, i_2, \ldots, i_m}\right| \quad = \quad (-1)^s |A|^{m-1} \cdot \left|A^{j_{m+1}, j_{m+2}, \ldots, j_n}_{i_{m+1}, i_{m+2}, \ldots, i_n}\right|$$

When $m = 2$, **(6.7)** becomes

(6.8)
$$\begin{vmatrix} \alpha_{i_1, j_1} & \alpha_{i_2, j_1} \\ \alpha_{i_1, j_2} & \alpha_{i_2, j_2} \end{vmatrix} \quad = \quad (-1)^{i_1+i_2+j_1+j_2} |A| \cdot \left|A^{j_3, j_4, \ldots, j_n}_{i_3, i_4, \ldots, i_n}\right|$$

$$= \quad |A| \cdot \text{ algebraic complement of } \left|A^{j_1, j_2}_{i_1, i_2}\right|$$

When $m = n-1$, **(6.7)** becomes

(6.9)
$$\left|M^{j_1, j_2, \ldots, j_{n-1}}_{i_1, i_2, \ldots, i_{n-1}}\right| \quad = \quad (-1)^{i_n + j_n} |A|^{n-2} a_{i_n, j_n}$$

When $m = n$, **(6.7)** becomes **(6.4)**.

SOLVED PROBLEMS

1. The adjoint of $A = \begin{bmatrix} a & b \\ c & d \end{bmatrix}$ is $\begin{bmatrix} \alpha_{11} & \alpha_{21} \\ \alpha_{12} & \alpha_{22} \end{bmatrix} = \begin{bmatrix} d & -b \\ -c & a \end{bmatrix}$.

2. The adjoint of $A = \begin{bmatrix} 1 & 2 & 3 \\ 1 & 3 & 4 \\ 1 & 4 & 3 \end{bmatrix}$ is $\begin{bmatrix} \begin{vmatrix} 3 & 4 \\ 4 & 3 \end{vmatrix} & -\begin{vmatrix} 2 & 3 \\ 4 & 3 \end{vmatrix} & \begin{vmatrix} 2 & 3 \\ 3 & 4 \end{vmatrix} \\ -\begin{vmatrix} 1 & 4 \\ 1 & 3 \end{vmatrix} & \begin{vmatrix} 1 & 3 \\ 1 & 3 \end{vmatrix} & -\begin{vmatrix} 1 & 3 \\ 1 & 4 \end{vmatrix} \\ \begin{vmatrix} 1 & 3 \\ 1 & 4 \end{vmatrix} & -\begin{vmatrix} 1 & 2 \\ 1 & 4 \end{vmatrix} & \begin{vmatrix} 1 & 2 \\ 1 & 3 \end{vmatrix} \end{bmatrix} = \begin{bmatrix} -7 & 6 & -1 \\ 1 & 0 & -1 \\ 1 & -2 & 1 \end{bmatrix}$.

3. Prove: If A is of order n and rank $n-1$, then adj A is of rank 1.

First we note that, since A is of rank $n-1$, there is at least one non-zero cofactor and the rank of adj A is at least one. By Theorem **X**, Chapter 5, the rank of adj A is at most $n-(n-1) = 1$. Hence, the rank is exactly one.

4. Prove: $\text{adj}\, AB = \text{adj}\, B \cdot \text{adj}\, A$.

By **(6.2)** $$AB\, \text{adj}\, AB = |AB| \cdot I = (\text{adj}\, AB)AB$$

Since $$AB \cdot \text{adj}\, B \cdot \text{adj}\, A = A(B \cdot \text{adj}\, B)\,\text{adj}\, A = A(|B| \cdot I)\,\text{adj}\, A = |B|(A\, \text{adj}\, A) = |B| \cdot |A| \cdot I = |AB| \cdot I$$

and $$(\text{adj}\, B \cdot \text{adj}\, A)AB = \text{adj}\, B\{(\text{adj}\, A)A\}B = \text{adj}\, B \cdot |A| \cdot I \cdot B = |A|\{(\text{adj}\, B)B\} = |AB| \cdot I$$

we conclude that $$\text{adj}\, AB = \text{adj}\, B \cdot \text{adj}\, A$$

5. Show that $\text{adj}(\text{adj}\, A) = |A|^{n-2} \cdot A$, if $|A| \neq 0$.

By **(6.2)** and **(6.4)**,

$$\begin{aligned} \text{adj}\, A \cdot \text{adj}(\text{adj}\, A) &= \text{diag}(|\text{adj}\, A|, |\text{adj}\, A|, \ldots, |\text{adj}\, A|) \\ &= \text{diag}(|A|^{n-1}, |A|^{n-1}, \ldots, |A|^{n-1}) \end{aligned}$$

Then

$$\begin{aligned} A \cdot \text{adj}\, A \cdot \text{adj}(\text{adj}\, A) &= |A|^{n-1} \cdot A \\ |A| \cdot \text{adj}(\text{adj}\, A) &= |A|^{n-1} \cdot A \end{aligned}$$

and $$\text{adj}(\text{adj}\, A) = |A|^{n-2} \cdot A$$

6. Prove: Let $\left|A^{j_1, j_2, \ldots, j_m}_{i_1, i_2, \ldots, i_m}\right|$ be an m-square minor of the n-square matrix $A = [a_{ij}]$,

let $\left|A^{j_{m+1}, j_{m+2}, \ldots, j_n}_{i_{m+1}, i_{m+2}, \ldots, i_n}\right|$ be its complement in A, and

let $\left|M^{j_1, j_2, \ldots, j_m}_{i_1, i_2, \ldots, i_m}\right|$ denote the m-square minor of adj A whose elements occupy the same

position in adj A as those of $\left|A^{j_1, j_2, \ldots, j_m}_{i_1, i_2, \ldots, i_m}\right|$ occupy in A. Then

$$|A| \cdot \left| M_{i_1,\, i_2,\, \ldots,\, i_m}^{j_1,\, j_2,\, \ldots,\, j_m} \right| \quad = \quad (-1)^s\, |A|^m \cdot \left| A_{i_{m+1},\, i_{m+2},\, \ldots,\, i_n}^{j_{m+1},\, j_{m+2},\, \ldots,\, j_n} \right|$$

where $s = i_1 + i_2 + \cdots + i_m + j_1 + j_2 + \cdots + j_m$.

From

$$\left[\begin{array}{cccc|ccc} a_{i_1,j_1} & a_{i_1,j_2} & \cdots & a_{i_1,j_m} & a_{i_1,j_{m+1}} & \cdots & a_{i_1,j_n} \\ a_{i_2,j_1} & a_{i_2,j_2} & \cdots & a_{i_2,j_m} & a_{i_2,j_{m+1}} & \cdots & a_{i_2,j_n} \\ \cdots & \cdots & \cdots & \cdots & \cdots & \cdots & \cdots \\ a_{i_m,j_1} & a_{i_m,j_2} & \cdots & a_{i_m,j_m} & a_{i_m,j_{m+1}} & \cdots & a_{i_m,j_n} \\ \hline a_{i_{m+1},j_1} & a_{i_{m+1},j_2} & \cdots & a_{i_{m+1},j_m} & a_{i_{m+1},j_{m+1}} & \cdots & a_{i_{m+1},j_n} \\ \cdots & \cdots & \cdots & \cdots & \cdots & \cdots & \cdots \\ \cdots & \cdots & \cdots & \cdots & \cdots & \cdots & \cdots \\ a_{i_n,j_1} & a_{i_n,j_2} & \cdots & a_{i_n,j_m} & a_{i_n,j_{m+1}} & \cdots & a_{i_n,j_n} \end{array}\right] \left[\begin{array}{cccc|cccc} \alpha_{i_1,j_1} & \alpha_{i_2,j_1} & \cdots & \alpha_{i_m,j_1} & 0 & 0 & \cdots & 0 \\ \alpha_{i_1,j_2} & \alpha_{i_2,j_2} & \cdots & \alpha_{i_m,j_2} & 0 & 0 & \cdots & 0 \\ \cdots & \cdots & \cdots & \cdots & \cdots & \cdots & \cdots & \cdots \\ \alpha_{i_1,j_m} & \alpha_{i_2,j_m} & \cdots & \alpha_{i_m,j_m} & 0 & 0 & \cdots & 0 \\ \hline \alpha_{i_1,j_{m+1}} & \alpha_{i_2,j_{m+1}} & \cdots & \alpha_{i_m,j_{m+1}} & 1 & 0 & \cdots & 0 \\ \cdots & \cdots & \cdots & \cdots & \cdots & \cdots & \cdots & \cdots \\ \cdots & \cdots & \cdots & \cdots & \cdots & \cdots & \cdots & \cdots \\ \alpha_{i_1,j_n} & \alpha_{i_2,j_n} & \cdots & \alpha_{i_m,j_n} & 0 & 0 & \cdots & 1 \end{array}\right]$$

$$= \left[\begin{array}{cccc|ccc} |A| & 0 & \cdots & 0 & a_{i_1,j_{m+1}} & \cdots & a_{i_1,j_n} \\ 0 & |A| & \cdots & 0 & a_{i_2,j_{m+1}} & \cdots & a_{i_2,j_n} \\ \cdots & \cdots & \cdots & \cdots & \cdots & \cdots & \cdots \\ 0 & 0 & \cdots & |A| & a_{i_m,j_{m+1}} & \cdots & a_{i_m,j_n} \\ \hline 0 & 0 & \cdots & 0 & a_{i_{m+1},j_{m+1}} & \cdots & a_{i_{m+1},j_n} \\ \cdots & \cdots & \cdots & \cdots & \cdots & \cdots & \cdots \\ \cdots & \cdots & \cdots & \cdots & \cdots & \cdots & \cdots \\ 0 & 0 & \cdots & 0 & a_{i_n,j_{m+1}} & \cdots & a_{i_n,j_n} \end{array}\right]$$

by taking determinants of both sides, we have

$$(-1)^s |A| \cdot \left| M_{i_1,\, i_2,\, \ldots,\, i_m}^{j_1,\, j_2,\, \ldots,\, j_m} \right| \quad = \quad |A|^m \cdot \left| A_{i_{m+1},\, i_{m+2},\, \ldots,\, i_n}^{j_{m+1},\, j_{m+2},\, \ldots,\, j_n} \right|$$

where s is as defined in the theorem. From this, the required form follows immediately.

7. Prove: If A is a skew-symmetric of order $2n$, then $|A|$ is the square of a polynomial in the elements of A.

By its definition, $|A|$ is a polynomial in its elements; we are to show that under the conditions given above this polynomial is a perfect square.

The theorem is true for $n = 1$ since, when $A = \begin{bmatrix} 0 & a \\ -a & 0 \end{bmatrix}$, $|A| = a^2$.

Assume now that the theorem is true when $n = k$ and consider the skew-symmetric matrix $A = [a_{ij}]$ of order $2k+2$. By partitioning, write $A = \begin{bmatrix} B & C \\ D & E \end{bmatrix}$ where $E = \begin{bmatrix} 0 & a_{2k+1,\,2k+2} \\ a_{2k+2,\,2k+1} & 0 \end{bmatrix}$. Then B is skew-symmetric of order $2k$ and, by assumption, $|B| = f^2$ where f is a polynomial in the elements of B.

If α_{ij} denotes the cofactor of a_{ij} in A, we have by Problem 6, Chapter 3, and (**6.8**)

$$\begin{vmatrix} \alpha_{2k+1,\,2k+1} & \alpha_{2k+2,\,2k+1} \\ \alpha_{2k+1,\,2k+2} & \alpha_{2k+2,\,2k+2} \end{vmatrix} = \begin{vmatrix} 0 & \alpha_{2k+2,\,2k+1} \\ \alpha_{2k+1,\,2k+2} & 0 \end{vmatrix} = |A| \cdot |B|$$

Moreover, $\alpha_{2k+2,\,2k+1} = -\alpha_{2k+1,\,2k+2}$; hence,

$$|A| \cdot f^2 = \alpha^2_{2k+1,\,2k+2} \qquad \text{and} \qquad |A| = \left\{\frac{\alpha_{2k+1,\,2k+2}}{f}\right\}^2$$

a perfect square.

SUPPLEMENTARY PROBLEMS

8. Compute the adjoint of:

(a) $\begin{bmatrix} 1 & 2 & 3 \\ 0 & 1 & 2 \\ 0 & 0 & 0 \end{bmatrix}$, (b) $\begin{bmatrix} 1 & 2 & 3 \\ 0 & 1 & 2 \\ 0 & 0 & 1 \end{bmatrix}$, (c) $\begin{bmatrix} 1 & 0 & 2 \\ 2 & 1 & 0 \\ 3 & 2 & 1 \end{bmatrix}$, (d) $\begin{bmatrix} 5 & 0 & 0 & 2 \\ 1 & 1 & 0 & 2 \\ 0 & 0 & 2 & 1 \\ 1 & 0 & 0 & 1 \end{bmatrix}$

Ans. (a) $\begin{bmatrix} 0 & 0 & 1 \\ 0 & 0 & -2 \\ 0 & 0 & 1 \end{bmatrix}$, (b) $\begin{bmatrix} 1 & -2 & 1 \\ 0 & 1 & -2 \\ 0 & 0 & 1 \end{bmatrix}$, (c) $\begin{bmatrix} 1 & 4 & -2 \\ -2 & -5 & 4 \\ 1 & -2 & 1 \end{bmatrix}$, (d) $\begin{bmatrix} 2 & 0 & 0 & -4 \\ 2 & 6 & 0 & -16 \\ 1 & 0 & 3 & -5 \\ -2 & 0 & 0 & 10 \end{bmatrix}$

9. Verify:

(a) The adjoint of a scalar matrix is a scalar matrix.

(b) The adjoint of a diagonal matrix is a diagonal matrix.

(c) The adjoint of a triangular matrix is a triangular matrix.

10. Write a matrix $A \neq 0$ of order 3 such that $\text{adj}\,A = 0$.

11. If A is a 2-square matrix, show that $\text{adj}(\text{adj}\,A) = A$.

12. Show that the adjoint of $A = \begin{bmatrix} -1 & -2 & -2 \\ 2 & 1 & -2 \\ 2 & -2 & 1 \end{bmatrix}$ is $3A'$ and the adjoint of $A = \begin{bmatrix} -4 & -3 & -3 \\ 1 & 0 & 1 \\ 4 & 4 & 3 \end{bmatrix}$ is A itself.

13. Prove: If an n-square matrix A is of rank $< n-1$, then $\text{adj}\,A = 0$.

14. Prove: If A is symmetric, so also is $\text{adj}\,A$.

15. Prove: If A is Hermitian, so also is $\text{adj}\,A$.

16. Prove: If A is skew-symmetric of order n, then $\text{adj}\,A$ is symmetric or skew-symmetric according as n is odd or even.

17. Is there a theorem similar to that of Problem 16 for skew-Hermitian matrices?

18. For the elementary matrices, show that

(*a*) adj $H_{ij}^{-1} = -H_{ij}$

(*b*) adj $H_i^{-1}(k) = \text{diag}(1/k, 1/k, \ldots, 1/k, 1, 1/k, \ldots, 1/k)$, where the element 1 stands in the ith row

(*c*) adj $H_{ij}^{-1}(k) = H_{ij}(k)$, with similar results for the K's.

19. Prove: If A is an n-square matrix of rank n or $n-1$ and if $H_s \ldots H_2 \cdot H_1 \cdot A \cdot K_1 \cdot K_2 \ldots K_t = \lambda$ where λ is I_n or $\begin{bmatrix} I_{n-1} & 0 \\ 0 & 0 \end{bmatrix}$, then

$$\text{adj } A = \text{adj } K_1^{-1} \cdot \text{adj } K_2^{-1} \ldots \text{adj } K_t^{-1} \cdot \text{adj } \lambda \cdot \text{adj } H_s^{-1} \ldots \text{adj } H_2^{-1} \cdot \text{adj } H_1^{-1}$$

20. Use the method of Problem 19 to compute the adjoint of

(*a*) A of Problem 7, Chapter 5 (*b*) $\begin{bmatrix} 1 & 1 & 1 & 0 \\ 2 & 3 & 3 & 2 \\ 1 & 2 & 3 & 2 \\ 4 & 6 & 7 & 4 \end{bmatrix}$

Ans. (*a*) $\begin{bmatrix} 7 & -3 & -3 \\ -1 & 1 & 0 \\ -1 & 0 & 1 \end{bmatrix}$, (*b*) $\begin{bmatrix} -14 & 2 & -2 & 2 \\ 14 & -2 & 2 & -2 \\ 0 & 0 & 0 & 0 \\ -7 & 1 & -1 & 1 \end{bmatrix}$

21. Let $A = [a_{ij}]$ and $B = [k - a_{ij}]$ be 3-square matrices. If $S(C)$ = sum of elements of matrix C, show that

$$S(\text{adj } A) = S(\text{adj } B) \quad \text{and} \quad |B| = k \cdot S(\text{adj } A) - |A|$$

22. Prove: If A is n-square then $|\text{adj}(\text{adj } A)| = |A|^{(n-1)^2}$.

23. Let $A_n = [a_{ij}]$ $(i, j = 1, 2, \ldots, n)$ be the lower triangular matrix whose triangle is the Pascal triangle; for example,

$$A_4 = \begin{bmatrix} 1 & 0 & 0 & 0 \\ 1 & 1 & 0 & 0 \\ 1 & 2 & 1 & 0 \\ 1 & 3 & 3 & 1 \end{bmatrix}$$

Define $b_{ij} = (-1)^{i+j} a_{ij}$ and verify for $n = 2, 3, 4$ that

(i) $$\text{adj } A_n = [b_{ij}] = A_n^{-1}$$

24. Let B be obtained from A by deleting its ith and pth rows and jth and qth columns. Show that

$$\begin{vmatrix} \alpha_{ij} & \alpha_{pj} \\ \alpha_{iq} & \alpha_{pq} \end{vmatrix} = (-1)^{i+p+j+q} |B| \cdot |A|$$

where α_{ij} is the cofactor of a_{ij} in $|A|$.

Chapter 7

The Inverse of a Matrix

IF A **AND** B are n-square matrices such that $AB = BA = I$, B is called the inverse of A, $(B = A^{-1})$ and A is called the inverse of B, $(A = B^{-1})$.

In Problem 1, we prove

I. An n-square matrix A has an inverse if and only if it is non-singular.

The inverse of a non-singular n-square matrix is unique. (See Problem 7, Chapter 2.)

II. If A is non-singular, then $AB = AC$ implies $B = C$.

THE INVERSE of a non-singular diagonal matrix $\text{diag}(k_1, k_2, \ldots, k_n)$ is the diagonal matrix

$$\text{diag}(1/k_1, 1/k_2, \ldots, 1/k_n)$$

If $A_1, A_2, \ldots, A_s$ are non-singular matrices, then the inverse of the direct sum $\text{diag}(A_1, A_2, \ldots, A_s)$ is

$$\text{diag}(A_1^{-1}, A_2^{-1}, \ldots, A_s^{-1})$$

Procedures for computing the inverse of a general non-singular matrix are given below.

INVERSE FROM THE ADJOINT. From **(6.2)** $A \text{ adj } A = |A| \cdot I$. If A is non-singular

$$(7.1) \qquad A^{-1} = \frac{\text{adj } A}{|A|} = \begin{bmatrix} \alpha_{11}/|A| & \alpha_{21}/|A| & \ldots & \alpha_{n1}/|A| \\ \alpha_{12}/|A| & \alpha_{22}/|A| & \ldots & \alpha_{n2}/|A| \\ \ldots & \ldots & \ldots & \ldots \\ \ldots & \ldots & \ldots & \ldots \\ \alpha_{1n}/|A| & \alpha_{2n}/|A| & \ldots & \alpha_{nn}/|A| \end{bmatrix}$$

Example 1. From Problem 2, Chapter 6, the adjoint of $A = \begin{bmatrix} 1 & 2 & 3 \\ 1 & 3 & 4 \\ 1 & 4 & 3 \end{bmatrix}$ is $\begin{bmatrix} -7 & 6 & -1 \\ 1 & 0 & -1 \\ 1 & -2 & 1 \end{bmatrix}$.

Since $|A| = -2$, $A^{-1} = \dfrac{\text{adj } A}{|A|} = \begin{bmatrix} 7/2 & -3 & \frac{1}{2} \\ -\frac{1}{2} & 0 & \frac{1}{2} \\ -\frac{1}{2} & 1 & -\frac{1}{2} \end{bmatrix}$.

See Problem 2.

INVERSE FROM ELEMENTARY MATRICES. Let the non-singular n-square matrix A be reduced to I by elementary transformations so that

$$H_s \dots H_2 \cdot H_1 \cdot A \cdot K_1 \cdot K_2 \dots K_t = PAQ = I$$

Then $A = P^{-1} \cdot Q^{-1}$ by (5.5) and, since $(B^{-1})^{-1} = B$,

$$\text{(7.2)} \qquad A^{-1} = (P^{-1} \cdot Q^{-1})^{-1} = Q \cdot P = K_1 \cdot K_2 \dots K_t \cdot H_s \dots H_2 \cdot H_1$$

Example 2. From Problem 7, Chapter 5,

$$H_2 H_1 A K_1 K_2 = \begin{bmatrix} 1 & 0 & 0 \\ 0 & 1 & 0 \\ -1 & 0 & 1 \end{bmatrix} \begin{bmatrix} 1 & 0 & 0 \\ -1 & 1 & 0 \\ 0 & 0 & 1 \end{bmatrix} \cdot A \cdot \begin{bmatrix} 1 & -3 & 0 \\ 0 & 1 & 0 \\ 0 & 0 & 1 \end{bmatrix} \begin{bmatrix} 1 & 0 & -3 \\ 0 & 1 & 0 \\ 0 & 0 & 1 \end{bmatrix} = I$$

Then $$A^{-1} = K_1 K_2 H_2 H_1 = \begin{bmatrix} 1 & -3 & 0 \\ 0 & 1 & 0 \\ 0 & 0 & 1 \end{bmatrix} \begin{bmatrix} 1 & 0 & -3 \\ 0 & 1 & 0 \\ 0 & 0 & 1 \end{bmatrix} \begin{bmatrix} 1 & 0 & 0 \\ -1 & 1 & 0 \\ 0 & 0 & 1 \end{bmatrix} \begin{bmatrix} 1 & 0 & 0 \\ 0 & 1 & 0 \\ -1 & 0 & 1 \end{bmatrix} = \begin{bmatrix} 7 & -3 & -3 \\ -1 & 1 & 0 \\ -1 & 0 & 1 \end{bmatrix}.$$

In Chapter 5 it was shown that a non-singular matrix can be reduced to normal form by row transformations alone. Then, from (7.2) with $Q = I$, we have

$$\text{(7.3)} \qquad A^{-1} = P = H_s \dots H_2 \cdot H_1$$

That is,

III. If A is reduced to I by a sequence of row transformations alone, then A^{-1} is equal to the product in reverse order of the corresponding elementary matrices.

Example 3. Find the inverse of $A = \begin{bmatrix} 1 & 3 & 3 \\ 1 & 4 & 3 \\ 1 & 3 & 4 \end{bmatrix}$ of Example 2 using only row transformations to reduce A to I.

Write the matrix $[A\ I_3]$ and perform the sequence of row transformations which carry A into I_3 on the rows of six elements. We have

$$[A\ I_3] = \left[\begin{array}{ccc|ccc} 1 & 3 & 3 & 1 & 0 & 0 \\ 1 & 4 & 3 & 0 & 1 & 0 \\ 1 & 3 & 4 & 0 & 0 & 1 \end{array}\right] \sim \left[\begin{array}{ccc|ccc} 1 & 3 & 3 & 1 & 0 & 0 \\ 0 & 1 & 0 & -1 & 1 & 0 \\ 0 & 0 & 1 & -1 & 0 & 1 \end{array}\right] \sim \left[\begin{array}{ccc|ccc} 1 & 0 & 3 & 4 & -3 & 0 \\ 0 & 1 & 0 & -1 & 1 & 0 \\ 0 & 0 & 1 & -1 & 0 & 1 \end{array}\right] \sim \left[\begin{array}{ccc|ccc} 1 & 0 & 0 & 7 & -3 & -3 \\ 0 & 1 & 0 & -1 & 1 & 0 \\ 0 & 0 & 1 & -1 & 0 & 1 \end{array}\right]$$

$$= [I_3\ A^{-1}]$$

by (7.3). Thus, as A is reduced to I_3, I_3 is carried into $A^{-1} = \begin{bmatrix} 7 & -3 & -3 \\ -1 & 1 & 0 \\ -1 & 0 & 1 \end{bmatrix}$.

See Problem 3.

INVERSE BY PARTITIONING. Let the matrix $A = [a_{ij}]$ of order n and its inverse $B = [b_{ij}]$ be partitioned into submatrices of indicated orders:

$$\left[\begin{array}{c|c} A_{11} & A_{12} \\ (p \times p) & (p \times q) \\ \hline A_{21} & A_{22} \\ (q \times p) & (q \times q) \end{array}\right] \quad \text{and} \quad \left[\begin{array}{c|c} B_{11} & B_{12} \\ (p \times p) & (p \times q) \\ \hline B_{21} & B_{22} \\ (q \times p) & (q \times q) \end{array}\right] \quad \text{where } p + q = n$$

Since $AB = BA = I_n$, we have

$$(7.4)\quad \begin{cases} \text{(i)}\ A_{11}B_{11} + A_{12}B_{21} = I_p & \text{(iii)}\ B_{21}A_{11} + B_{22}A_{21} = 0 \\ \text{(ii)}\ A_{11}B_{12} + A_{12}B_{22} = 0 & \text{(iv)}\ B_{21}A_{12} + B_{22}A_{22} = I_q \end{cases}$$

Then, provided A_{11} is non-singular,

$$(7.5)\quad \begin{cases} B_{11} = A_{11}^{-1} + (A_{11}^{-1}A_{12})\xi^{-1}(A_{21}A_{11}^{-1}) & B_{21} = -\xi^{-1}(A_{21}A_{11}^{-1}) \\ B_{12} = -(A_{11}^{-1}A_{12})\xi^{-1} & B_{22} = \xi^{-1} \end{cases}$$

where $\xi = A_{22} - A_{21}(A_{11}^{-1}A_{12})$.

See Problem 4.

In practice, A_{11} is usually taken of order $n-1$. To obtain A_{11}^{-1}, the following procedure is used. Let

$$G_2 = \begin{bmatrix} a_{11} & a_{12} \\ a_{21} & a_{22} \end{bmatrix}, \quad G_3 = \begin{bmatrix} a_{11} & a_{12} & a_{13} \\ a_{21} & a_{22} & a_{23} \\ a_{31} & a_{32} & a_{33} \end{bmatrix}, \quad G_4 = \begin{bmatrix} a_{11} & a_{12} & a_{13} & a_{14} \\ a_{21} & a_{22} & a_{23} & a_{24} \\ a_{31} & a_{32} & a_{33} & a_{34} \\ a_{41} & a_{42} & a_{43} & a_{44} \end{bmatrix}, \quad \ldots$$

After computing G_2^{-1}, partition G_3 so that $A_{22} = [a_{33}]$ and use **(7.5)** to obtain G_3^{-1}. Repeat the process on G_4 after partitioning it so that $A_{22} = [a_{44}]$, and so on.

Example 4. Find the inverse of $A = \begin{bmatrix} 1 & 3 & 3 \\ 1 & 4 & 3 \\ 1 & 3 & 4 \end{bmatrix}$, using partitioning.

Take $A_{11} = \begin{bmatrix} 1 & 3 \\ 1 & 4 \end{bmatrix}$, $A_{12} = \begin{bmatrix} 3 \\ 3 \end{bmatrix}$, $A_{21} = [1\ 3]$, and $A_{22} = [4]$. Now

$$A_{11}^{-1} = \begin{bmatrix} 4 & -3 \\ -1 & 1 \end{bmatrix}, \quad A_{11}^{-1}A_{12} = \begin{bmatrix} 4 & -3 \\ -1 & 1 \end{bmatrix}\begin{bmatrix} 3 \\ 3 \end{bmatrix} = \begin{bmatrix} 3 \\ 0 \end{bmatrix}, \quad A_{21}A_{11}^{-1} = [1\ 3]\begin{bmatrix} 4 & -3 \\ -1 & 1 \end{bmatrix} = [1\ 0],$$

$$\xi = A_{22} - A_{21}(A_{11}^{-1}A_{12}) = [4] - [1\ 3]\begin{bmatrix} 3 \\ 0 \end{bmatrix} = [1], \quad \text{and} \quad \xi^{-1} = [1]$$

Then

$$B_{11} = A_{11}^{-1} + (A_{11}^{-1}A_{12})\xi^{-1}(A_{21}A_{11}^{-1}) = \begin{bmatrix} 4 & -3 \\ -1 & 1 \end{bmatrix} + \begin{bmatrix} 3 \\ 0 \end{bmatrix}[1]\cdot[1\ 0]$$

$$= \begin{bmatrix} 4 & -3 \\ -1 & 1 \end{bmatrix} + \begin{bmatrix} 3 & 0 \\ 0 & 0 \end{bmatrix} = \begin{bmatrix} 7 & -3 \\ -1 & 1 \end{bmatrix}$$

$$B_{12} = -(A_{11}^{-1}A_{12})\xi^{-1} = \begin{bmatrix} -3 \\ 0 \end{bmatrix}$$

$$B_{21} = -\xi^{-1}(A_{21}A_{11}^{-1}) = [-1, 0]$$

$$B_{22} = \xi^{-1} = [1]$$

and

$$A^{-1} = \begin{bmatrix} B_{11} & B_{12} \\ B_{21} & B_{22} \end{bmatrix} = \begin{bmatrix} 7 & -3 & -3 \\ -1 & 1 & 0 \\ -1 & 0 & 1 \end{bmatrix}$$

See Problems 5-6.

THE INVERSE OF A SYMMETRIC MATRIX. When A is symmetric, $\alpha_{ij} = \alpha_{ji}$ and only $\frac{1}{2}n(n+1)$ cofactors need be computed instead of the usual n^2 in obtaining A^{-1} from adj A.

If there is to be any gain in computing A^{-1} as the product of elementary matrices, the elementary transformations must be performed so that the property of being symmetric is preserved. This requires that the transformations occur in pairs, a row transformation followed immediately by the same column transformation. For example,

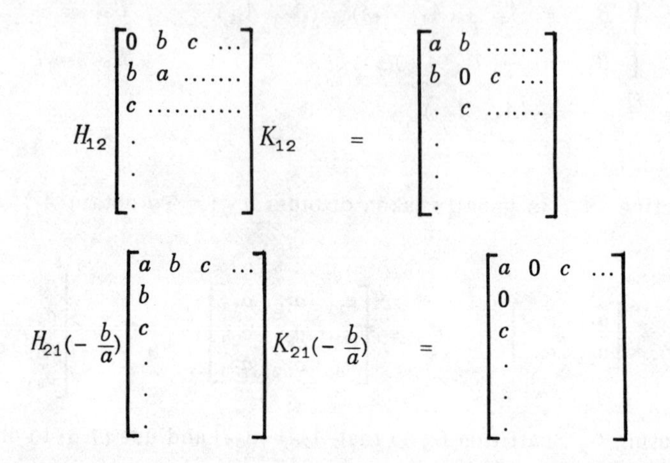

However, when the element a in the diagonal is replaced by 1, the pair of transformations are $H_1(1/\sqrt{a})$ and $K_1(1/\sqrt{a})$. In general, $\sqrt{a}$ is either irrational or imaginary; hence, this procedure is not recommended.

The maximum gain occurs when the method of partitioning is used since then **(7.5)** reduces to

$$\textbf{(7.6)}\qquad \begin{aligned} B_{11} &= A_{11}^{-1} + (A_{11}^{-1}A_{12})\xi^{-1}(A_{11}^{-1}A_{12})', \qquad & B_{21} &= B_{12}' \\ B_{12} &= -(A_{11}^{-1}A_{12})\xi^{-1}, & B_{22} &= \xi^{-1} \end{aligned}$$

where $\xi = A_{22} - A_{21}(A_{11}^{-1}A_{12})$.

See Problem 7.

When A is not symmetric, the above procedure may be used to find the inverse of $A'A$, which is symmetric, and then the inverse of A is found by

$$\textbf{(7.7)}\qquad A^{-1} = (A'A)^{-1}A'$$

SOLVED PROBLEMS

1. Prove: An n-square matrix A has an inverse if and only if it is non-singular.

   Suppose A is non-singular. By Theorem IV, Chapter 5, there exist non-singular matrices P and Q such that $PAQ = I$. Then $A = P^{-1}\cdot Q^{-1}$ and $A^{-1} = Q\cdot P$ exists.

   Suppose A^{-1} exists. The $A\cdot A^{-1} = I$ is of rank n. If A were singular, AA^{-1} would be of rank $< n$; hence, A is non-singular.

2. (a) When $A = \begin{bmatrix} 2 & 3 \\ 1 & 4 \end{bmatrix}$, then $|A| = 5$, $\operatorname{adj} A = \begin{bmatrix} 4 & -3 \\ -1 & 2 \end{bmatrix}$, and $A^{-1} = \begin{bmatrix} 4/5 & -3/5 \\ -1/5 & 2/5 \end{bmatrix}$.

(b) When $A = \begin{bmatrix} 2 & 3 & 1 \\ 1 & 2 & 3 \\ 3 & 1 & 2 \end{bmatrix}$, then $|A| = 18$, $\operatorname{adj} A = \begin{bmatrix} 1 & -5 & 7 \\ 7 & 1 & -5 \\ -5 & 7 & 1 \end{bmatrix}$, and $A^{-1} = \dfrac{1}{18}\begin{bmatrix} 1 & -5 & 7 \\ 7 & 1 & -5 \\ -5 & 7 & 1 \end{bmatrix}$.

3. Find the inverse of $A = \begin{bmatrix} 2 & 4 & 3 & 2 \\ 3 & 6 & 5 & 2 \\ 2 & 5 & 2 & -3 \\ 4 & 5 & 14 & 14 \end{bmatrix}$.

$$[A\ I_4] = \left[\begin{array}{cccc|cccc} 2 & 4 & 3 & 2 & 1 & 0 & 0 & 0 \\ 3 & 6 & 5 & 2 & 0 & 1 & 0 & 0 \\ 2 & 5 & 2 & -3 & 0 & 0 & 1 & 0 \\ 4 & 5 & 14 & 14 & 0 & 0 & 0 & 1 \end{array}\right] \sim \left[\begin{array}{cccc|cccc} 1 & 2 & 3/2 & 1 & 1/2 & 0 & 0 & 0 \\ 3 & 6 & 5 & 2 & 0 & 1 & 0 & 0 \\ 2 & 5 & 2 & -3 & 0 & 0 & 1 & 0 \\ 4 & 5 & 14 & 14 & 0 & 0 & 0 & 1 \end{array}\right] \sim \left[\begin{array}{cccc|cccc} 1 & 2 & 3/2 & 1 & 1/2 & 0 & 0 & 0 \\ 0 & 0 & 1/2 & -1 & -3/2 & 1 & 0 & 0 \\ 0 & 1 & -1 & -5 & -1 & 0 & 1 & 0 \\ 0 & -3 & 8 & 10 & -2 & 0 & 0 & 1 \end{array}\right]$$

$$\sim \left[\begin{array}{cccc|cccc} 1 & 2 & 3/2 & 1 & 1/2 & 0 & 0 & 0 \\ 0 & 1 & -1 & -5 & -1 & 0 & 1 & 0 \\ 0 & 0 & 1/2 & -1 & -3/2 & 1 & 0 & 0 \\ 0 & -3 & 8 & 10 & -2 & 0 & 0 & 1 \end{array}\right] \sim \left[\begin{array}{cccc|cccc} 1 & 0 & 7/2 & 11 & 5/2 & 0 & -2 & 0 \\ 0 & 1 & -1 & -5 & -1 & 0 & 1 & 0 \\ 0 & 0 & 1 & -2 & -3 & 2 & 0 & 0 \\ 0 & 0 & 5 & -5 & -5 & 0 & 3 & 1 \end{array}\right]$$

$$\sim \left[\begin{array}{cccc|cccc} 1 & 0 & 0 & 18 & 13 & -7 & -2 & 0 \\ 0 & 1 & 0 & -7 & -4 & 2 & 1 & 0 \\ 0 & 0 & 1 & -2 & -3 & 2 & 0 & 0 \\ 0 & 0 & 0 & 5 & 10 & -10 & 3 & 1 \end{array}\right] \sim \left[\begin{array}{cccc|cccc} 1 & 0 & 0 & 18 & 13 & -7 & -2 & 0 \\ 0 & 1 & 0 & -7 & -4 & 2 & 1 & 0 \\ 0 & 0 & 1 & -2 & -3 & 2 & 0 & 0 \\ 0 & 0 & 0 & 1 & 2 & -2 & 3/5 & 1/5 \end{array}\right]$$

$$\sim \left[\begin{array}{cccc|cccc} 1 & 0 & 0 & 0 & -23 & 29 & -64/5 & -18/5 \\ 0 & 1 & 0 & 0 & 10 & -12 & 26/5 & 7/5 \\ 0 & 0 & 1 & 0 & 1 & -2 & 6/5 & 2/5 \\ 0 & 0 & 0 & 1 & 2 & -2 & 3/5 & 1/5 \end{array}\right]$$

$$= [I_4\ A^{-1}]$$

The inverse is $A^{-1} = \begin{bmatrix} -23 & 29 & -64/5 & -18/5 \\ 10 & -12 & 26/5 & 7/5 \\ 1 & -2 & 6/5 & 2/5 \\ 2 & -2 & 3/5 & 1/5 \end{bmatrix}$.

4. Solve $\begin{cases} \textbf{(i)}\ \ A_{11}B_{11} + A_{12}B_{21} = I & \textbf{(iii)}\ \ B_{21}A_{11} + B_{22}A_{21} = 0 \\ \textbf{(ii)}\ \ A_{11}B_{12} + A_{12}B_{22} = 0 & \textbf{(iv)}\ \ B_{21}A_{12} + B_{22}A_{22} = I \end{cases}$ for B_{11}, B_{12}, B_{21}, and B_{22}.

Set $B_{22} = \xi^{-1}$. From **(ii)**, $B_{12} = -(A_{11}^{-1}A_{12})\xi^{-1}$; from **(iii)**, $B_{21} = -\xi^{-1}(A_{21}A_{11}^{-1})$; and, from **(i)**, $B_{11} = A_{11}^{-1} - A_{11}^{-1}A_{12}B_{21} = A_{11}^{-1} + (A_{11}^{-1}A_{12})\xi^{-1}(A_{21}A_{11}^{-1})$.

Finally, substituting in **(iv)**,

$$-\xi^{-1}(A_{21}A_{11}^{-1})A_{12} + \xi^{-1}A_{22} = I \quad \text{and} \quad \xi = A_{22} - (A_{21}A_{11}^{-1})A_{12}$$

5. Find the inverse of $A = \begin{bmatrix} 1 & 2 & 3 & 1 \\ 1 & 3 & 3 & 2 \\ 2 & 4 & 3 & 3 \\ 1 & 1 & 1 & 1 \end{bmatrix}$ by partitioning.

(*a*) Take $G_3 = \begin{bmatrix} 1 & 2 & 3 \\ 1 & 3 & 3 \\ 2 & 4 & 3 \end{bmatrix}$ and partition so that

$$A_{11} = \begin{bmatrix} 1 & 2 \\ 1 & 3 \end{bmatrix}, \quad A_{12} = \begin{bmatrix} 3 \\ 3 \end{bmatrix}, \quad A_{21} = [2 \;\; 4], \quad \text{and} \quad A_{22} = [3]$$

Now $$A_{11}^{-1} = \begin{bmatrix} 3 & -2 \\ -1 & 1 \end{bmatrix}, \quad A_{11}^{-1}A_{12} = \begin{bmatrix} 3 & -2 \\ -1 & 1 \end{bmatrix}\begin{bmatrix} 3 \\ 3 \end{bmatrix} = \begin{bmatrix} 3 \\ 0 \end{bmatrix}, \quad A_{21}A_{11}^{-1} = [2 \;\; 4]\begin{bmatrix} 3 & -2 \\ -1 & 1 \end{bmatrix} = [2 \;\; 0],$$

$$\xi = A_{22} - A_{21}(A_{11}^{-1}A_{12}) = [3] - [2 \;\; 4]\begin{bmatrix} 3 \\ 0 \end{bmatrix} = [-3], \quad \text{and} \quad \xi^{-1} = [-1/3]$$

Then $$B_{11} = A_{11}^{-1} + (A_{11}^{-1}A_{12})\xi^{-1}(A_{21}A_{11}^{-1}) = \begin{bmatrix} 3 & -2 \\ -1 & 1 \end{bmatrix} + \begin{bmatrix} 3 \\ 0 \end{bmatrix}\left[-\frac{1}{3}\right][2 \;\; 0] = \begin{bmatrix} 3 & -2 \\ -1 & 1 \end{bmatrix} - \begin{bmatrix} 2 & 0 \\ 0 & 0 \end{bmatrix}$$

$$= \frac{1}{3}\begin{bmatrix} 3 & -6 \\ -3 & 3 \end{bmatrix}$$

$$B_{12} = -(A_{11}^{-1}A_{12})\xi^{-1} = \frac{1}{3}\begin{bmatrix} 3 \\ 0 \end{bmatrix}, \quad B_{21} = -\xi^{-1}(A_{21}A_{11}^{-1}) = \frac{1}{3}[2 \;\; 0], \quad B_{22} = \xi^{-1} = \left[-\frac{1}{3}\right]$$

and $$G_3^{-1} = \begin{bmatrix} B_{11} & B_{12} \\ B_{21} & B_{22} \end{bmatrix} = \frac{1}{3}\begin{bmatrix} 3 & -6 & 3 \\ -3 & 3 & 0 \\ 2 & 0 & -1 \end{bmatrix}$$

(*b*) Partition A so that $A_{11} = \begin{bmatrix} 1 & 2 & 3 \\ 1 & 3 & 3 \\ 2 & 4 & 3 \end{bmatrix}$, $A_{12} = \begin{bmatrix} 1 \\ 2 \\ 3 \end{bmatrix}$, $A_{21} = [1 \;\; 1 \;\; 1]$, and $A_{22} = [1]$.

Now $$A_{11}^{-1} = \frac{1}{3}\begin{bmatrix} 3 & -6 & 3 \\ -3 & 3 & 0 \\ 2 & 0 & -1 \end{bmatrix}, \quad A_{11}^{-1}A_{12} = \frac{1}{3}\begin{bmatrix} 0 \\ 3 \\ -1 \end{bmatrix}, \quad A_{21}A_{11}^{-1} = \frac{1}{3}[2 \;\; -3 \;\; 2],$$

$$\xi = [1] - [1 \;\; 1 \;\; 1]\left(\frac{1}{3}\right)\begin{bmatrix} 0 \\ 3 \\ -1 \end{bmatrix} = \left[\frac{1}{3}\right], \quad \text{and} \quad \xi^{-1} = [3]$$

Then $$B_{11} = \frac{1}{3}\begin{bmatrix} 3 & -6 & 3 \\ -3 & 3 & 0 \\ 2 & 0 & -1 \end{bmatrix} + \frac{1}{3}\begin{bmatrix} 0 \\ 3 \\ -1 \end{bmatrix}[3]\frac{1}{3}[2 \;\; -3 \;\; 2] = \frac{1}{3}\begin{bmatrix} 3 & -6 & 3 \\ -3 & 3 & 0 \\ 2 & 0 & -1 \end{bmatrix} + \frac{1}{3}\begin{bmatrix} 0 & 0 & 0 \\ 6 & -9 & 6 \\ -2 & 3 & -2 \end{bmatrix} = \begin{bmatrix} 1 & -2 & 1 \\ 1 & -2 & 2 \\ 0 & 1 & -1 \end{bmatrix},$$

$$B_{12} = \begin{bmatrix} 0 \\ -3 \\ 1 \end{bmatrix}, \quad B_{21} = [-2 \;\; 3 \;\; -2], \quad B_{22} = [3]$$

and $$A^{-1} = \begin{bmatrix} B_{11} & B_{12} \\ B_{21} & B_{22} \end{bmatrix} = \begin{bmatrix} 1 & -2 & 1 & 0 \\ 1 & -2 & 2 & -3 \\ 0 & 1 & -1 & 1 \\ -2 & 3 & -2 & 3 \end{bmatrix}$$

6. Find the inverse of $A = \begin{bmatrix} 1 & 3 & 3 \\ 1 & 3 & 4 \\ 1 & 4 & 3 \end{bmatrix}$ by partitioning.

We cannot take $A_{11} = \begin{bmatrix} 1 & 3 \\ 1 & 3 \end{bmatrix}$ since this is singular.

By Example 3, the inverse of $H_{23}A = \begin{bmatrix} 1 & 0 & 0 \\ 0 & 0 & 1 \\ 0 & 1 & 0 \end{bmatrix} A = \begin{bmatrix} 1 & 3 & 3 \\ 1 & 4 & 3 \\ 1 & 3 & 4 \end{bmatrix} = B$ is $B^{-1} = \begin{bmatrix} 7 & -3 & -3 \\ -1 & 1 & 0 \\ -1 & 0 & 1 \end{bmatrix}$. Then

$$A^{-1} = B^{-1}H_{23} = \begin{bmatrix} 7 & -3 & -3 \\ -1 & 1 & 0 \\ -1 & 0 & 1 \end{bmatrix} \cdot \begin{bmatrix} 1 & 0 & 0 \\ 0 & 0 & 1 \\ 0 & 1 & 0 \end{bmatrix} = \begin{bmatrix} 7 & -3 & -3 \\ -1 & 0 & 1 \\ -1 & 1 & 0 \end{bmatrix}$$

Thus, if the $(n-1)$-square minor A_{11} of the n-square non-singular matrix A is singular, we first bring a non-singular $(n-1)$-square matrix into the upper left corner to obtain B, find the inverse of B, and by the proper transformation on B^{-1} obtain A^{-1}.

7. Compute the inverse of the symmetric matrix $A = \begin{bmatrix} 2 & 1 & -1 & 2 \\ 1 & 3 & 2 & -3 \\ -1 & 2 & 1 & -1 \\ 2 & -3 & -1 & 4 \end{bmatrix}$.

Consider first the submatrix $G_3 = \begin{bmatrix} 2 & 1 & -1 \\ 1 & 3 & 2 \\ -1 & 2 & 1 \end{bmatrix}$ partitioned so that

$$A_{11} = \begin{bmatrix} 2 & 1 \\ 1 & 3 \end{bmatrix}, \quad A_{12} = \begin{bmatrix} -1 \\ 2 \end{bmatrix}, \quad A_{21} = [-1 \;\; 2], \quad A_{22} = [1]$$

Now
$$A_{11}^{-1} = \begin{bmatrix} 3/5 & -1/5 \\ -1/5 & 2/5 \end{bmatrix}, \quad A_{11}^{-1}A_{12} = \begin{bmatrix} 3/5 & -1/5 \\ -1/5 & 2/5 \end{bmatrix}\begin{bmatrix} -1 \\ 2 \end{bmatrix} = \begin{bmatrix} -1 \\ 1 \end{bmatrix},$$

$$\xi = A_{22} - A_{21}(A_{11}^{-1}A_{12}) = [1] - [-1 \;\; 2]\begin{bmatrix} -1 \\ 1 \end{bmatrix} = [-2] \quad \text{and} \quad \xi^{-1} = [-\tfrac{1}{2}]$$

Then
$$B_{11} = \begin{bmatrix} 3/5 & -1/5 \\ -1/5 & 2/5 \end{bmatrix} + \begin{bmatrix} -1 \\ 1 \end{bmatrix}[-\tfrac{1}{2}][-1 \;\; 1] = \begin{bmatrix} 3/5 & -1/5 \\ -1/5 & 2/5 \end{bmatrix} + \begin{bmatrix} -\frac{1}{2} & \frac{1}{2} \\ \frac{1}{2} & -\frac{1}{2} \end{bmatrix} = \frac{1}{10}\begin{bmatrix} 1 & 3 \\ 3 & -1 \end{bmatrix},$$

$$B_{12} = \begin{bmatrix} -\frac{1}{2} \\ \frac{1}{2} \end{bmatrix}, \quad B_{21} = [-\tfrac{1}{2} \;\; \tfrac{1}{2}], \quad B_{22} = [-\tfrac{1}{2}]$$

and
$$G_3^{-1} = \frac{1}{10}\begin{bmatrix} 1 & 3 & -5 \\ 3 & -1 & 5 \\ -5 & 5 & -5 \end{bmatrix}$$

Consider now the matrix A partitioned so that

$$A_{11} = \begin{bmatrix} 2 & 1 & -1 \\ 1 & 3 & 2 \\ -1 & 2 & 1 \end{bmatrix}, \quad A_{12} = \begin{bmatrix} 2 \\ -3 \\ -1 \end{bmatrix}, \quad A_{21} = [2 \;\; -3 \;\; -1], \quad A_{22} = [4]$$

Now $A_{11}^{-1} = \dfrac{1}{10}\begin{bmatrix} 1 & 3 & -5 \\ 3 & -1 & 5 \\ -5 & 5 & -5 \end{bmatrix}$, $A_{11}^{-1}A_{12} = \begin{bmatrix} -1/5 \\ 2/5 \\ -2 \end{bmatrix}$, $\xi = [18/5]$, $\xi^{-1} = [5/18]$.

Then $B_{11} = \frac{1}{18}\begin{bmatrix} 2 & 5 & -7 \\ 5 & -1 & 5 \\ -7 & 5 & 11 \end{bmatrix}$, $B_{12} = \frac{1}{18}\begin{bmatrix} 1 \\ -2 \\ 10 \end{bmatrix}$, $B_{21} = \frac{1}{18}[1\ -2\ 10]$, $B_{22} = [5/18]$

and
$$A^{-1} = \frac{1}{18}\begin{bmatrix} 2 & 5 & -7 & 1 \\ 5 & -1 & 5 & -2 \\ -7 & 5 & 11 & 10 \\ 1 & -2 & 10 & 5 \end{bmatrix}$$

SUPPLEMENTARY PROBLEMS

8. Find the adjoint and inverse of each of the following:

$(a)\ \begin{bmatrix} 1 & 2 & -1 \\ -1 & 1 & 2 \\ 2 & -1 & 1 \end{bmatrix}$, $(b)\ \begin{bmatrix} 2 & 3 & 4 \\ 4 & 3 & 1 \\ 1 & 2 & 4 \end{bmatrix}$, $(c)\ \begin{bmatrix} 1 & 2 & 3 \\ 2 & 4 & 5 \\ 3 & 5 & 6 \end{bmatrix}$, $(d)\ \begin{bmatrix} 1 & 2 & 0 & 0 \\ 0 & 3 & 0 & 0 \\ 0 & 0 & 2 & 1 \\ 0 & 0 & 0 & 3 \end{bmatrix}$

Ans. Inverses $(a)\ \frac{1}{14}\begin{bmatrix} 3 & -1 & 5 \\ 5 & 3 & -1 \\ -1 & 5 & 3 \end{bmatrix}$, $(b)\ \frac{1}{5}\begin{bmatrix} -10 & 4 & 9 \\ 15 & -4 & -14 \\ -5 & 1 & 6 \end{bmatrix}$, $(c)\ \begin{bmatrix} 1 & -3 & 2 \\ -3 & 3 & -1 \\ 2 & -1 & 0 \end{bmatrix}$, $(d)\ \begin{bmatrix} 1 & -2/3 & 0 & 0 \\ 0 & 1/3 & 0 & 0 \\ 0 & 0 & 1/2 & -1/6 \\ 0 & 0 & 0 & 1/3 \end{bmatrix}$

9. Find the inverse of the matrix of Problem 8(d) as a direct sum.

10. Obtain the inverses of the matrices of Problem 8 using the method of Problem 3.

11. Same, for the matrices $(a)\ \begin{bmatrix} 1 & 1 & 1 & 1 \\ 1 & 2 & 3 & -4 \\ 2 & 3 & 5 & -5 \\ 3 & -4 & -5 & 8 \end{bmatrix}$, $(b)\ \begin{bmatrix} 3 & 4 & 2 & 7 \\ 2 & 3 & 3 & 2 \\ 5 & 7 & 3 & 9 \\ 2 & 3 & 2 & 3 \end{bmatrix}$, $(c)\ \begin{bmatrix} 2 & 5 & 2 & 3 \\ 2 & 3 & 3 & 4 \\ 3 & 6 & 3 & 2 \\ 4 & 12 & 0 & 8 \end{bmatrix}$, $(d)\ \begin{bmatrix} 1 & 3 & 3 & 2 & 1 \\ 1 & 4 & 3 & 3 & -1 \\ 1 & 3 & 4 & 1 & 1 \\ 1 & 1 & 1 & 1 & -1 \\ 1 & -2 & -1 & 2 & 2 \end{bmatrix}$

Ans. $(a)\ \frac{1}{18}\begin{bmatrix} 2 & 16 & -6 & 4 \\ 22 & 41 & -30 & -1 \\ -10 & -44 & 30 & -2 \\ 4 & -13 & 6 & -1 \end{bmatrix}$ $\quad (c)\ \frac{1}{48}\begin{bmatrix} -144 & 36 & 60 & 21 \\ 48 & -20 & -12 & -5 \\ 48 & -4 & -12 & -13 \\ 0 & 12 & -12 & 3 \end{bmatrix}$

$(b)\ \frac{1}{2}\begin{bmatrix} -1 & 11 & 7 & -26 \\ -1 & -7 & -3 & 16 \\ 1 & 1 & -1 & 0 \\ 1 & -1 & -1 & 2 \end{bmatrix}$ $\quad (d)\ \frac{1}{15}\begin{bmatrix} 30 & -20 & -15 & 25 & -5 \\ 30 & -11 & -18 & 7 & -8 \\ -30 & 12 & 21 & -9 & 6 \\ -15 & 12 & 6 & -9 & 6 \\ 15 & -7 & -6 & -1 & -1 \end{bmatrix}$

12. Use the result of Example 4 to obtain the inverse of the matrix of Problem 11(d) by partitioning.

13. Obtain by partitioning the inverses of the matrices of Problems 8(a), 8(b), 11(a) – 11(c).

14. Obtain by partitioning the inverses of the symmetric matrices (a) $\begin{bmatrix} 1 & 2 & -1 & 2 \\ 2 & 2 & -1 & 1 \\ -1 & -1 & 1 & -1 \\ 2 & 1 & -1 & 2 \end{bmatrix}$, (b) $\begin{bmatrix} 0 & 1 & 2 & 2 \\ 1 & 1 & 2 & 3 \\ 2 & 2 & 2 & 3 \\ 2 & 3 & 3 & 3 \end{bmatrix}$.

Ans. (a) $-\frac{1}{2}\begin{bmatrix} 1 & -1 & -1 & -1 \\ -1 & -1 & -1 & 1 \\ -1 & -1 & -5 & -1 \\ -1 & 1 & -1 & -1 \end{bmatrix}$, (b) $\begin{bmatrix} -3 & 3 & -3 & 2 \\ 3 & -4 & 4 & -2 \\ -3 & 4 & -5 & 3 \\ 2 & -2 & 3 & -2 \end{bmatrix}$

15. Prove: If A is non-singular, then $AB = AC$ implies $B = C$.

16. Show that if the non-singular matrices A and B commute, so also do
(a) A^{-1} and B, (b) A and B^{-1}, (c) A^{-1} and B^{-1}. *Hint.* (a) $A^{-1}(AB)A^{-1} = A^{-1}(BA)A^{-1}$.

17. Show that if the non-singular matrix A is symmetric, so also is A^{-1}.
Hint: $A^{-1}A = I = (AA^{-1})' = (A^{-1})'A$.

18. Show that if the non-singular symmetric matrices A and B commute, then (a) $A^{-1}B$, (b) AB^{-1}, and (c) $A^{-1}B^{-1}$ are symmetric. *Hint*: (a) $(A^{-1}B)' = (BA^{-1})' = (A^{-1})'B' = A^{-1}B$.

19. An $m \times n$ matrix A is said to have a *right inverse* B if $AB = I$ and a *left inverse* C if $CA = I$. Show that A has a right inverse if and only if A is of rank m and has a left inverse if and only if the rank of A is n.

20. Find a right inverse of $A = \begin{bmatrix} 1 & 3 & 2 & 3 \\ 1 & 4 & 1 & 3 \\ 1 & 3 & 5 & 4 \end{bmatrix}$ if one exists.

Hint. The rank of A is 3 and the submatrix $S = \begin{bmatrix} 1 & 3 & 2 \\ 1 & 4 & 1 \\ 1 & 3 & 5 \end{bmatrix}$ is non-singular with inverse S^{-1}. A right inverse of

A is the 4×3 matrix $B = \begin{bmatrix} S^{-1} \\ 0 \end{bmatrix} = \frac{1}{3}\begin{bmatrix} 17 & -9 & -5 \\ -4 & 3 & 1 \\ -1 & 0 & 1 \\ 0 & 0 & 0 \end{bmatrix}$.

21. Show that the submatrix $T = \begin{bmatrix} 1 & 3 & 3 \\ 1 & 4 & 3 \\ 1 & 3 & 4 \end{bmatrix}$ of A of Problem 20 is non-singular and obtain $\begin{bmatrix} 7 & -3 & -3 \\ -1 & 1 & 0 \\ 0 & 0 & 0 \\ -1 & 0 & 1 \end{bmatrix}$ as another right inverse of A.

22. Obtain $\begin{bmatrix} 7 & -1 & -1 & a \\ -3 & 1 & 0 & b \\ -3 & 0 & 1 & c \end{bmatrix}$ as a left inverse of $\begin{bmatrix} 1 & 1 & 1 \\ 3 & 4 & 3 \\ 3 & 3 & 4 \\ 0 & 0 & 0 \end{bmatrix}$, where a, b, and c are arbitrary.

23. Show that $A = \begin{bmatrix} 1 & 3 & 4 & 7 \\ 1 & 4 & 5 & 9 \\ 2 & 3 & 5 & 8 \end{bmatrix}$ has neither a right nor a left inverse.

24. Prove: If $|A_{11}| \neq 0$, then $\begin{vmatrix} A_{11} & A_{12} \\ A_{21} & A_{22} \end{vmatrix} = |A_{11}| \cdot |A_{22} - A_{21}A_{11}^{-1}A_{12}|$.

25. If $|I + A| \neq 0$, then $(I+A)^{-1}$ and $(I-A)$ commute.

26. Prove:(i) of Problem 23, Chapter 6.

Chapter 8

Fields

NUMBER FIELDS. A collection or set S of real or complex numbers, consisting of more than the element 0, is called a **number field** provided the operations of addition, subtraction, multiplication, and division (except by 0) on any two of the numbers yield a number of S.

Examples of number fields are:

(*a*) the set of all rational numbers,
(*b*) the set of all real numbers,
(*c*) the set of all numbers of the form $a+b\sqrt{3}$, where a and b are rational numbers,
(*d*) the set of all complex numbers $a+bi$, where a and b are real numbers.

The set of all integers and the set of all numbers of the form $b\sqrt{3}$, where b is a rational number, are **not** number fields.

GENERAL FIELDS. A collection or set S of two or more elements, together with two operations called addition (+) and multiplication (·), is called a field F provided that ($a, b, c, \ldots$ are elements of F, i.e. scalars),

A_1: $a+b$ is a unique element of F
A_2: $a+b = b+a$
A_3: $a+(b+c) = (a+b)+c$
A_4: For every element a in F there exists an element 0 in F such that $a+0 = 0+a = a$.
A_5: For each element a in F there exists a unique element $-a$ in F such that $a+(-a) = 0$
M_1: $ab = a \cdot b$ is a unique element of F
M_2: $ab = ba$
M_3: $(ab)c = a(bc)$
M_4: For every element a in F there exists an element $1 \neq 0$ such that $1 \cdot a = a \cdot 1 = a$.
M_5: For each element $a \neq 0$ in F there exists a unique element a^{-1} in F such that $a \cdot a^{-1} = a^{-1} \cdot a = 1$.
D_1: $a(b+c) = ab+ac$
D_2: $(a+b)c = ac+bc$

In addition to the number fields listed above, other examples of fields are:

(*e*) the set of all quotients $\dfrac{P(x)}{Q(x)}$ of polynomials in x with real coefficients,

(*f*) the set of all 2×2 matrices of the form $\begin{bmatrix} a & -b \\ b & a \end{bmatrix}$ where a and b are real numbers.

(*g*) the set in which $a+a = 0$. This field, called **of characteristic 2**, will be **excluded** hereafter. In this field, for example, the customary proof that a determinant having two rows identical is 0 is not valid. By interchanging the two identical rows, we are led to $D = -D$ or $2D = 0$; but D is not necessarily 0.

SUBFIELDS. If S and T are two sets and if every member of S is also a member of T, then S is called a **subset** of T.

If S and T are fields and if S is a subset of T, then S is called a **subfield** of T. For example, the field of all real numbers is a subfield of the field of all complex numbers; the field of all rational numbers is a subfield of the field of all real numbers and the field of all complex numbers.

MATRICES OVER A FIELD. When all of the elements of a matrix A are in a field F, we say that "**A is over F**". For example,

$$A = \begin{bmatrix} 1 & 1/2 \\ 1/4 & 2/3 \end{bmatrix} \text{ is over the rational field and } B = \begin{bmatrix} 1 & 1+i \\ 2 & 1-3i \end{bmatrix} \text{ is over the complex field}$$

Here, A is also over the real field while B is not; also A is over the complex field.

Let $A, B, C, \ldots$ be matrices over the same field F and let F be the smallest field which contains all the elements; that is, if all the elements are rational numbers, the field F is the rational field and not the real or complex field. An examination of the various operations defined on these matrices, individually or collectively, in the previous chapters shows that no elements other than those in F are ever required. For example:

The sum, difference, and product are matrices over F.

If A is non-singular, its inverse is over F.

If $A \sim I$ then there exist matrices P and Q over F such that $PAQ = I$ and I is over F.

If A is over the rational field and is of rank r, its rank is unchanged when considered over the real or the complex field.

Hereafter when A is said to be over F it will be assumed that F is the smallest field containing all of its elements.

In later chapters it will at times be necessary to restrict the field, say, to the real field. At other times, the field of the elements will be extended, say, from the rational field to the real field. Otherwise, the statement "A over F" implies no restriction on the field, **except for the excluded field of characteristic two**.

SOLVED PROBLEM

1. Verify that the set of all complex numbers constitutes a field.

To do this we simply check the properties $A_1 - A_5$, $M_1 - M_5$, and $D_1 - D_2$. The zero element (A_4) is 0 and the unit element (M_4) is 1. If $a+bi$ and $c+di$ are two elements, the negative (A_5) of $a+bi$ is $-a-bi$, the product (M_1) is $(a+bi)(c+di) = (ac-bd) + (ad+bc)i$; the inverse ($M_5$) of $a+bi \neq 0$ is

$$\frac{1}{a+bi} = \frac{a-bi}{a^2+b^2} = \frac{a}{a^2+b^2} - \frac{bi}{a^2+b^2}$$

Verification of the remaining properties is left as an exercise for the reader.

SUPPLEMENTARY PROBLEMS

2. Verify (*a*) the set of all real numbers of the form $a+b\sqrt{5}$ where a and b are rational numbers and

(*b*) the set of all quotients $\dfrac{P(x)}{Q(x)}$ of polynomials in x with real coefficients constitute fields.

3. Verify (*a*) the set of all rational numbers,

(*b*) the set of all numbers $a+b\sqrt{3}$, where a and b are rational numbers, and

(*c*) the set of all numbers $a+bi$, where a and b are rational numbers are subfields of the complex field.

4. Verify that the set of all 2×2 matrices of the form $\begin{bmatrix} a & -b \\ b & a \end{bmatrix}$, where a and b are rational numbers, forms a field.

Show that this is a subfield of the field of all 2×2 matrices of the form $\begin{bmatrix} a & -b \\ b & a \end{bmatrix}$ where a and b are real numbers.

5. Why does not the set of all 2×2 matrices with real elements form a field?

6. A set R of elements $a,b,c,\ldots$ satisfying the conditions (A_1, A_2, A_3, A_4, A_5; M_1, M_3; D_1, D_2) of Page 64 is called a **ring**. To emphasize the fact that multiplication is not commutative, R may be called a **non-commutative ring**. When a ring R satisfies M_2, it is called **commutative**. When a ring R satisfies M_4, it is spoken of as a ring **with unit element**.

Verify:

(*a*) the set of even integers $0, \pm 2, \pm 4, \ldots$ is an example of a commutative ring without unit element.

(*b*) the set of all integers $0, \pm 1, \pm 2, \pm 3, \ldots$ is an example of a commutative ring with unit element.

(*c*) the set of all n-square matrices over F is an example of a non-commutative ring with unit element.

(*d*) the set of all 2×2 matrices of the form $\begin{bmatrix} a & -b \\ b & a \end{bmatrix}$, where a and b are real numbers, is an example of a commutative ring with unit element.

7. Can the set (*a*) of Problem 6 be turned into a commutative ring with unit element by simply adjoining the elements ± 1 to the set?

8. By Problem 4, the set (*d*) of Problem 6 is a field. Is every field a ring? Is every commutative ring with unit element a field?

9. Describe the ring of all 2×2 matrices $\begin{bmatrix} 0 & 0 \\ a & b \end{bmatrix}$, where a and b are in F. If A is any matrix of the ring and $L = \begin{bmatrix} 0 & 0 \\ 1 & 1 \end{bmatrix}$, show that $LA = A$. Call L a **left unit element**. Is there a right unit element?

10. Let C be the field of all complex numbers $p+qi$ and K be the field of all 2×2 matrices $\begin{bmatrix} u & -v \\ v & u \end{bmatrix}$ where p, q, u, v are real numbers. Take the complex number $a+bi$ and the matrix $\begin{bmatrix} a & -b \\ b & a \end{bmatrix}$ as corresponding elements of the two sets and call each the **image** of the other.

(*a*) Write the image of $\begin{bmatrix} 2 & -3 \\ 3 & 2 \end{bmatrix}$, $\begin{bmatrix} 0 & -4 \\ 4 & 0 \end{bmatrix}$; $3+2i$, 5.

(*b*) Show that the image of the sum (product) of two elements of K is the sum (product) of their images in C.

(*c*) Show that the image of the identity element of K is the identity element of C.

(*d*) What is the image of the conjugate of $a+bi$?

(*e*) What is the image of the inverse of $\begin{bmatrix} a & -b \\ b & a \end{bmatrix}$?

This is an example of an **isomorphism** between two sets.

Chapter 9

Linear Dependence of Vectors and Forms

THE ORDERED PAIR of real numbers (x_1, x_2) is used to denote a point X in a plane. The same pair of numbers, written as $[x_1, x_2]$, will be used here to denote the **two-dimensional vector** or **2-vector** OX (see Fig. 9-1).

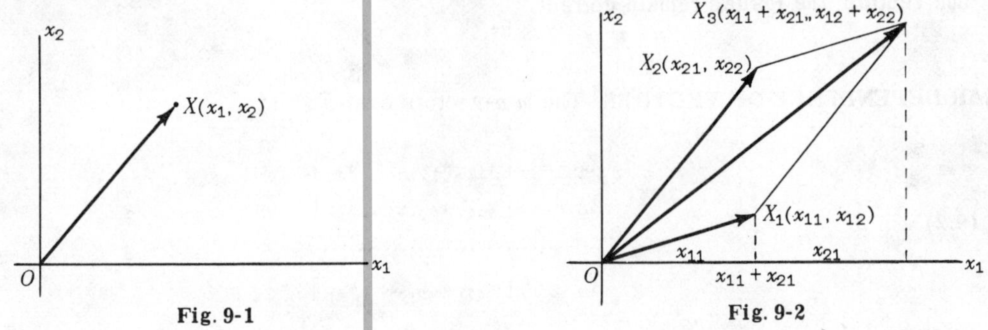

Fig. 9-1 Fig. 9-2

If $X_1 = [x_{11}, x_{12}]$ and $X_2 = [x_{21}, x_{22}]$ are distinct 2-vectors, the parallelogram law for their sum (see Fig. 9-2) yields

$$X_3 \quad = \quad X_1 + X_2 \quad = \quad [x_{11}+x_{21},\ x_{12}+x_{22}]$$

Treating X_1 and X_2 as 1×2 matrices, we see that this is merely the rule for adding matrices given in Chapter 1. Moreover, if k is any scalar,

$$kX_1 \quad = \quad [kx_{11},\ kx_{12}]$$

is the familiar multiplication of a vector by a real number of physics.

VECTORS. By an **n-dimensional vector** or **n-vector** X over F is meant an ordered set of n elements x_i of F, as

$$\textbf{(9.1)} \qquad X \quad = \quad [x_1, x_2, \ldots, x_n]$$

The elements $x_1, x_2, \ldots, x_n$ are called respectively the first, second, ..., nth **components** of X.

Later we shall find it more convenient to write the components of a vector in a column, as

$$\textbf{(9.1}'\textbf{)} \qquad X \quad = \quad [x_1, x_2, \ldots, x_n]' \quad = \quad \begin{bmatrix} x_1 \\ x_2 \\ \cdot \\ \cdot \\ \cdot \\ x_n \end{bmatrix}$$

Now **(9.1)** and **(9.1′)** denote the same vector; however, we shall speak of **(9.1)** as a **row vector** and **(9.1′)** as a **column vector**. We may, then, consider the $p \times q$ matrix A as defining p row vectors (the elements of a row being the components of a q-vector) or as defining q column vectors.

The vector, all of whose components are zero, is called the **zero vector** and is denoted by 0.

The sum and difference of two row (column) vectors and the product of a scalar and a vector are formed by the rules governing matrices.

Example 1. Consider the 3-vectors

$$X_1 = [3,1,-4],\quad X_2 = [2,2,-3],\quad X_3 = [0,-4,1], \quad \text{and} \quad X_4 = [-4,-4,6]$$

(a) $2X_1 - 5X_2 = 2[3,1,-4] - 5[2,2,-3] = [6,2,-8] - [10,10,-15] = [-4,-8,7]$

(b) $2X_2 + X_4 = 2[2,2,-3] + [-4,-4,6] = [0,0,0] = 0$

(c) $2X_1 - 3X_2 - X_3 = 0$

(d) $2X_1 - X_2 - X_3 + X_4 = 0$

The vectors used here are row vectors. Note that if each bracket is primed to denote column vectors, the results remain correct.

LINEAR DEPENDENCE OF VECTORS. The m n-vectors over F

$$(9.2)\qquad \begin{aligned} X_1 &= [x_{11}, x_{12}, \ldots, x_{1n}] \\ X_2 &= [x_{21}, x_{22}, \ldots, x_{2n}] \\ &\cdots\cdots\cdots\cdots\cdots \\ X_m &= [x_{m1}, x_{m2}, \ldots, x_{mn}] \end{aligned}$$

are said to be **linearly dependent** over F provided there exist m elements $k_1, k_2, \ldots, k_m$ of F, not all zero, such that

$$(9.3)\qquad k_1X_1 + k_2X_2 + \cdots + k_mX_m = 0$$

Otherwise, the m vectors are said to be **linearly independent**.

Example 2. Consider the four vectors of Example 1. By (b) the vectors X_2 and X_4 are linearly dependent; so also are X_1, X_2, and X_3 by (c) and the entire set by (d).

The vectors X_1 and X_2, however, are linearly independent. For, assume the contrary so that

$$k_1X_1 + k_2X_2 = [3k_1 + 2k_2,\ k_1 + 2k_2,\ -4k_1 - 3k_2] = [0,0,0]$$

Then $3k_1 + 2k_2 = 0$, $k_1 + 2k_2 = 0$, and $-4k_1 - 3k_2 = 0$. From the first two relations $k_1 = 0$ and then $k_2 = 0$.

Any n-vector X and the n-zero vector 0 are linearly dependent.

A vector X_{m+1} is said to be expressible as a **linear combination** of the vectors $X_1, X_2, \ldots, X_m$ if there exist elements $k_1, k_2, \ldots, k_m$ of F such that

$$X_{m+1} = k_1X_1 + k_2X_2 + \cdots + k_mX_m$$

BASIC THEOREMS. If in (9.3), $k_i \neq 0$, we may solve for

$$X_i = -\frac{1}{k_i}\{k_1X_1 + \cdots + k_{i-1}X_{i-1} + k_{i+1}X_{i+1} + \cdots + k_mX_m\} \qquad \text{or}$$

$$(9.4)\qquad X_i = s_1X_1 + \cdots + s_{i-1}X_{i-1} + s_{i+1}X_{i+1} + \cdots + s_mX_m$$

Thus,

I. If m vectors are linearly dependent, some one of them may always be expressed as a linear combination of the others.

II. If m vectors $X_1, X_2, \ldots, X_m$ are linearly independent while the set obtained by adding another vector X_{m+1} is linearly dependent, then X_{m+1} can be expressed as a linear combination of $X_1, X_2, \ldots, X_m$.

Example 3. From Example 2, the vectors X_1, and X_2 are linearly independent while X_1, X_2, and X_3 are linearly dependent, satisfying the relations $2X_1 - 3X_2 - X_3 = 0$. Clearly, $X_3 = 2X_1 - 3X_2$.

III. If among the m vectors $X_1, X_2, \ldots, X_m$ there is a subset of $r<m$ vectors which are linearly dependent, the vectors of the entire set are linearly dependent.

Example 4. By (*b*) of Example 1, the vectors X_2 and X_4 are linearly dependent; by (*d*), the set of four vectors is linearly dependent.
See Problem 1.

IV. If the rank of the matrix

(9.5)
$$A = \begin{bmatrix} x_{11} & x_{12} & \ldots & x_{1n} \\ x_{21} & x_{22} & \ldots & x_{2n} \\ \ldots & \ldots & \ldots & \ldots \\ x_{m1} & x_{m2} & \ldots & x_{mn} \end{bmatrix}, \qquad m \leq n,$$

associated with the m vectors **(9.2)** is $r<m$, there are exactly r vectors of the set which are linearly independent while each of the remaining m-r vectors can be expressed as a linear combination of these r vectors.
See Problems 2-3.

V. A necessary and sufficient condition that the vectors **(9.2)** be linearly dependent is that the matrix **(9.5)** of the vectors be of rank $r<m$. If the rank is m, the vectors are linearly independent.

The set of vectors **(9.2)** is necessarily linearly dependent if $m>n$.

If the set of vectors **(9.2)** is linearly independent so also is every subset of them.

A LINEAR FORM over F in n variables $x_1, x_2, \ldots, x_n$ is a polynomial of the type

(9.6)
$$\sum_{i=1}^{n} a_i x_i = a_1x_1 + a_2x_2 + \cdots + a_nx_n$$

where the coefficients are in F.

Consider a system of m linear forms in n variables

(9.7)
$$\begin{cases} f_1 = a_{11}x_1 + a_{12}x_2 + \cdots + a_{1n}x_n \\ f_2 = a_{21}x_1 + a_{22}x_2 + \cdots + a_{2n}x_n \\ \ldots\ldots\ldots\ldots\ldots\ldots\ldots\ldots \\ f_m = a_{m1}x_1 + a_{m2}x_2 + \cdots + a_{mn}x_n \end{cases}$$

and the associated matrix

$$A = \begin{bmatrix} a_{11} & a_{12} & \ldots & a_{1n} \\ a_{21} & a_{22} & \ldots & a_{2n} \\ \ldots & \ldots & \ldots & \ldots \\ a_{m1} & a_{m2} & \ldots & a_{mn} \end{bmatrix}$$

If there exist elements $k_1, k_2, \ldots, k_m$, not all zero, in F such that

$$k_1f_1 + k_2f_2 + \ldots + k_m f_m = 0$$

the forms (**9.7**) are said to be **linearly dependent**; otherwise the forms are said to be **linearly independent**. Thus, the linear dependence or independence of the forms of (**9.7**) is equivalent to the linear dependence or independence of the row vectors of A.

Example 5. The forms $f_1 = 2x_1 - x_2 + 3x_3$, $f_2 = x_1 + 2x_2 + 4x_3$, $f_3 = 4x_1 - 7x_2 + x_3$ are linearly dependent since $A = \begin{bmatrix} 2 & -1 & 3 \\ 1 & 2 & 4 \\ 4 & -7 & 1 \end{bmatrix}$ is of rank 2. Here, $3f_1 - 2f_2 - f_3 = 0$.

The system (**9.7**) is necessarily dependent if $m>n$. Why?

SOLVED PROBLEMS

1. Prove: If among the m vectors $X_1, X_2, \ldots, X_m$ there is a subset, say, $X_1, X_2, \ldots, X_r$, $r<m$, which is linearly dependent, so also are the m vectors.

Since, by hypothesis, $k_1X_1 + k_2X_2 + \cdots + k_rX_r = 0$ with not all of the k's equal to zero, then

$$k_1X_1 + k_2X_2 + \cdots + k_rX_r + 0\cdot X_{r+1} + \cdots + 0\cdot X_m = 0$$

with not all of the k's equal to zero and the entire set of vectors is linearly dependent.

2. Prove: If the rank of the matrix associated with a set of m n-vectors is $r<m$, there are exactly r vectors which are linearly independent while each of the remaining m-r vectors can be expressed as a linear combination of these r vectors.

Let (**9.5**) be the matrix and suppose first that $m \leq n$. If the r-rowed minor in the upper left hand corner is equal to zero, we interchange rows and columns as are necessary to bring a non-vanishing r-rowed minor into this position and then renumber all rows and columns in natural order. Thus, we have

$$\Delta = \begin{vmatrix} x_{11} & x_{12} & \ldots & x_{1r} \\ x_{21} & x_{22} & \ldots & x_{2r} \\ \ldots & \ldots & \ldots & \ldots \\ x_{r1} & x_{r2} & \ldots & x_{rr} \end{vmatrix} \neq 0$$

Consider now an $(r+1)$-rowed minor

$$\nabla = \begin{vmatrix} x_{11} & x_{12} & \ldots & x_{1r} & x_{1q} \\ x_{21} & x_{22} & \ldots & x_{2r} & x_{2q} \\ \ldots & \ldots & \ldots & \ldots & \ldots \\ x_{r1} & x_{r2} & \ldots & x_{rr} & x_{rq} \\ x_{p1} & x_{p2} & \ldots & x_{pr} & x_{pq} \end{vmatrix} = 0$$

where the elements x_{pj} and x_{iq} are respectively from any row and any column not included in Δ. Let $k_1, k_2, \ldots, k_{r+1} = \Delta$ be the respective cofactors of the elements $x_{1q}, x_{2q}, \ldots, x_{rq}, x_{pq}$ of the last column of ∇. Then, by (**3.10**)

$$k_1x_{1i} + k_2x_{2i} + \cdots + k_rx_{ri} + k_{r+1}x_{pi} = 0 \qquad (i = 1, 2, \ldots, r)$$

and by hypothesis $\quad k_1x_{1q} + k_2x_{2q} + \cdots + k_rx_{rq} + k_{r+1}x_{pq} = \nabla = 0$

Now let the last column of ∇ be replaced by another of the remaining columns, say the column numbered u, not appearing in Δ. The cofactors of the elements of this column are precisely the k's obtained above so that

$$k_1x_{1u} + k_2x_{2u} + \cdots + k_rx_{ru} + k_{r+1}x_{pu} = 0$$

Thus,

$$k_1x_{1t} + k_2x_{2t} + \cdots + k_rx_{rt} + k_{r+1}x_{pt} = 0 \qquad (t = 1, 2, \ldots, n)$$

and, summing over all values of t,

$$k_1X_1 + k_2X_2 + \cdots + k_rX_r + k_{r+1}X_p = 0$$

Since $k_{r+1} = \Delta \neq 0$, X_p is a linear combination of the r linearly independent vectors $X_1, X_2, \ldots, X_r$. But X_p was any one of the m-r vectors $X_{r+1}, X_{r+2}, \ldots, X_m$; hence, each of these may be expressed as a linear combination of $X_1, X_2, \ldots, X_r$.

For the case $m > n$, consider the matrix when to each of the given m vectors m-n additional zero components are added. This matrix is $[A \mid 0]$. Clearly the linear dependence or independence of the vectors and also the rank of A have not been changed.

Thus, in either case, the vectors $X_{r+1}, \ldots, X_m$ are linear combinations of the linearly independent vectors $X_1, X_2, \ldots, X_r$ as was to be proved.

3. Show, using a matrix, that each triple of vectors

$$(a)\quad \begin{aligned} X_1 &= [1, 2, -3, 4] \\ X_2 &= [3, -1, 2, 1] \\ X_3 &= [1, -5, 8, -7] \end{aligned} \qquad \text{and} \qquad (b)\quad \begin{aligned} X_1 &= [2, 3, 1, -1] \\ X_2 &= [2, 3, 1, -2] \\ X_3 &= [4, 6, 2, -3] \end{aligned}$$

is linearly dependent. In each determine a maximum subset of linearly independent vectors and express the others as linear combinations of these.

(*a*) Here, $\begin{bmatrix} 1 & 2 & -3 & 4 \\ 3 & -1 & 2 & 1 \\ 1 & -5 & 8 & -7 \end{bmatrix}$ is of rank 2; there are two linearly independent vectors, say X_1 and X_2. The minor $\begin{vmatrix} 1 & 2 \\ 3 & -1 \end{vmatrix} \neq 0$. Consider then the minor $\begin{vmatrix} 1 & 2 & -3 \\ 3 & -1 & 2 \\ 1 & -5 & 8 \end{vmatrix}$. The cofactors of the elements of the third column are respectively -14, 7, and -7. Then $-14X_1 + 7X_2 - 7X_3 = 0$ and $X_3 = -2X_1 + X_2$.

(*b*) Here $\begin{bmatrix} 2 & 3 & 1 & -1 \\ 2 & 3 & 1 & -2 \\ 4 & 6 & 2 & -3 \end{bmatrix}$ is of rank 2; there are two linearly independent vectors, say X_1 and X_2. Now the minor $\begin{vmatrix} 2 & 3 \\ 2 & 3 \end{vmatrix} = 0$; we interchange the 2nd and 4th columns to obtain $\begin{bmatrix} 2 & -1 & 1 & 3 \\ 2 & -2 & 1 & 3 \\ 4 & -3 & 2 & 6 \end{bmatrix}$ for which $\begin{vmatrix} 2 & -1 \\ 2 & -2 \end{vmatrix} \neq 0$.

The cofactors of the elements of the last column of $\begin{vmatrix} 2 & -1 & 1 \\ 2 & -2 & 1 \\ 4 & -3 & 2 \end{vmatrix}$ are $2, 2, -2$ respectively. Then

$$2X_1 + 2X_2 - 2X_3 = 0 \qquad \text{and} \qquad X_3 = X_1 + X_2$$

4. Let $P_1(1,1,1)$, $P_2(1,2,3)$, $P_3(3,1,2)$, and $P_4(2,3,4)$ be points in ordinary space. The points P_1, P_2 and the origin of coordinates determine a plane π of equation

$$\text{(i)} \qquad \begin{vmatrix} x & y & z & 1 \\ 1 & 1 & 1 & 1 \\ 1 & 2 & 3 & 1 \\ 0 & 0 & 0 & 1 \end{vmatrix} = x - 2y + z = 0$$

Substituting the coordinates of P_4 into the left member of **(i)**, we have

$$\begin{vmatrix} 2 & 3 & 4 & 1 \\ 1 & 1 & 1 & 1 \\ 1 & 2 & 3 & 1 \\ 0 & 0 & 0 & 1 \end{vmatrix} = \begin{vmatrix} 2 & 3 & 4 & 0 \\ 1 & 1 & 1 & 0 \\ 1 & 2 & 3 & 0 \\ 0 & 0 & 0 & 1 \end{vmatrix} = \begin{vmatrix} 2 & 3 & 4 \\ 1 & 1 & 1 \\ 1 & 2 & 3 \end{vmatrix} = 0$$

Thus, P_4 lies in π. The significant fact here is that $[P_4, P_1, P_2]' = \begin{bmatrix} 2 & 3 & 4 \\ 1 & 1 & 1 \\ 1 & 2 & 3 \end{bmatrix}$ is of rank 2.

We have verified: Any three points of ordinary space lie in a plane through the origin provided the matrix of their coordinates is of rank 2.

Show that P_3 does not lie in π.

SUPPLEMENTARY PROBLEMS

5. Prove: If m vectors $X_1, X_2, \ldots, X_m$ are linearly independent while the set obtained by adding another vector X_{m+1} is linearly dependent, then X_{m+1} can be expressed as a linear combination of $X_1, X_2, \ldots, X_m$.

6. Show that the representation of X_{m+1} in Problem 5 is unique.

   Hint: Suppose $X_{m+1} = \sum_{i=1}^{m} k_i X_i = \sum_{i=1}^{n} s_i X_i$ and consider $\sum_{i=1}^{m} (k_i - s_i) X_i$.

7. Prove: A necessary and sufficient condition that the vectors **(9.2)** be linearly dependent is that the matrix **(9.5)** of the vectors be of rank $r<m$.

   Hint: Suppose the m vectors are linearly dependent so that **(9.4)** holds. In **(9.5)** subtract from the ith row the product of the first row by s_1, the product of the second row by s_2, ... as indicated in **(9.4)**. For the converse, see Problem 2.

8. Examine each of the following sets of vectors over the real field for linear dependence or independence. In each dependent set select a maximum linearly independent subset and express each of the remaining vectors as a linear combination of these.

   (*a*) $X_1 = [2, -1, 3, 2]$, $X_2 = [1, 3, 4, 2]$, $X_3 = [3, -5, 2, 2]$

   (*b*) $X_1 = [1, 2, 1]$, $X_2 = [2, 1, 4]$, $X_3 = [4, 5, 6]$, $X_4 = [1, 8, -3]$

   (*c*) $X_1 = [2, 1, 3, 2, -1]$, $X_2 = [4, 2, 1, -2, 3]$, $X_3 = [0, 0, 5, 6, -5]$, $X_4 = [6, 3, -1, -6, 7]$

   Ans. (*a*) $X_3 = 2X_1 - X_2$ (*b*) $X_3 = 2X_1 + X_2$, $X_4 = 5X_1 - 2X_2$ (*c*) $X_3 = 2X_1 - X_2$, $X_4 = 2X_2 - X_1$

9. Why can there be no more than n linearly independent n-vectors over F?

10. Show that if in (9.2) either $X_i = X_j$ or $X_i = aX_j$, a in F, the set of vectors is linearly dependent. Is the converse true?

11. Show that any n-vector X and the n-zero vector are linearly dependent; hence, X and 0 are considered proportional.
Hint: Consider $k_1X + k_2 \cdot 0 = 0$ where $k_1 = 0$ and $k_2 \neq 0$.

12. (a) Show that $X_1 = [1, 1+i, i]$, $X_2 = [i, -i, 1-i]$ and $X_3 = [1+2i, 1-i, 2-i]$ are linearly dependent over the rational field and, hence, over the complex field.
(b) Show that $X_1 = [1, 1+i, i]$, $X_2 = [i, -i, 1-i]$, and $X_3 = [0, 1-2i, 2-i]$ are linearly independent over the real field but are linearly dependent over the complex field.

13. Investigate the linear dependence or independence of the linear forms:

(a) $f_1 = 3x_1 - x_2 + 2x_3 + x_4$, $f_2 = 2x_1 + 3x_2 - x_3 + 2x_4$, $f_3 = 5x_1 - 9x_2 + 8x_3 - x_4$

(b) $f_1 = 2x_1 - 3x_2 + 4x_3 - 2x_4$, $f_2 = 3x_1 + 2x_2 - 2x_3 + 5x_4$, $f_3 = 5x_1 - x_2 + 2x_3 + x_4$

Ans. (a) $3f_1 - 2f_2 - f_3 = 0$

14. Consider the linear dependence or independence of a system of polynomials

$$P_i = a_{i0}x^n + a_{i1}x^{n-1} + \cdots + a_{i\,n-1}x + a_{in} \qquad (i = 1, 2, \ldots, m)$$

and show that the system is linearly dependent or independent according as the row vectors of the coefficient matrix

$$A = \begin{bmatrix} a_{10} & a_{11} & \cdots & a_{1n} \\ a_{20} & a_{21} & \cdots & a_{2n} \\ \cdots & \cdots & \cdots & \cdots \\ a_{m0} & a_{m1} & \cdots & a_{mn} \end{bmatrix}$$

are linearly dependent or independent, that is, according as the rank r of A is less than or equal to m.

15. If the polynomials of either system are linearly dependent, find a linear combination which is identically zero.

(a) $P_1 = x^3 - 3x^2 + 4x - 2$, $P_2 = 2x^2 - 6x + 4$, $P_3 = x^3 - 2x^2 + x$

(b) $P_1 = 2x^4 + 3x^3 - 4x^2 + 5x + 3$, $P_2 = x^3 + 2x^2 - 3x + 1$, $P_3 = x^4 + 2x^3 - x^2 + x + 2$

Ans. (a) $2P_1 + P_2 - 2P_3 = 0$ (b) $P_1 + P_2 - 2P_3 = 0$

16. Consider the linear dependence or independence of a set of 2×2 matrices $M_1 = \begin{bmatrix} a & b \\ c & d \end{bmatrix}$, $M_2 = \begin{bmatrix} e & f \\ g & h \end{bmatrix}$, $M_3 = \begin{bmatrix} p & q \\ s & t \end{bmatrix}$ over F.

Show that $k_1M_1 + k_2M_2 + k_3M_3 = 0$, when not all the k's (in F) are zero, requires that the rank of the matrix $\begin{bmatrix} a & b & c & d \\ e & f & g & h \\ p & q & s & t \end{bmatrix}$ be < 3. (Note that the matrices M_1, M_2, M_3 are considered as defining vectors of four components.)

Extend the result to a set of $m \times n$ matrices.

17. Show that $\begin{bmatrix} 1 & 2 & 3 \\ 3 & 2 & 4 \\ 1 & 3 & 2 \end{bmatrix}$, $\begin{bmatrix} 2 & 1 & 3 \\ 3 & 4 & 2 \\ 2 & 2 & 1 \end{bmatrix}$, and $\begin{bmatrix} 0 & 3 & 3 \\ 3 & 0 & 6 \\ 0 & 4 & 3 \end{bmatrix}$ are linearly dependent.

18. Show that any 2×2 matrix can be written as a linear combination of the matrices $\begin{bmatrix} 1 & 0 \\ 0 & 0 \end{bmatrix}$, $\begin{bmatrix} 0 & 1 \\ 0 & 0 \end{bmatrix}$, $\begin{bmatrix} 0 & 0 \\ 1 & 0 \end{bmatrix}$, an $\begin{bmatrix} 0 & 0 \\ 0 & 1 \end{bmatrix}$. Generalize to $n \times n$ matrices.

19. If the n-vectors $X_1, X_2, \ldots, X_n$ are linearly independent, show that the vectors $Y_1, Y_2, \ldots, Y_n$, where $Y_i = \sum_{j=1}^{n} a_{ij}X_j$, are linearly independent if and only if $A = [a_{ij}]$ is non-singular.

20. If A is of rank r, show how to construct a non-singular matrix B such that $AB = [C_1, C_2, \ldots, C_r, 0, \ldots, 0$ where $C_1, C_2, \ldots, C_r$ are a given set of linearly independent columns of A.

21. Given the points $P_1(1,1,1,1)$, $P_2(1,2,3,4)$, $P_3(2,2,2,2)$, and $P_4(3,4,5,6)$ of four-dimensional space.
(*a*) Show that the rank of $[P_1, P_3]'$ is 1 so that the points lie on a line through the origin.
(*b*) Show that $[P_1, P_2, P_3, P_4]'$ is of rank 2 so that these points lie in a plane through the origin.
(*c*) Does $P_5(2,3,2,5)$ lie in the plane of (*b*)?

22. Show that every n-square matrix A over F satisfies an equation of the form

$$A^p + k_1 A^{p-1} + k_2 A^{p-2} + \ldots + k_{p-1}A + k_p I = 0$$

where the k_i are scalars of F.

Hint: Consider $I, A, A^2, A^3, \ldots, A^{n^2}$ in the light of Problem 16.

23. Find the equation of minimum degree (see Problem 22) which is satisfied by

$$(a)\ A = \begin{bmatrix} 1 & 1 \\ 1 & 1 \end{bmatrix}, \quad (b)\ A = \begin{bmatrix} 1 & -1 \\ 1 & 1 \end{bmatrix}, \quad (c)\ A = \begin{bmatrix} 1 & 0 \\ 1 & 1 \end{bmatrix}$$

Ans. (*a*) $A^2 - 2A = 0$, (*b*) $A^2 - 2A + 2I = 0$, (*c*) $A^2 - 2A + I = 0$

24. In Problem 23(*b*) and (*c*), multiply each equation by A^{-1} to obtain (*b*) $A^{-1} = I - \frac{1}{2}A$, (*c*) $A^{-1} = 2I - A$, and thus verify: If A over F is non-singular, then A^{-1} can be expressed as a polynomial in A whose coefficients are scalars of F.

Chapter 10

Linear Equations

DEFINITIONS. Consider a system of m linear equations in the n unknowns $x_1, x_2, \ldots, x_n$

$$
(10.1) \qquad \begin{cases} a_{11}x_1 + a_{12}x_2 + \ldots + a_{1n}x_n = h_1 \\ a_{21}x_1 + a_{22}x_2 + \ldots + a_{2n}x_n = h_2 \\ \ldots\ldots\ldots\ldots\ldots\ldots\ldots\ldots \\ a_{m1}x_1 + a_{m2}x_2 + \ldots + a_{mn}x_n = h_m \end{cases}
$$

in which the coefficients (a's) and the constant terms (h's) are in F.

By a solution in F of the system is meant any set of values of $x_1, x_2, \ldots, x_n$ in F which satisfy simultaneously the m equations. When the system has a solution, it is said to be **consistent**; otherwise, the system is said to be **inconsistent**. A consistent system has either just one solution or infinitely many solutions.

Two systems of linear equations over F in the same number of unknowns are called **equivalent** if every solution of either system is a solution of the other. A system of equations equivalent to **(10.1)** may be obtained from it by applying one or more of the transformations: (a) interchanging any two of the equations, (b) multiplying any equation by any non-zero constant in F, or (c) adding to any equation a constant multiple of another equation. Solving a system of consistent equations consists in replacing the given system by an equivalent system of prescribed form.

SOLUTION USING A MATRIX. In matrix notation the system of linear equations **(10.1)** may be written as

$$
(10.2) \qquad \begin{bmatrix} a_{11} & a_{12} & \ldots & a_{1n} \\ a_{21} & a_{22} & \ldots & a_{2n} \\ \ldots & \ldots & \ldots & \ldots \\ a_{m1} & a_{m2} & \ldots & a_{mn} \end{bmatrix} \begin{bmatrix} x_1 \\ x_2 \\ \cdot \\ x_n \end{bmatrix} = \begin{bmatrix} h_1 \\ h_2 \\ \cdot \\ h_m \end{bmatrix}
$$

or, more compactly, as

$$
(10.3) \qquad AX = H
$$

where $A = [a_{ij}]$ is the coefficient matrix, $X = [x_1, x_2, \ldots, x_n]'$, and $H = [h_1, h_2, \ldots, h_m]'$.

Consider now for the system **(10.1)** the **augmented matrix**

$$
(10.4) \qquad \begin{bmatrix} a_{11} & a_{12} & \ldots & a_{1n} & h_1 \\ a_{21} & a_{22} & \ldots & a_{2n} & h_2 \\ \ldots & \ldots & \ldots & \ldots & \ldots \\ a_{m1} & a_{m2} & \ldots & a_{mn} & h_m \end{bmatrix} = [A \;\; H]
$$

(Each row of **(10.4)** is simply an abbreviation of a corresponding equation of **(10.1)**; to read the equation from the row, we simply supply the unknowns and the + and = signs properly.)

To solve the system **(10.1)** by means of **(10.4)**, we proceed by elementary row transformations to replace A by the row equivalent canonical matrix of Chapter 5. In doing this, we operate on the entire rows of **(10.4)**.

Example 1. Solve the system $\begin{cases} x_1 + 2x_2 + x_3 = 2 \\ 3x_1 + x_2 - 2x_3 = 1 \\ 4x_1 - 3x_2 - x_3 = 3 \\ 2x_1 + 4x_2 + 2x_3 = 4 \end{cases}$.

The augmented matrix $[A\ \ H] = \begin{bmatrix} 1 & 2 & 1 & 2 \\ 3 & 1 & -2 & 1 \\ 4 & -3 & -1 & 3 \\ 2 & 4 & 2 & 4 \end{bmatrix} \sim \begin{bmatrix} 1 & 2 & 1 & 2 \\ 0 & -5 & -5 & -5 \\ 0 & -11 & -5 & -5 \\ 0 & 0 & 0 & 0 \end{bmatrix} \sim \begin{bmatrix} 1 & 2 & 1 & 2 \\ 0 & 1 & 1 & 1 \\ 0 & -11 & -5 & -5 \\ 0 & 0 & 0 & 0 \end{bmatrix}$

$$\sim \begin{bmatrix} 1 & 0 & -1 & 0 \\ 0 & 1 & 1 & 1 \\ 0 & 0 & 1 & 1 \\ 0 & 0 & 0 & 0 \end{bmatrix} \sim \begin{bmatrix} 1 & 0 & 0 & 1 \\ 0 & 1 & 0 & 0 \\ 0 & 0 & 1 & 1 \\ 0 & 0 & 0 & 0 \end{bmatrix}.$$

Thus, the solution is the equivalent system of equations: $x_1 = 1,\ x_2 = 0,\ x_3 = 1$. Expressed in vector form, we have $X = [1, 0, 1]'$.

FUNDAMENTAL THEOREMS. When the coefficient matrix A of the system **(10.1)** is reduced to the row equivalent canonical form C, suppose $[A\ \ H]$ is reduced to $[C\ \ K]$, where $K = [k_1, k_2, \ldots, k_m]'$. If A is of rank r, the first r rows of C contain one or more non-zero elements. The first non-zero element in each of these rows is 1 and the column in which that 1 stands has zeros elsewhere. The remaining rows consist of zeros. From the first r rows of $[C\ \ K]$, we may obtain each of the variables $x_{j_1}, x_{j_2}, \ldots, x_{j_r}$ (the notation is that of Chapter 5) in terms of the remaining variables $x_{j_{r+1}}, x_{j_{r+2}}, \ldots, x_{j_n}$ and one of the $k_1, k_2, \ldots, k_r$.

If $k_{r+1} = k_{r+2} = \ldots = k_m = 0$, then **(10.1)** is consistent and an arbitrarily selected set of values for $x_{j_{r+1}}, x_{j_{r+2}}, \ldots, x_{j_n}$ together with the resulting values of $x_{j_1}, x_{j_2}, \ldots, x_{j_r}$ constitute a solution. On the other hand, if at least one of $k_{r+1}, k_{r+2}, \ldots, k_m$ is different from zero, say $k_t \neq 0$, the corresponding equation reads

$$0x_1 + 0x_2 + \ldots + 0x_n = k_t \neq 0$$

and **(10.1)** is inconsistent.

In the consistent case, A and $[A\ \ H]$ have the same rank; in the inconsistent case, they have different ranks. Thus

I. A system $AX = H$ of m linear equations in n unknowns is consistent if and only if the coefficient matrix and the augmented matrix of the system have the same rank.

II. In a consistent system **(10.1)** of rank $r<n$, $n-r$ of the unknowns may be chosen so that the coefficient matrix of the remaining r unknowns is of rank r. When these $n-r$ unknowns are assigned any values whatever, the other r unknowns are uniquely determined.

Example 2. For the system $\begin{cases} x_1 + 2x_2 - 3x_3 - 4x_4 = 6 \\ x_1 + 3x_2 + x_3 - 2x_4 = 4 \\ 2x_1 + 5x_2 - 2x_3 - 5x_4 = 10 \end{cases}$

$$[A\ H] = \begin{bmatrix} 1 & 2 & -3 & -4 & 6 \\ 1 & 3 & 1 & -2 & 4 \\ 2 & 5 & -2 & -5 & 10 \end{bmatrix} \sim \begin{bmatrix} 1 & 2 & -3 & -4 & 6 \\ 0 & 1 & 4 & 2 & -2 \\ 0 & 1 & 4 & 3 & -2 \end{bmatrix} \sim \begin{bmatrix} 1 & 0 & -11 & -8 & 10 \\ 0 & 1 & 4 & 2 & -2 \\ 0 & 0 & 0 & 1 & 0 \end{bmatrix}$$

$$\sim \begin{bmatrix} 1 & 0 & -11 & 0 & 10 \\ 0 & 1 & 4 & 0 & -2 \\ 0 & 0 & 0 & 1 & 0 \end{bmatrix} = [C\ K]$$

Since A and $[A\ H]$ are each of rank $r = 3$, the given system is consistent; moreover, the general solution contains $n-r = 4-3 = 1$ arbitrary constants. From the last row of $[C\ K]$, $x_4 = 0$. Let $x_3 = a$, where a is arbitrary; then $x_1 = 10+11a$ and $x_2 = -2-4a$. The solution of the system is given by $x_1 = 10+11a$, $x_2 = -2-4a$, $x_3 = a$, $x_4 = 0$ or $X = [10+11a,\ -2-4a,\ a,\ 0]'$.

If a consistent system of equations over F has a unique solution (Example 1) that solution is over F. If the system has infinitely many solutions (Example 2) it has infinitely many solutions over F when the arbitrary values to be assigned are over F. However, the system has infinitely many solutions over any field $\mathfrak{F}$ of which F is a subfield. For example, the system of Example 2 has infinitely many solutions over F (the rational field) if a is restricted to rational numbers, it has infinitely many real solutions if a 's restricted to real numbers, it has infinitely many complex solutions if a is any complex number whatever.

See Problems 1-2.

NON-HOMOGENEOUS EQUATIONS. A linear equation

$$a_1x_1 + a_2x_2 + \dots + a_nx_n = h$$

is called **non-homogeneous** if $h \neq 0$. A system $AX = H$ is called a system of non-homogeneous equations provided H is not a zero vector. The systems of Examples 1 and 2 are non-homogeneous systems.

In Problem 3 we prove

III. A system of n non-homogeneous equations in n unknowns has a unique solution provided the rank of its coefficient matrix A is n, that is, provided $|A| \neq 0$.

In addition to the method above, two additional procedures for solving a consistent system of n non-homogeneous equations in as many unknowns $AX = H$ are given below. The first of these is the familiar solution by determinants.

(*a*) Solution by Cramer's Rule. Denote by A_i, $(i = 1, 2, \dots, n)$ the matrix obtained from A by replacing its ith column with the column of constants (the h's). Then, if $|A| \neq 0$, the system $AX = H$ has the unique solution

$$\textbf{(10.5)} \qquad x_1 = \frac{|A_1|}{|A|}, \quad x_2 = \frac{|A_2|}{|A|}, \quad \dots, \quad x_n = \frac{|A_n|}{|A|}$$

See Problem 4.

Example 3. Solve the system
$$\begin{cases} 2x_1 + x_2 + 5x_3 + x_4 = 5 \\ x_1 + x_2 - 3x_3 - 4x_4 = -1 \\ 3x_1 + 6x_2 - 2x_3 + x_4 = 8 \\ 2x_1 + 2x_2 + 2x_3 - 3x_4 = 2 \end{cases}$$
using Cramer's Rule.

We find

$$|A| = \begin{vmatrix} 2 & 1 & 5 & 1 \\ 1 & 1 & -3 & -4 \\ 3 & 6 & -2 & 1 \\ 2 & 2 & 2 & -3 \end{vmatrix} = -120, \qquad |A_1| = \begin{vmatrix} 5 & 1 & 5 & 1 \\ -1 & 1 & -3 & -4 \\ 8 & 6 & -2 & 1 \\ 2 & 2 & 2 & -3 \end{vmatrix} = -240$$

$$|A_2| = \begin{vmatrix} 2 & 5 & 5 & 1 \\ 1 & -1 & -3 & -4 \\ 3 & 8 & -2 & 1 \\ 2 & 2 & 2 & -3 \end{vmatrix} = -24, \qquad |A_3| = \begin{vmatrix} 2 & 1 & 5 & 1 \\ 1 & 1 & -1 & -4 \\ 3 & 6 & 8 & 1 \\ 2 & 2 & 2 & -3 \end{vmatrix} = 0$$

and

$$|A_4| = \begin{vmatrix} 2 & 1 & 5 & 5 \\ 1 & 1 & -3 & -1 \\ 3 & 6 & -2 & 8 \\ 2 & 2 & 2 & 2 \end{vmatrix} = -96$$

Then $x_1 = \dfrac{|A_1|}{|A|} = \dfrac{-240}{-120} = 2,\quad x_2 = \dfrac{|A_2|}{|A|} = \dfrac{-24}{-120} = \dfrac{1}{5},\quad x_3 = \dfrac{|A_3|}{|A|} = \dfrac{0}{-120} = 0$, and

$$x_4 = \frac{|A_4|}{|A|} = \frac{-96}{-120} = \frac{4}{5}.$$

(*b*) Solution using A^{-1}. If $|A| \neq 0$, A^{-1} exists and the solution of the system $AX = H$ is given by

(10.6) $$A^{-1}\cdot AX = A^{-1}H \quad \text{or} \quad X = A^{-1}H$$

Example 4. The coefficient matrix of the system $\begin{cases} 2x_1 + 3x_2 + x_3 = 9 \\ x_1 + 2x_2 + 3x_3 = 6 \\ 3x_1 + x_2 + 2x_3 = 8 \end{cases}$ is $A = \begin{bmatrix} 2 & 3 & 1 \\ 1 & 2 & 3 \\ 3 & 1 & 2 \end{bmatrix}$.

From Problem 2(*b*), Chapter 7, $A^{-1} = \dfrac{1}{18}\begin{bmatrix} 1 & -5 & 7 \\ 7 & 1 & -5 \\ -5 & 7 & 1 \end{bmatrix}$. Then

$$A^{-1}\cdot AX = X = A^{-1}H = \frac{1}{18}\begin{bmatrix} 1 & -5 & 7 \\ 7 & 1 & -5 \\ -5 & 7 & 1 \end{bmatrix}\begin{bmatrix} 9 \\ 6 \\ 8 \end{bmatrix} = \frac{1}{18}\begin{bmatrix} 35 \\ 29 \\ 5 \end{bmatrix}$$

The solution of the system is $x_1 = 35/18$, $x_2 = 29/18$, $x_3 = 5/18$.

See Problem 5.

HOMOGENEOUS EQUATIONS. A linear equation

(10.7) $$a_1x_1 + a_2x_2 + \ldots + a_nx_n = 0$$

is called **homogeneous.** A system of linear equations

(10.8) $$AX = 0$$

in n unknowns is called a **system of homogeneous equations.** For the system **(10.8)** the rank of the coefficient matrix A and the augmented matrix $[A\ 0]$ are the same; thus, the system is always consistent. Note that $X = 0$, that is, $x_1 = x_2 = \ldots = x_n = 0$ is always a solution; it is called **the trivial solution.**

If the rank of A is n, then n of the equations of **(10.8)** can be solved by Cramer's rule for the unique solution $x_1 = x_2 = \ldots = x_n = 0$ and the system has only the trivial solution. If the rank of A is $r<n$, Theorem **II** assures the existence of non-trivial solutions. Thus,

IV. A necessary and sufficient condition for **(10.8)** to have a solution other than the trivial solution is that the rank of A be $r < n$.

V. A necessary and sufficient condition that a system of n homogeneous equations in n unknowns has a solution other than the trivial solution is $|A| = 0$.

VI. If the rank of **(10.8)** is $r < n$, the system has exactly n-r linearly independent solutions such that every solution is a linear combination of these n-r and every such linear combination is a solution.

See Problem 6.

LET X_1 and X_2 be two distinct solutions of $AX = H$. Then $AX_1 = H$, $AX_2 = H$, and $A(X_1 - X_2) = AY = 0$. Thus, $Y = X_1 - X_2$ is a non-trivial solution of $AX = 0$.

Conversely, if Z is any non-trivial solution of $AX = 0$ and if X_p is any solution of $AX = H$, then $X = X_p + Z$ is also a solution of $AX = H$. As Z ranges over the complete solution of $AX = 0$, $X_p + Z$ ranges over the complete solution of $AX = H$. Thus,

VII. If the system of non-homogeneous equations $AX = H$ is consistent, a complete solution of the system is given by the complete solution of $AX = 0$ plus any particular solution of $AX = H$.

Example 5. In the system $\begin{cases} x_1 - 2x_2 + 3x_3 = 4 \\ x_1 + x_2 + 2x_3 = 5 \end{cases}$ set $x_1 = 0$; then $x_3 = 2$ and $x_2 = 1$. A particular solution is $X_p = [0, 1, 2]'$. The complete solution of $\begin{cases} x_1 - 2x_2 + 3x_3 = 0 \\ x_1 + x_2 + 2x_3 = 0 \end{cases}$ is $[-7a, a, 3a]'$, where a is arbitrary. Then the complete solution of the given system is

$$X = [-7a, a, 3a]' + [0, 1, 2]' = [-7a, 1+a, 2+3a]'$$

Note. The above procedure may be extended to larger systems. However, it is first necessary to show that the system is consistent. This is a long step in solving the system by the augmented matrix method given earlier.

SOLVED PROBLEMS

1. Solve $\begin{cases} x_1 + x_2 - 2x_3 + x_4 + 3x_5 = 1 \\ 2x_1 - x_2 + 2x_3 + 2x_4 + 6x_5 = 2 \\ 3x_1 + 2x_2 - 4x_3 - 3x_4 - 9x_5 = 3 \end{cases}$.

Solution:

The augmented matrix

$$[A\ H] = \begin{bmatrix} 1 & 1 & -2 & 1 & 3 & 1 \\ 2 & -1 & 2 & 2 & 6 & 2 \\ 3 & 2 & -4 & -3 & -9 & 3 \end{bmatrix} \sim \begin{bmatrix} 1 & 1 & -2 & 1 & 3 & 1 \\ 0 & -3 & 6 & 0 & 0 & 0 \\ 0 & -1 & 2 & -6 & -18 & 0 \end{bmatrix} \sim \begin{bmatrix} 1 & 1 & -2 & 1 & 3 & 1 \\ 0 & 1 & -2 & 0 & 0 & 0 \\ 0 & -1 & 2 & -6 & -18 & 0 \end{bmatrix}$$

$$\sim \begin{bmatrix} 1 & 0 & 0 & 1 & 3 & 1 \\ 0 & 1 & -2 & 0 & 0 & 0 \\ 0 & 0 & 0 & -6 & -18 & 0 \end{bmatrix} \sim \begin{bmatrix} 1 & 0 & 0 & 1 & 3 & 1 \\ 0 & 1 & -2 & 0 & 0 & 0 \\ 0 & 0 & 0 & 1 & 3 & 0 \end{bmatrix}$$

$$\sim \begin{bmatrix} 1 & 0 & 0 & 0 & 0 & 1 \\ 0 & 1 & -2 & 0 & 0 & 0 \\ 0 & 0 & 0 & 1 & 3 & 0 \end{bmatrix}$$

Then $x_1 = 1$, $x_2 - 2x_3 = 0$, and $x_4 + 3x_5 = 0$. Take $x_3 = a$ and $x_5 = b$, where a and b are arbitrary; the complete solution may be given as $x_1 = 1$, $x_2 = 2a$, $x_3 = a$, $x_4 = -3b$, $x_5 = b$ or as $X = [1, 2a, a, -3b, b]'$.

2. Solve $\begin{cases} x_1 + x_2 + 2x_3 + x_4 = 5 \\ 2x_1 + 3x_2 - x_3 - 2x_4 = 2 \\ 4x_1 + 5x_2 + 3x_3 = 7 \end{cases}$.

Solution:

$$[A\ H] = \begin{bmatrix} 1 & 1 & 2 & 1 & 5 \\ 2 & 3 & -1 & -2 & 2 \\ 4 & 5 & 3 & 0 & 7 \end{bmatrix} \sim \begin{bmatrix} 1 & 1 & 2 & 1 & 5 \\ 0 & 1 & -5 & -4 & -8 \\ 0 & 1 & -5 & -4 & -13 \end{bmatrix} \sim \begin{bmatrix} 1 & 0 & 7 & 5 & 13 \\ 0 & 1 & -5 & -4 & -8 \\ 0 & 0 & 0 & 0 & -5 \end{bmatrix}.$$

The last row reads $0\cdot x_1 + 0\cdot x_2 + 0\cdot x_3 + 0\cdot x_4 = -5$; thus the given system is inconsistent and has no solution.

3. Prove: A system $AX = H$ of n non-homogeneous equations in n unknowns has a unique solution provided $|A| \neq 0$.

If A is non-singular, it is equivalent to I. When A is reduced by row transformations only to I, suppose $[A\ H]$ is reduced to $[I\ K]$. Then $X = K$ is a solution of the system.

Suppose next that $X = L$ is a second solution of the system; then $AK = H$ and $AL = H$, and $AK = AL$. Since A is non-singular, $K = L$, and the solution is unique.

4. Derive Cramer's Rule.

Let the system of non-homogeneous equations be

$$(1)\qquad \begin{cases} a_{11}x_1 + a_{12}x_2 + \dots + a_{1n}x_n = h_1 \\ a_{21}x_1 + a_{22}x_2 + \dots + a_{2n}x_n = h_2 \\ \dots\dots\dots\dots\dots\dots \\ a_{n1}x_1 + a_{n2}x_2 + \dots + a_{nn}x_n = h_n \end{cases}$$

Denote by A the coefficient matrix $[a_{ij}]$ and let α_{ij} be the cofactor of a_{ij} in A. Multiply the first equation of (1) by α_{11}, the second equation by α_{21}, ..., the last equation by α_{n1}, and add. We have

$$\sum_{i=1}^{n} a_{i1}\alpha_{i1}x_1 + \sum_{i=1}^{n} a_{i2}\alpha_{i1}x_2 + \dots + \sum_{i=1}^{n} a_{in}\alpha_{i1}x_n = \sum_{i=1}^{n} h_i\alpha_i$$

which by Theorems **X** and **XI**, and Problem 10, Chapter 3, reduces to

$$|A|\cdot x_1 = \begin{vmatrix} h_1 & a_{12} & \dots & a_{1n} \\ h_2 & a_{22} & \dots & a_{2n} \\ \dots & \dots & \dots & \dots \\ h_n & a_{n2} & \dots & a_{nn} \end{vmatrix} = |A_1| \quad \text{so that} \quad x_1 = \frac{|A_1|}{|A|}$$

Next, multiply the equations of (1) respectively by $\alpha_{12}, \alpha_{22}, \dots, \alpha_{n2}$ and sum to obtain

$$|A|\cdot x_2 = \begin{vmatrix} a_{11} & h_1 & a_{13} & \dots & a_{1n} \\ a_{21} & h_2 & a_{23} & \dots & a_{2n} \\ \dots & \dots & \dots & \dots & \dots \\ a_{n1} & h_n & a_{n3} & \dots & a_{nn} \end{vmatrix} = |A_2| \quad \text{so that} \quad x_2 = \frac{|A_2|}{|A|}$$

Continuing in this manner, we finally multiply the equations of (1) respectively by $\alpha_{1n}, \alpha_{2n}, \dots, \alpha_{nn}$ and sum to obtain

$$|A|\cdot x_n = \begin{vmatrix} a_{11} & \dots & a_{1,n-1} & h_1 \\ a_{21} & \dots & a_{2,n-1} & h_2 \\ \dots & \dots & \dots & \dots \\ a_{n1} & \dots & a_{n,n-1} & h_n \end{vmatrix} = |A_n| \quad \text{so that} \quad x_n = \frac{|A_n|}{|A|}$$

5. Solve the system $\begin{cases} 2x_1 + x_2 + 5x_3 + x_4 = 5 \\ x_1 + x_2 - 3x_3 - 4x_4 = -1 \\ 3x_1 + 6x_2 - 2x_3 + x_4 = 8 \\ 2x_1 + 2x_2 + 2x_3 - 3x_4 = 2 \end{cases}$ using the inverse of the coefficient matrix.

Solution:

The inverse of $A = \begin{bmatrix} 2 & 1 & 5 & 1 \\ 1 & 1 & -3 & -4 \\ 3 & 6 & -2 & 1 \\ 2 & 2 & 2 & -3 \end{bmatrix}$ is $\dfrac{1}{120}\begin{bmatrix} 120 & 120 & 0 & -120 \\ -69 & -73 & 17 & 80 \\ -15 & -35 & -5 & 40 \\ 24 & 8 & 8 & -40 \end{bmatrix}$. Then

$$X = \frac{1}{120}\begin{bmatrix}120 & 120 & 0 & -120\\ -69 & -73 & 17 & 80\\ -15 & -35 & -5 & 40\\ 24 & 8 & 8 & -40\end{bmatrix}\begin{bmatrix}5\\-1\\8\\2\end{bmatrix} = \begin{bmatrix}2\\1/5\\0\\4/5\end{bmatrix}$$

(See Example 3.)

6. Solve $\begin{cases} x_1 + x_2 + x_3 + x_4 = 0 \\ x_1 + 3x_2 + 2x_3 + 4x_4 = 0 \\ 2x_1 + x_3 - x_4 = 0 \end{cases}$.

Solution:

$$[A\ H] = \begin{bmatrix}1&1&1&1&0\\1&3&2&4&0\\2&0&1&-1&0\end{bmatrix} \sim \begin{bmatrix}1&1&1&1&0\\0&2&1&3&0\\0&-2&-1&-3&0\end{bmatrix} \sim \begin{bmatrix}1&1&1&1&0\\0&2&1&3&0\\0&0&0&0&0\end{bmatrix}$$

$$\sim \begin{bmatrix}1&1&1&1&0\\0&1&\frac{1}{2}&\frac{3}{2}&0\\0&0&0&0&0\end{bmatrix} \sim \begin{bmatrix}1&0&\frac{1}{2}&-\frac{1}{2}&0\\0&1&\frac{1}{2}&\frac{3}{2}&0\\0&0&0&0&0\end{bmatrix}$$

The complete solution of the system is $x_1 = -\frac{1}{2}a + \frac{1}{2}b$, $x_2 = -\frac{1}{2}a - \frac{3}{2}b$, $x_3 = a$, $x_4 = b$. Since the rank of A is 2, we may obtain exactly $n - r = 4 - 2 = 2$ linearly independent solutions. One such pair, obtained by first taking $a = 1$, $b = 1$ and then $a = 3$, $b = 1$ is

$$x_1 = 0,\ x_2 = -2,\ x_3 = 1,\ x_4 = 1 \quad \text{and} \quad x_1 = -1,\ x_2 = -3,\ x_3 = 3,\ x_4 = 1$$

What can be said of the pair of solutions obtained by taking $a = b = 1$ and $a = b = 3$?

7. Prove: In a square matrix A of order n and rank n-1, the cofactors of the elements of any two rows (columns) are proportional.

Since $|A| = 0$, the cofactors of the elements of any row (column) of A are a solution X_1 of the system $AX = 0$ $(A'X = 0)$.

Now the system has but one linearly independent solution since A (A') is of rank n-1. Hence, for the cofactors of another row (column) of A (another solution X_2 of the system), we have $X_2 = kX_1$.

8. Prove: If $f_1, f_2, \ldots, f_m$ are $m<n$ linearly independent linear forms over F in n variables, then the p linear forms

$$g_j = \sum_{i=1}^{m} s_{ij} f_i, \qquad (j = 1, 2, \ldots, p)$$

are linearly dependent if and only if the $m \times p$ matrix $[s_{ij}]$ is of rank $r<p$.

The g's are linearly dependent if and only if there exist scalars $a_1, a_2, \ldots, a_p$ in F, not all zero, such that

$$\begin{aligned} a_1 g_1 + a_2 g_2 + \ldots + a_p g_p &= a_1 \sum_{i=1}^{m} s_{i1} f_i + a_2 \sum_{i=1}^{m} s_{i2} f_i + \ldots + a_p \sum_{i=1}^{m} s_{ip} f_i \\ &= (\sum_{j=1}^{p} a_j s_{1j}) f_1 + (\sum_{j=1}^{p} a_j s_{2j}) f_2 + \ldots + (\sum_{j=1}^{p} a_j s_{mj}) f_m \\ &= \sum_{i=1}^{m} (\sum_{j=1}^{p} a_j s_{ij}) f_i = 0 \end{aligned}$$

Since the f's are linearly independent, this requires

$$\sum_{j=1}^{p} a_j s_{ij} = a_1 s_{i1} + a_2 s_{i2} + \ldots + a_p s_{ip} = 0 \qquad (i = 1, 2, \ldots, m)$$

Now, by Theorem **IV**, the system of m homogeneous equations in p unknowns $\sum_{j=1}^{p} s_{ij} x_j = 0$ has a non-trivial solution $X = [a_1, a_2, \ldots, a_p]'$ if and only if $[s_{ij}]$ is of rank $r<p$.

9. Suppose $A = [a_{ij}]$ of order n is singular. Show that there always exists a matrix $B = [b_{ij}] \neq 0$ of order n such that $AB = 0$.

Let $B_1, B_2, \ldots, B_n$ be the column vectors of B. Then, by hypothesis, $AB_1 = AB_2 = \ldots = AB_n = 0$. Consider any one of these, say $AB_t = 0$, or

$$\begin{cases} a_{11}b_{1t} + a_{12}b_{2t} + \ldots + a_{1n}b_{nt} = 0 \\ a_{21}b_{1t} + a_{22}b_{2t} + \ldots + a_{2n}b_{nt} = 0 \\ \ldots\ldots\ldots\ldots\ldots\ldots\ldots\ldots\ldots \\ a_{n1}b_{1t} + a_{n2}b_{2t} + \ldots + a_{nn}b_{nt} = 0 \end{cases}$$

Since the coefficient matrix A is singular, the system in the unknowns $b_{1t}, b_{2t}, \ldots, b_{nt}$ has solutions other than the trivial solution. Similarly, $AB_1 = 0$, $AB_2 = 0$, ... have solutions, each being a column of B.

SUPPLEMENTARY PROBLEMS

10. Find all solutions of:

(a) $x_1 - 2x_2 + x_3 - 3x_4 = 1$

(b) $\begin{cases} x_1 + x_2 + x_3 = 4 \\ 2x_1 + 5x_2 - 2x_3 = 3 \end{cases}$

(c) $\begin{cases} x_1 + x_2 + x_3 = 4 \\ 2x_1 + 5x_2 - 2x_3 = 3 \\ x_1 + 7x_2 - 7x_3 = 5 \end{cases}$

(d) $\begin{cases} x_1 + x_2 + x_3 + x_4 = 0 \\ x_1 + x_2 + x_3 - x_4 = 4 \\ x_1 + x_2 - x_3 + x_4 = -4 \\ x_1 - x_2 + x_3 + x_4 = 2 \end{cases}$

Ans. (a) $x_1 = 1 + 2a - b + 3c$, $x_2 = a$, $x_3 = b$, $x_4 = c$

(b) $x_1 = -7a/3 + 17/3$, $x_2 = 4a/3 - 5/3$, $x_3 = a$

(d) $x_1 = -x_2 = 1$, $x_3 = -x_4 = 2$

11. Find all non-trivial solutions of:

(a) $\begin{cases} x_1 - 2x_2 + 3x_3 = 0 \\ 2x_1 + 5x_2 + 6x_3 = 0 \end{cases}$

(b) $\begin{cases} 2x_1 - x_2 + 3x_3 = 0 \\ 3x_1 + 2x_2 + x_3 = 0 \\ x_1 - 4x_2 + 5x_3 = 0 \end{cases}$

(c) $\begin{cases} x_1 + 2x_2 + 3x_3 = 0 \\ 2x_1 + x_2 + 3x_3 = 0 \\ 3x_1 + 2x_2 + x_3 = 0 \end{cases}$

(d) $\begin{cases} 4x_1 - x_2 + 2x_3 + x_4 = 0 \\ 2x_1 + 3x_2 - x_3 - 2x_4 = 0 \\ 7x_2 - 4x_3 - 5x_4 = 0 \\ 2x_1 - 11x_2 + 7x_3 + 8x_4 = 0 \end{cases}$

Ans. (a) $x_1 = -3a$, $x_2 = 0$, $x_3 = a$

(b) $x_1 = -x_2 = -x_3 = a$

(d) $x_1 = -\frac{5}{8}a + \frac{3}{8}b$, $x_2 = a$, $x_3 = \frac{7}{4}a - \frac{5}{4}b$, $x_4 = b$

12. Reconcile the solution of 10(d) with another $x_1 = c,\ x_2 = d,\ x_3 = -\frac{10}{3}c - \frac{d}{3},\ x_4 = \frac{8}{3}c + \frac{5}{3}d$.

13. Given $A = \begin{bmatrix} 1 & 1 & 2 \\ 2 & 2 & 4 \\ 3 & 3 & 6 \end{bmatrix}$, find a matrix B of rank 2 such that $AB = 0$. Hint. Select the columns of B from the solutions of $AX = 0$.

14. Show that a square matrix is singular if and only if its rows (columns) are linearly dependent.

15. Let $AX = 0$ be a system of n homogeneous equations in n unknowns and suppose A of rank $r = n-1$. Show that any non-zero vector of cofactors $[\alpha_{i1}, \alpha_{i2}, \ldots, \alpha_{in}]'$ of a row of A is a solution of $AX = 0$.

16. Use Problem 15 to solve:

$$(a)\begin{cases} x_1 - 2x_2 + 3x_3 = 0 \\ 2x_1 + 5x_2 + 6x_3 = 0 \end{cases},\qquad (b)\begin{cases} 2x_1 + 3x_2 - x_3 = 0 \\ 3x_1 - 4x_2 + 2x_3 = 0 \end{cases},\qquad (c)\begin{cases} 2x_1 + 3x_2 + 4x_3 = 0 \\ 2x_1 - x_2 + 6x_3 = 0 \end{cases}$$

Hint. To the equations of (a) adjoin $0x_1 + 0x_2 + 0x_3 = 0$ and find the cofactors of the elements of the third row of $\begin{bmatrix} 1 & -2 & 3 \\ 2 & 5 & 6 \\ 0 & 0 & 0 \end{bmatrix}$.

Ans. (a) $x_1 = -27a,\ x_2 = 0,\ x_3 = 9a$ or $[3a, 0, -a]'$, (b) $[2a, -7a, -17a]'$, (c) $[11a, -2a, -4a]'$

17. Let the coefficient and the augmented matrix of the system of 3 non-homogeneous equations in 5 unknowns $AX = H$ be of rank 2 and assume the canonical form of the augmented matrix to be

$$\begin{bmatrix} 1 & 0 & b_{13} & b_{14} & b_{15} & c_1 \\ 0 & 1 & b_{23} & b_{24} & b_{25} & c_2 \\ 0 & 0 & 0 & 0 & 0 & 0 \end{bmatrix}$$

with not both of c_1, c_2 equal to 0. First choose $x_3 = x_4 = x_5 = 0$ and obtain $X_1 = [c_1, c_2, 0, 0, 0]'$ as a solution of $AX = H$. Then choose $x_3 = 1,\ x_4 = x_5 = 0$, also $x_3 = x_5 = 0,\ x_4 = 1$ and $x_3 = x_4 = 0,\ x_5 = 1$ to obtain other solutions X_2, X_3, and X_4. Show that these $5-2+1 = 4$ solutions are linearly independent.

18. Consider the linear combination $Y = s_1X_1 + s_2X_2 + s_3X_3 + s_4X_4$ of the solutions of Problem 17. Show that Y is a solution of $AX = H$ if and only if (i) $s_1 + s_2 + s_3 + s_4 = 1$. Thus, with s_1, s_2, s_3, s_4 arbitrary except for (i), Y is a complete solution of $AX = H$.

19. Prove: Theorem **VI**. Hint. Follow Problem 17 with $c_1 = c_2 = 0$.

20. Prove: If A is an $m \times p$ matrix of rank r_1 and B is a $p \times n$ matrix of rank r_2 such that $AB = 0$, then $r_1 + r_2 \leq p$. Hint. Use Theorem **VI**.

21. Using the 4×5 matrix $A = [a_{ij}]$ of rank 2, verify: In an $m\times n$ matrix A of rank r, the r-square determinants formed from the columns of a submatrix consisting of any r rows of A are proportional to the r-square determinants formed from any other submatrix consisting of r rows of A.
Hint. Suppose the first two rows are linearly independent so that $a_{3j} = p_{31}a_{1j} + p_{32}a_{2j},\ a_{4j} = p_{41}a_{1j} + p_{42}a_{2j}$, $(j = 1, 2, \ldots, 5)$. Evaluate the 2-square determinants

$$\begin{vmatrix} a_{1q} & a_{1s} \\ a_{2q} & a_{2s} \end{vmatrix},\qquad \begin{vmatrix} a_{1q} & a_{1s} \\ a_{3q} & a_{3s} \end{vmatrix},\qquad \text{and} \qquad \begin{vmatrix} a_{3q} & a_{3s} \\ a_{4q} & a_{4s} \end{vmatrix}$$

22. Write a proof of the theorem of Problem 21.

23. From Problem 7, obtain: If the n-square matrix A is of rank n-1, then the following relations among its cofactors hold

$$(a)\ \ \alpha_{ij}\,\alpha_{hk} = \alpha_{ik}\,\alpha_{hj}, \qquad (b)\ \ \alpha_{ii}\,\alpha_{jj} = \alpha_{ij}\,\alpha_{ji}$$

where $(h, i, j, k = 1, 2, \ldots, n)$.

24. Show that $B = \begin{bmatrix} 1 & 1 & 1 & 1 & 4 \\ 1 & 2 & 3 & -4 & 2 \\ 2 & 1 & 1 & 2 & 6 \\ 3 & 2 & -1 & -1 & 3 \\ 1 & 2 & 2 & -2 & 4 \\ 2 & 3 & -3 & 1 & 1 \end{bmatrix}$ is row equivalent to $\begin{bmatrix} 1 & 0 & 0 & 0 & 0 \\ 0 & 1 & 0 & 0 & 0 \\ 0 & 0 & 1 & 0 & 0 \\ 0 & 0 & 0 & 1 & 0 \\ 0 & 0 & 0 & 0 & 1 \\ 0 & 0 & 0 & 0 & 0 \end{bmatrix}$. From $B = [A \;\; H]$ infer that the system of 6 linear equations in 4 unknowns has 5 linearly independent equations. Show that a system of $m > n$ linear equations in n unknowns can have at most $n+1$ linearly independent equations. Show that when there are $n+1$, the system is inconsistent.

25. If $AX = H$ is consistent and of rank r, for what set of r variables can one solve?

26. Generalize the results of Problems 17 and 18 to m non-homogeneous equations in n unknowns with coefficient and augmented matrix of the same rank r to prove: If the coefficient and the augmented matrix of the system $AX = H$ of m non-homogeneous equations in n unknowns have rank r and if $X_1, X_2, \ldots, X_{n-r+1}$ are linearly independent solutions of the system, then

$$X = s_1 X_1 + s_1 X_2 + \ldots + s_{n-r+1} X_{n-r+1}$$

where $\sum_{i=1}^{n-r+1} s_i = 1$, is a complete solution.

27. In a four-pole electrical network, the input quantities E_1 and I_1 are given in terms of the output quantities E_2 and I_2 by

$$\begin{aligned} E_1 &= aE_2 + bI_2 \\ I_1 &= cE_2 + dI_2 \end{aligned} \quad \text{or} \quad \begin{bmatrix} E_1 \\ I_1 \end{bmatrix} = \begin{bmatrix} a & b \\ c & d \end{bmatrix} \begin{bmatrix} E_2 \\ I_2 \end{bmatrix} = A \begin{bmatrix} E_2 \\ I_2 \end{bmatrix}$$

Show that $\begin{bmatrix} E_1 \\ E_2 \end{bmatrix} = \dfrac{1}{c} \begin{bmatrix} a & -|A| \\ 1 & -d \end{bmatrix} \begin{bmatrix} I_1 \\ I_2 \end{bmatrix}$ and $\begin{bmatrix} E_1 \\ I_2 \end{bmatrix} = \dfrac{1}{d} \begin{bmatrix} b & |A| \\ 1 & -c \end{bmatrix} \begin{bmatrix} I_1 \\ E_2 \end{bmatrix}$.

Solve also for E_2 and I_2, I_1 and I_2, I_1 and E_2.

28. Let the system of n linear equations in n unknowns $AX = H$, $H \neq 0$, have a unique solution. Show that the system $AX = K$ has a unique solution for any n-vector $K \neq 0$.

29. Solve the set of linear forms $AX = \begin{bmatrix} 1 & -1 & 1 \\ 2 & 1 & 3 \\ 1 & 2 & 3 \end{bmatrix} \begin{bmatrix} x_1 \\ x_2 \\ x_3 \end{bmatrix} = Y = \begin{bmatrix} y_1 \\ y_2 \\ y_3 \end{bmatrix}$ for the x_i as linear forms in the y's.

Now write down the solution of $A'X = Y$.

30. Let A be n-square and non-singular, and let S_i be the solution of $AX = E_i$, $(i = 1, 2, \ldots, n)$, where E_i is the n-vector whose ith component is 1 and whose other components are 0. Identify the matrix $[S_1, S_2, \ldots, S_n]$.

31. Let A be an $m \times n$ matrix with $m < n$ and let S_i be a solution of $AX = E_i$, $(i = 1, 2, \ldots, m)$, where E_i is the m-vector whose ith component is 1 and whose other components are 0. If $K = [k_1, k_2, \ldots, k_m]'$, show that

$$k_1 S_1 + k_2 S_2 + \ldots + k_m S_m$$

is a solution of $AX = K$.

Chapter 11

Vector Spaces

UNLESS STATED OTHERWISE, all vectors will now be column vectors. When components are displayed, we shall write $[x_1, x_2, ..., x_n]'$. The transpose mark ($'$) indicates that the elements are to be written in a column.

A set of such n-vectors over F is said to be **closed under addition** if the sum of any two of them is a vector of the set. Similarly, the set is said to be **closed under scalar multiplication** if every scalar multiple of a vector of the set is a vector of the set.

Example 1. (a) The set of all vectors $[x_1, x_2, x_3]'$ of ordinary space having equal components ($x_1 = x_2 = x_3$) is closed under both addition and scalar multiplication. For, the sum of any two of the vectors and k times any vector (k real) are again vectors having equal components.

(b) The set of **all** vectors $[x_1, x_2, x_3]'$ of ordinary space is closed under addition and scalar multiplication.

VECTOR SPACES. Any set of n-vectors over F which is closed under both addition and scalar multiplication is called a **vector space**. Thus, if $X_1, X_2, ..., X_m$ are n-vectors over F, the set of all linear combinations

$$k_1 X_1 + k_2 X_2 + \cdots + k_m X_m \qquad (k_i \text{ in } F) \tag{11.1}$$

is a vector space over F. For example, both of the sets of vectors (a) and (b) of Example 1 are vector spaces. Clearly, every vector space **(11.1)** contains the zero n-vector while the zero n-vector alone is a vector space. (The space **(11.1)** is also called a **linear vector space**.)

The totality $V_n(F)$ of **all** n-vectors over F is called the **n-dimensional vector space** over F.

SUBSPACES. A set V of the vectors of $V_n(F)$ is called a **subspace** of $V_n(F)$ provided V is closed under addition and scalar multiplication. Thus, the zero n-vector is a subspace of $V_n(F)$; so also is $V_n(F)$ itself. The set (a) of Example 1 is a subspace (a line) of ordinary space. In general, if $X_1, X_2, ..., X_m$ belong to $V_n(F)$, the space of all linear combinations **(11.1)** is a subspace of $V_n(F)$.

A vector space V is said to be **spanned** or generated by the n-vectors $X_1, X_2, ..., X_m$ provided (a) the X_i lie in V and (b) every vector of V is a linear combination **(11.1)**. Note that the vectors $X_1, X_2, ..., X_m$ are not restricted to be linearly independent.

Example 2. Let F be the field R of real numbers so that the 3-vectors $X_1 = [1, 1, 1]'$, $X_2 = [1, 2, 3]'$, $X_3 = [1, 3, 2]'$, and $X_4 = [3, 2, 1]'$ lie in ordinary space $S = V_3(R)$. Any vector $[a, b, c]'$ of S can be expressed as

$$y_1X_1 + y_2X_2 + y_3X_3 + y_4X_4 \quad = \quad \begin{bmatrix} y_1 + \ \ y_2 + \ \ y_3 + 3y_4 \\ y_1 + 2y_2 + 3y_3 + 2y_4 \\ y_1 + 3y_2 + 2y_3 + \ \ y_4 \end{bmatrix}$$

since the resulting system of equations

$$\text{(i)} \qquad \begin{aligned} y_1 + \ \ y_2 + \ \ y_3 + 3y_4 &= a \\ y_1 + 2y_2 + 3y_3 + 2y_4 &= b \\ y_1 + 3y_2 + 2y_3 + \ \ y_4 &= c \end{aligned}$$

is consistent. Thus, the vectors X_1, X_2, X_3, X_4 span S.

The vectors X_1 and X_2 are linearly independent. They span a subspace (the plane π) of S which contains every vector $hX_1 + kX_2$, where h and k are real numbers.

The vector X_4 spans a subspace (the line L) of S which contains every vector hX_4, where h is a real number.

See Problem 1.

BASIS AND DIMENSION. By the dimension of a vector space V is meant the maximum number of linearly independent vectors in V or, what is the same thing, the minimum number of linearly independent vectors required to span V. In elementary geometry, ordinary space is considered as a 3-space (space of dimension three) of points (a, b, c). Here we have been considering it as a 3-space of vectors $[a, b, c]'$. The plane π of Example 2 is of dimension 2 and the line L is of dimension 1.

A vector space of dimension r consisting of n-vectors will be denoted by $V_n^r(F)$. When $r = n$, we shall agree to write $V_n(F)$ for $V_n^n(F)$.

A set of r linearly independent vectors of $V_n^r(F)$ is called a **basis** of the space. Each vector of the space is then a **unique** linear combination of the vectors of this basis. All bases of $V_n^r(F)$ have exactly the same number of vectors but any r linearly independent vectors of the space will serve as a basis.

Example 3. The vectors X_1, X_2, X_3 of Example 2 span S since any vector $[a, b, c]'$ of S can be expressed as

$$y_1X_1 + y_2X_2 + y_3X_3 \quad = \quad \begin{bmatrix} y_1 + \ \ y_2 + \ \ y_3 \\ y_1 + 2y_2 + 3y_3 \\ y_1 + 3y_2 + 2y_3 \end{bmatrix}$$

The resulting system of equations $\begin{cases} y_1 + \ \ y_2 + \ \ y_3 = a \\ y_1 + 2y_2 + 3y_3 = b, \\ y_1 + 3y_2 + 2y_3 = c \end{cases}$ unlike the system (i), has a unique solution. The vectors X_1, X_2, X_3 are a basis of S. The vectors X_1, X_2, X_4 are not a basis of S. (Show this.) They span the subspace π of Example 2, whose basis is the set X_1, X_2.

Theorems **I-V** of Chapter 9 apply here, of course. In particular, Theorem **IV** may be restated as:

I. If $X_1, X_2, \ldots, X_m$ are a set of n-vectors over F and if r is the rank of the $n \times m$ matrix of their components, then from the set r linearly independent vectors may be selected. These r vectors span a $V_n^r(F)$ in which the remaining m-r vectors lie.

See Problems 2-3.

Of considerable importance are:

II. If $X_1, X_2, \ldots, X_m$ are $m<n$ linearly independent n-vectors of $V_n(F)$ and if $X_{m+1}, X_{m+2}, \ldots, X_n$ are any $n-m$ vectors of $V_n(F)$ which together with $X_1, X_2, \ldots, X_m$ form a linearly independent set, then the set $X_1, X_2, \ldots, X_n$ is a basis of $V_n(F)$.

See Problem 4.

III. If $X_1, X_2, \ldots, X_m$ are $m<n$ linearly independent n-vectors over F, then the p vectors

$$Y_j = \sum_{i=1}^{m} s_{ij} X_i \qquad (j = 1, 2, \ldots, p)$$

are linearly dependent if $p>m$ or, when $p \leq m$, if $[s_{ij}]$ is of rank $r<p$.

IV. If $X_1, X_2, \ldots, X_n$ are linearly independent n-vectors over F, then the vectors

$$Y_i = \sum_{j=1}^{n} a_{ij} X_j \qquad (i = 1, 2, \ldots, n)$$

are linearly independent if and only if $[a_{ij}]$ is nonsingular.

IDENTICAL SUBSPACES. If ${}_1V_n^r(F)$ and ${}_2V_n^r(F)$ are two subspaces of $V_n(F)$, they are identical if and only if each vector of ${}_1V_n^r(F)$ is a vector of ${}_2V_n^r(F)$ and conversely, that is, if and only if each is a subspace of the other.

See Problem 5.

SUM AND INTERSECTION OF TWO SPACES. Let $V_n^h(F)$ and $V_n^k(F)$ be two vector spaces. By their **sum** is meant the totality of vectors $X+Y$ where X is in $V_n^h(F)$ and Y is in $V_n^k(F)$. Clearly, this is a vector space; we call it the **sum space** $V_n^s(F)$. The dimension s of the sum space of two vector spaces does not exceed the sum of their dimensions.

By the **intersection** of the two vector spaces is meant the totality of vectors common to the two spaces. Now if X is a vector common to the two spaces, so also is aX; likewise, if X and Y are common to the two spaces so also is $aX+bY$. Thus, the intersection of two spaces is a vector space; we call it the **intersection space** $V_n^t(F)$. The dimension of the intersection space of two vector spaces cannot exceed the smaller of the dimensions of the two spaces.

V. If two vector spaces $V_n^h(F)$ and $V_n^k(F)$ have $V_n^s(F)$ as sum space and $V_n^t(F)$ as intersection space, then $h+k = s+t$.

Example 4. Consider the subspace π_1 spanned by X_1 and X_2 of Example 2 and the subspace π_2 spanned by X_3 and X_4. Since π_1 and π_2 are not identical (prove this) and since the four vectors span S, the sum space of π_1 and π_2 is S.

Now $4X_1 - X_2 = X_4$; thus, X_4 lies in both π_1 and π_2. The subspace (line L) spanned by X_4 is then the intersection space of π_1 and π_2. Note that π_1 and π_2 are each of dimension 2, S is of dimension 3, and L is of dimension 1. This agrees with Theorem **V**.

See Problems 6-8.

NULLITY OF A MATRIX. For a system of homogeneous equations $AX = 0$, the solution vectors X constitute a vector space called the **null space** of A. The dimension of this space, denoted by N_A, is called the **nullity** of A.

Restating Theorem **VI**, Chapter 10, we have

VI. If A has nullity N_A, then $AX = 0$ has N_A linearly independent solutions $X_1, X_2, \ldots,$

X_{N_A} such that every solution of $AX = 0$ is a linear combination of them and every such linear combination is a solution.

A basis for the null space of A is any set of N_A linearly independent solutions of $AX = 0$.

See Problem 9.

VII. For an $m \times n$ matrix A of rank r_A and nullity N_A,

$$r_A + N_A = n \tag{11.2}$$

SYLVESTER'S LAWS OF NULLITY. If A and B are of order n and respective ranks r_A and r_B, the rank and nullity of their product AB satisfy the inequalities

$$\begin{aligned} r_{AB} &\geq r_A + r_B - n \\ N_{AB} &> N_A, \quad N_{AB} > N_B \\ N_{AB} &\leq N_A + N_B \end{aligned} \tag{11.3}$$

See Problem 10.

BASES AND COORDINATES. The n-vectors

$$E_1 = [1, 0, 0, \ldots, 0]', \quad E_2 = [0, 1, 0, \ldots, 0]', \quad \ldots, \quad E_n = [0, 0, 0, \ldots, 1]'$$

are called **elementary** or **unit** vectors over F. The elementary vector E_j, whose jth component is 1, is called the jth elementary vector. The elementary vectors $E_1, E_2, \ldots, E_n$ constitute an important basis for $V_n(F)$.

Every vector $X = [x_1, x_2, \ldots, x_n]'$ of $V_n(F)$ can be expressed uniquely as the sum

$$X = \sum_{i=1}^{n} x_i E_i = x_1 E_1 + x_2 E_2 + \cdots + x_n E_n$$

of the elementary vectors. The components $x_1, x_2, \ldots, x_n$ of X are now called the **coordinates** of X relative to the E-basis. Hereafter, unless otherwise specified, we shall assume that a vector X is given relative to this basis.

Let $Z_1, Z_2, \ldots, Z_n$ be another basis of $V_n(F)$. Then there exist unique scalars $a_1, a_2, \ldots, a_n$ in F such that

$$X = \sum_{i=1}^{n} a_i Z_i = a_1 Z_1 + a_2 Z_2 + \cdots + a_n Z_n$$

These scalars $a_1, a_2, \ldots, a_n$ are called the coordinates of X relative to the Z-basis. Writing $X_Z = [a_1, a_2, \ldots, a_n]'$, we have

$$X = [Z_1, Z_2, \ldots, Z_n] X_Z = Z \cdot X_Z \tag{11.4}$$

where Z is the matrix whose columns are the basis vectors $Z_1, Z_2, \ldots, Z_n$.

Example 5. If $Z_1 = [2, -1, 3]'$, $Z_2 = [1, 2, -1]'$, $Z_3 = [1, -1, -1]'$ is a basis of $V_3(F)$ and $X_Z = [1, 2, 3]'$ is a vector of $V_3(F)$ relative to that basis, then

$$X = [Z_1, Z_2, Z_3] X_Z = \begin{bmatrix} 2 & 1 & 1 \\ -1 & 2 & -1 \\ 3 & -1 & -1 \end{bmatrix} \begin{bmatrix} 1 \\ 2 \\ 3 \end{bmatrix} = \begin{bmatrix} 7 \\ 0 \\ -2 \end{bmatrix} = [7, 0, -2]'$$

relative to the E-basis.

See Problem 11.

Let $W_1, W_2, \ldots, W_n$ be yet another basis of $V_n(F)$. Suppose $X_W = [b_1, b_2, \ldots, b_n]'$ so that

$$X = [W_1, W_2, \ldots, W_n] X_W = W \cdot X_W \tag{11.5}$$

From **(11.4)** and **(11.5)**, $X = Z \cdot X_Z = W \cdot X_W$ and

$$X_W = W^{-1} \cdot Z \cdot X_Z = P X_Z \tag{11.6}$$

where $P = W^{-1} Z$.

Thus,

VIII. If a vector of $V_n(F)$ has coordinates X_Z and X_W respectively relative to two bases of $V_n(F)$, then there exists a non-singular matrix P, determined solely by the two bases and given by **(11.6)** such that $X_W = P X_Z$.

See Problem 12.

SOLVED PROBLEMS

1. The set of all vectors $X = [x_1, x_2, x_3, x_4]'$, where $x_1 + x_2 + x_3 + x_4 = 0$ is a subspace V of $V_4(F)$ since the sum of any two vectors of the set and any scalar multiple of a vector of the set have components whose sum is zero, that is, are vectors of the set.

2. Since $\begin{bmatrix} 1 & 3 & 1 \\ 2 & 4 & 0 \\ 2 & 4 & 0 \\ 1 & 3 & 1 \end{bmatrix}$ is of rank 2, the vectors $X_1 = [1, 2, 2, 1]'$, $X_2 = [3, 4, 4, 3]'$, and $X_3 = [1, 0, 0, 1]'$ are linearly dependent and span a vector space $V_4^2(F)$.

Now any two of these vectors are linearly independent; hence, we may take X_1 and X_2, X_1 and X_3, or X_2 and X_3 as a basis of the $V_4^2(F)$.

3. Since $\begin{bmatrix} 1 & 4 & 2 & 4 \\ 1 & 3 & 1 & 2 \\ 1 & 2 & 0 & 0 \\ 0 & -1 & -1 & -2 \end{bmatrix}$ is of rank 2, the vectors $X_1 = [1,1,1,0]'$, $X_2 = [4,3,2,-1]'$, $X_3 = [2,1,0,-1]'$, and $X_4 = [4,2,0,-2]'$ are linearly dependent and span a $V_4^2(F)$.

For a basis, we may take any two of the vectors except the pair X_3, X_4.

4. The vectors X_1, X_2, X_3 of Problem 2 lie in $V_4(F)$. Find a basis.

For a basis of this space we may take $X_1, X_2, X_4 = [1, 0, 0, 0]'$, and $X_5 = [0, 1, 0, 0]'$ or $X_1, X_2, X_6 = [1, 2, 3, 4]'$, and $X_7 = [1, 3, 6, 8]'$, ... since the matrices $[X_1, X_2, X_4, X_5]$ and $[X_1, X_2, X_6, X_7]$ are of rank 4.

5. Let $X_1 = [1,2,1]'$, $X_2 = [1,2,3]'$, $X_3 = [3,6,5]'$, $Y_1 = [0,0,1]'$, $Y_2 = [1,2,5]'$ be vectors of $V_3(F)$. Show that the space spanned by X_1, X_2, X_3 and the space spanned by Y_1, Y_2 are identical.

First, we note that X_1 and X_2 are linearly independent while $X_3 = 2X_1 + X_2$. Thus, the X_i span a space of dimension two, say ${}_1V_3^2(F)$. Also, the Y_i being linearly independent span a space of dimension two, say ${}_2V_3^2(F)$.

Next, $Y_1 = \frac{1}{2}X_2 - \frac{1}{2}X_1$, $Y_2 = 2X_2 - X_1$; $X_1 = Y_2 - 4Y_1$, $X_2 = Y_2 - 2Y_1$. Thus, any vector $aY_1 + bY_2$ of ${}_2V_3^2(F)$ is a vector $(\frac{1}{2}a + 2b)X_2 - (\frac{1}{2}a + b)X_1$ of ${}_1V_3^2(F)$ and any vector $cX_1 + dX_2$ of ${}_1V_3^2(F)$ is a vector $(c+d)Y_2 - (4c+2d)Y_1$ of ${}_2V_3^2(F)$. Hence, the two spaces are identical.

6. (a) If $X = [x_1, x_2, x_3]'$ lies in the $V_3^2(F)$ spanned by $X_1 = [1,-1,1]'$ and $X_2 = [3,4,-2]'$, then

$$\begin{vmatrix} x_1 & 1 & 3 \\ x_2 & -1 & 4 \\ x_3 & 1 & -2 \end{vmatrix} = -2x_1 + 5x_2 + 7x_3 = 0.$$

(b) If $X = [x_1, x_2, x_3, x_4]'$ lies in the $V_4^2(F)$ spanned by $X_1 = [1,1,2,3]'$ and $X_2 = [1,0,-2,1]'$, then

$$\begin{bmatrix} x_1 & 1 & 1 \\ x_2 & 1 & 0 \\ x_3 & 2 & -2 \\ x_4 & 3 & 1 \end{bmatrix} \text{ is of rank 2. Since } \begin{vmatrix} 1 & 1 \\ 1 & 0 \end{vmatrix} \neq 0, \text{ this requires } \begin{vmatrix} x_1 & 1 & 1 \\ x_2 & 1 & 0 \\ x_3 & 2 & -2 \end{vmatrix} = -2x_1 + 4x_2 - x_3 = 0 \text{ and}$$

$$\begin{vmatrix} x_1 & 1 & 1 \\ x_2 & 1 & 0 \\ x_4 & 3 & 1 \end{vmatrix} = x_1 + 2x_2 - x_4 = 0.$$

These problems illustrate that every $V_n^k(F)$ may be defined as the totality of solutions over F of a system of n-k linearly independent homogeneous linear equations over F in n unknowns.

7. Prove: If two vector spaces $V_n^h(F)$ and $V_n^k(F)$ have $V_n^s(F)$ as sum space and $V_n^t(F)$ as intersection space, then $h+k = s+t$.

Suppose $t = h$; then $V_n^h(F)$ is a subspace of $V_n^k(F)$ and their sum space is V_n^k itself. Thus, $s = k$, $t = h$ and $s+t = h+k$. The reader will show that the same is true if $t = k$.

Suppose next that $t<h$, $t<k$ and let $X_1, X_2, \ldots, X_t$ span $V_n^t(F)$. Then by Theorem **II** there exist vectors $Y_{t+1}, Y_{t+2}, \ldots, Y_h$ so that $X_1, X_2, \ldots, X_t, Y_{t+1}, \ldots, Y_h$ span $V_n^h(F)$ and vectors $Z_{t+1}, Z_{t+2}, \ldots, Z_k$ so that $X_1, X_2, \ldots, X_t, Z_{t+1}, \ldots, Z_k$ span $V_n^k(F)$.

Now suppose there exist scalars a's and b's such that

$$\textbf{(11.4)} \qquad \sum_{i=1}^{t} a_i X_i + \sum_{i=t+1}^{h} a_i Y_i + \sum_{i=t+1}^{k} b_i Z_i = 0 \qquad \text{or}$$

$$\sum_{i=1}^{t} a_i X_i + \sum_{i=t+1}^{h} a_i Y_i = \sum_{i=t+1}^{k} -b_i Z_i$$

The vector on the left belongs to $V_n^h(F)$, and from the right member, belongs also to $V_n^k(F)$; thus it belongs to $V_n^t(F)$. But $X_1, X_2, \ldots, X_t$ span $V_n^t(F)$; hence, $a_{t+1} = a_{t+2} = \ldots = a_h = 0$.

Now from **(11.4)**, $$\sum_{i=1}^{t} a_i X_i + \sum_{i=t+1}^{k} b_i Z_i = 0$$

But the X's and Z's are linearly independent so that $a_1 = a_2 = \ldots = a_t = b_{t+1} = b_{t+2} = \ldots = b_k = 0$; thus, the X's, Y's, and Z's are a linearly independent set and span $V_n^s(F)$. Then $s = h+k-t$ as was to be proved.

8. Consider ${}_1V_3^2(F)$ having $X_1 = [1,2,3]'$ and $X_2 = [1,1,1]'$ as basis and ${}_2V_3^2(F)$ having $Y_1 = [3,1,2]'$ and $Y_2 = [1,0,1]'$ as basis. Since the matrix of the components $\begin{bmatrix} 1 & 1 & 3 & 1 \\ 2 & 1 & 1 & 0 \\ 3 & 1 & 2 & 1 \end{bmatrix}$ is of rank 3, the sum space is $V_3(F)$. As a basis, we may take X_1, X_2, and Y_1.

From $h + k = s + t$, the intersection space is a $V_3^1(F)$. To find a basis, we equate linear combinations of the vectors of the bases of ${}_1V_3^2(F)$ and ${}_2V_3^2(F)$ as

$$aX_1 + bX_2 = cY_1 + dY_2$$

take $d = 1$ for convenience, and solve $\begin{cases} a + b - 3c = 1 \\ 2a + b - c = 0 \\ 3a + b - 2c = 1 \end{cases}$ obtaining $a = 1/3$, $b = -4/3$, $c = -2/3$. Then $aX_1 + bX_2 = [-1,-2/3,-1/3]'$ is a basis of the intersection space. The vector $[3,2,1]'$ is also a basis.

9. Determine a basis for the null space of $A = \begin{bmatrix} 1 & 1 & 3 & 3 \\ 0 & 2 & 2 & 4 \\ 1 & 0 & 2 & 1 \\ 1 & 1 & 3 & 3 \end{bmatrix}$.

Consider the system of equations $AX = 0$ which reduces to $\begin{cases} x_1 + 2x_3 + x_4 = 0 \\ x_2 + x_3 + 2x_4 = 0 \end{cases}$.

A basis for the null space of A is the pair of linearly independent solutions $[1,2,0,-1]'$ and $[2,1,-1,0]'$ of these equations.

10. Prove: $r_{AB} \geq r_A + r_B - n$.

Suppose first that A has the form $\begin{bmatrix} I_{r_A} & 0 \\ 0 & 0 \end{bmatrix}$. Then the first r_A rows of AB are the first r_A rows of B while the remaining rows are zeros. By Problem 10, Chapter 5, the rank of AB is $r_{AB} \geq r_A + r_B - n$.

Suppose next that A is not of the above form. Then there exist nonsingular matrices P and Q such that PAQ has that form while the rank of $PAQB$ is exactly that of AB (why?).

The reader may consider the special case when $B = \begin{bmatrix} I_{r_B} & 0 \\ 0 & 0 \end{bmatrix}$.

11. Let $X = [1,2,1]'$ relative to the E-basis. Find its coordinates relative to a new basis $Z_1 = [1,1,0]'$, $Z_2 = [1,0,1]'$, and $Z_3 = [1,1,1]'$.

Solution (a). Write

(i) $X = aZ_1 + bZ_2 + cZ_3$, that is, $\begin{bmatrix} 1 \\ 2 \\ 1 \end{bmatrix} = a\begin{bmatrix} 1 \\ 1 \\ 0 \end{bmatrix} + b\begin{bmatrix} 1 \\ 0 \\ 1 \end{bmatrix} + c\begin{bmatrix} 1 \\ 1 \\ 1 \end{bmatrix}$. Then $\begin{cases} a + b + c = 1 \\ a + c = 2 \\ b + c = 1 \end{cases}$ and $a = 0$, $b = -1$, $c = 2$. Thus relative to the Z-basis, we have $X_Z = [0,-1,2]'$.

Solution (*b*). Rewriting (i) as $X = [Z_1, Z_2, Z_3]X_Z = ZX_Z$, we have

$$X_Z = Z^{-1}X = \begin{bmatrix} 1 & 0 & -1 \\ 1 & -1 & 0 \\ -1 & 1 & 1 \end{bmatrix}\begin{bmatrix} 1 \\ 2 \\ 1 \end{bmatrix} = [0,-1,2]'$$

12. Let X_Z and X_W be the coordinates of a vector X with respect to the two bases $Z_1 = [1,1,0]'$, $Z_2 = [1,0,1]'$, $Z_3 = [1,1,1]'$ and $W_1 = [1,1,2]'$, $W_2 = [2,2,1]'$, $W_3 = [1,2,2]'$. Determine the matrix P such that $X_W = PX_Z$.

Here $Z = [Z_1, Z_2, Z_3] = \begin{bmatrix} 1 & 1 & 1 \\ 1 & 0 & 1 \\ 0 & 1 & 1 \end{bmatrix}$, $W = \begin{bmatrix} 1 & 2 & 1 \\ 1 & 2 & 2 \\ 2 & 1 & 2 \end{bmatrix}$, and $W^{-1} = \frac{1}{3}\begin{bmatrix} 2 & -3 & 2 \\ 2 & 0 & -1 \\ -3 & 3 & 0 \end{bmatrix}$.

Then $P = W^{-1}Z = \frac{1}{3}\begin{bmatrix} -1 & 4 & 1 \\ 2 & 1 & 1 \\ 0 & -3 & 0 \end{bmatrix}$ by (**11.6**).

SUPPLEMENTARY PROBLEMS

13. Let $[x_1, x_2, x_3, x_4]'$ be an arbitrary vector of $V_4(R)$, where R denotes the field of real numbers. Which of the following sets are subspaces of $V_4(R)$?

(*a*) All vectors with $x_1 = x_2 = x_3 = x_4$.

(*b*) All vectors with $x_1 = x_2$, $x_3 = 2x_4$.

(*c*) All vectors with $x_4 = 0$.

(*d*) All vectors with $x_1 = 1$.

(*e*) All vectors with x_1, x_2, x_3, x_4 integral.

Ans. All except (*d*) and (*e*).

14. Show that $[1,1,1,1]'$ and $[2,3,3,2]'$ are a basis of the $V_4^2(F)$ of Problem 2.

15. Determine the dimension of the vector space spanned by each set of vectors. Select a basis for each.

(*a*) $[1,2,3,4,5]'$, $[5,4,3,2,1]'$, $[1,1,1,1,1]'$

(*b*) $[1,1,0,-1]'$, $[1,2,3,4]'$, $[2,3,3,3]'$

(*c*) $[1,1,1,1]'$, $[3,4,5,6]'$, $[1,2,3,4]'$, $[1,0,-1,-2]'$

Ans. (*a*), (*b*), (*c*), $r = 2$

16. (*a*) Show that the vectors $X_1 = [1,-1,1]'$ and $X_2 = [3,4,-2]'$ span the same space as $Y_1 = [9,5,-1]'$ and $Y_2 = [-17,-11,3]'$.

(*b*) Show that the vectors $X_1 = [1,-1,1]'$ and $X_2 = [3,4,-2]'$ do not span the same space as $Y_1 = [-2,2,-2]'$ and $Y_2 = [4,3,1]'$.

17. Show that if the set $X_1, X_2, \ldots, X_k$ is a basis for $V_n^k(F)$, then any other vector Y of the space can be represented *uniquely* as a linear combination of $X_1, X_2, \ldots, X_k$.

Hint. Assume $Y = \sum_{i=1}^{k} a_i X_i = \sum_{i=1}^{k} b_i X_i$.

18. Consider the 4×4 matrix whose columns are the vectors of a basis of the $V_4^2(R)$ of Problem 2 and a basis of the $V_4^2(R)$ of Problem 3. Show that the rank of this matrix is 4; hence, $V_4(R)$ is the sum space and $V_4^0(R)$, the zero space, is the intersection space of the two given spaces.

19. Follow the proof given in Problem 8, Chapter 10, to prove Theorem **III**.

20. Show that the space spanned by $[1,0,0,0,0]'$, $[0,0,0,0,1]'$, $[1,0,1,0,0]'$, $[0,0,1,0,0]'$, $[1,0,0,1,1]'$ and the space spanned by $[1,0,0,0,1]'$, $[0,1,0,1,0]'$, $[0,1,-2,1,0]'$, $[1,0,-1,0,1]'$, $[0,1,1,1,0]'$ are of dimensions 4 and 3, respectively. Show that $[1,0,1,0,1]'$ and $[1,0,2,0,1]'$ are a basis for the intersection space.

21. Find, relative to the basis $Z_1 = [1,1,2]'$, $Z_2 = [2,2,1]'$, $Z_3 = [1,2,2]'$ the coordinates of the vectors (*a*) $[1,1,0]'$, (*b*) $[1,0,1]'$, (*c*) $[1,1,1]'$.

Ans. (*a*) $[-1/3, 2/3, 0]'$, (*b*) $[4/3, 1/3, -1]'$, (*c*) $[1/3, 1/3, 0]'$

22. Find, relative to the basis $Z_1 = [0,1,0]'$, $Z_2 = [1,1,1]'$, $Z_3 = [3,2,1]'$ the coordinates of the vectors (*a*) $[2,-1,0]'$, (*b*) $[1,-3,5]'$, (*c*) $[0,0,1]'$.

Ans. (*a*) $[-2,-1,1]'$, (*b*) $[-6,7,-2]'$, (*c*) $[-1/2, 3/2, -1/2]'$

23. Let X_Z and X_W be the coordinates of a vector X with respect to the given pair of bases. Determine the matrix P such that $X_W = PX_Z$.

(*a*) $Z_1 = [1,0,0]'$, $Z_2 = [1,0,1]'$, $Z_3 = [1,1,1]'$; $W_1 = [0,1,0]'$, $W_2 = [1,2,3]'$, $W_3 = [1,-1,1]'$

(*b*) $Z_1 = [0,1,0]'$, $Z_2 = [1,1,0]'$, $Z_3 = [1,2,3]'$; $W_1 = [1,1,0]'$, $W_2 = [1,1,1]'$, $W_3 = [1,2,1]'$

Ans. (*a*) $P = \frac{1}{2}\begin{bmatrix} 5 & 2 & 4 \\ -1 & 0 & 0 \\ 3 & 2 & 2 \end{bmatrix}$, (*b*) $P = \begin{bmatrix} 0 & 1 & -2 \\ -1 & 0 & 2 \\ 1 & 0 & 1 \end{bmatrix}$

24. Prove: If P_j is a solution of $AX = E_j$, $(j = 1,2,\ldots,n)$, then $\sum_{i=1}^{n} h_j P_j$ is a solution of $AX = H$, where $H = [h_1, h_2, \ldots, h_n]'$.

Hint. $H = h_1E_1 + h_2E_2 + \cdots + h_nE_n$.

25. The vector space defined by all linear combinations of the columns of a matrix A is called the **column space** of A. The vector space defined by all linear combinations of the rows of A is called the **row space** of A. Show that the columns of AB are in the column space of A and the rows of AB are in the row space of B.

26. Show that $AX = H$, a system of m non-homogeneous equations in n unknowns, is consistent if and only if the the vector H belongs to the column space of A.

27. Determine a basis for the null space of (*a*) $\begin{bmatrix} 1 & 1 & 0 \\ 0 & 1 & -1 \\ 1 & 0 & 1 \end{bmatrix}$, (*b*) $\begin{bmatrix} 1 & 1 & 1 & 1 \\ 1 & 2 & 1 & 2 \\ 3 & 4 & 3 & 4 \end{bmatrix}$.

Ans. (*a*) $[1,-1,-1]'$, (*b*) $[1,1,-1,-1]'$, $[1,2,-1,-2]'$

28. Prove: (*a*) $N_{AB} \geq N_A$, $N_{AB} \geq N_B$ (*b*) $N_{AB} \leq N_A + N_B$

Hint: (*a*) $N_{AB} = n - r_{AB}$; $r_{AB} \leq r_A$ and r_B.

(*b*) Consider $n - r_{AB}$, using the theorem of Problem 10.

29. Derive a procedure for Problem 16 using only column transformations on $A = [X_1, X_2, Y_1, Y_2]$. Then resolve Problem 5.

Chapter 12

Linear Transformations

DEFINITION. Let $X = [x_1, x_2, \ldots, x_n]'$ and $Y = [y_1, y_2, \ldots, y_n]'$ be two vectors of $V_n(F)$, their coordinates being relative to the **same** basis of the space. Suppose that the coordinates of X . Y are related by

$$(12.1)\qquad \begin{cases} y_1 = a_{11}x_1 + a_{12}x_2 + \cdots + a_{1n}x_n \\ y_2 = a_{21}x_1 + a_{22}x_2 + \cdots + a_{2n}x_n \\ \cdots\cdots\cdots\cdots\cdots\cdots\cdots\cdots\cdots \\ y_n = a_{n1}x_1 + a_{n2}x_2 + \cdots + a_{nn}x_n \end{cases}$$

or, briefly,

$$Y = AX$$

where $A = [a_{ij}]$ is over F. Then **(12.1)** is a transformation T which carries any vector X of $V_n(F)$ into (usually) another vector Y of the same space, called its **image**.

If **(12.1)** carries X_1 into Y_1 and X_2 into Y_2, then

(*a*) it carries kX_1 into kY_1, for every scalar k, and

(*b*) it carries $aX_1 + bX_2$ into $aY_1 + bY_2$, for every pair of scalars a and b. For this reason, the transformation is called **linear**.

Example 1. Consider the linear transformation $Y = AX = \begin{bmatrix} 1 & 1 & 2 \\ 1 & 2 & 5 \\ 1 & 3 & 3 \end{bmatrix} X$ in ordinary space $V_3(R)$.

(*a*) The image of $X = [2,0,5]'$ is $Y = \begin{bmatrix} 1 & 1 & 2 \\ 1 & 2 & 5 \\ 1 & 3 & 3 \end{bmatrix}\begin{bmatrix} 2 \\ 0 \\ 5 \end{bmatrix} = \begin{bmatrix} 12 \\ 27 \\ 17 \end{bmatrix} = [12,27,17]'$.

(*b*) The vector X whose image is $Y = [2,0,5]'$ is obtained by solving $\begin{bmatrix} 1 & 1 & 2 \\ 1 & 2 & 5 \\ 1 & 3 & 3 \end{bmatrix}\begin{bmatrix} x_1 \\ x_2 \\ x_3 \end{bmatrix} = \begin{bmatrix} 2 \\ 0 \\ 5 \end{bmatrix}$.

Since $\begin{bmatrix} 1 & 1 & 2 & 2 \\ 1 & 2 & 5 & 0 \\ 1 & 3 & 3 & 5 \end{bmatrix} \sim \begin{bmatrix} 1 & 0 & 0 & 13/5 \\ 0 & 1 & 0 & 11/5 \\ 0 & 0 & 1 & -7/5 \end{bmatrix}$, $\quad X = [13/5, 11/5, -7/5]'$.

BASIC THEOREMS. If in **(12.1)**, $X = [1,0,\ldots,0]' = E_1$ then $Y = [a_{11}, a_{21}, \ldots, a_{n1}]'$ and, in general, if $X = E_j$ then $Y = [a_{1j}, a_{2j}, \ldots, a_{nj}]'$. Hence,

I. A linear transformation **(12.1)** is uniquely determined when the images (Y's) of the basis vectors are known, the respective columns of A being the coordinates of the images of these vectors.

See Problem 1.

A linear transformation **(12.1)** is called **non-singular** if the images of distinct vectors X_i are distinct vectors Y_i. Otherwise the transformation is called **singular**.

II. A linear transformation **(12.1)** is non-singular if and only if A, the matrix of the transformation, is non-singular. See Problem 2.

III. A non-singular linear transformation carries linearly independent (dependent) vectors into linearly independent (dependent) vectors. See Problem 3.

From Theorem **III** follows

IV. Under a non-singular transformation **(12.1)** the image of a vector space $V_n^k(F)$ is a vector space $V_n^k(F)$, that is, the dimension of the vector space is preserved. In particular, the transformation is a mapping of $V_n(F)$ onto itself.

When A is non-singular, the inverse of **(12.1)**

$$X = A^{-1}Y$$

carries the set of vectors $Y_1, Y_2, \ldots, Y_n$ whose components are the columns of A into the basis vectors of the space. It is also a linear transformation.

V. The elementary vectors E_i of $V_n(F)$ may be transformed into any set of n linearly independent n-vectors by a non-singular linear transformation and conversely.

VI. If $Y = AX$ carries a vector X into a vector Y, if $Z = BY$ carries Y into Z, and if $W = CZ$ carries Z into W, then $Z = BY = (BA)X$ carries X into Z and $W = (CBA)X$ carries X into W.

VII. When any two sets of n linearly independent n-vectors are given, there exists a non-singular linear transformation which carries the vectors of one set into the vectors of the other.

CHANGE OF BASIS. Relative to a Z-basis, let $Y_Z = AX_Z$ be a linear transformation of $V_n(F)$. Suppose that the basis is changed and let X_W and Y_W be the coordinates of X_Z and Y_Z respectively relative to the new basis. By Theorem **VIII**, Chapter 11, there exists a non-singular matrix P such that $X_W = PX_Z$ and $Y_W = PY_Z$ or, setting $P^{-1} = Q$, such that

$$X_Z = QX_W \quad \text{and} \quad Y_Z = QY_W$$

Then

$$Y_W = Q^{-1}Y_Z = Q^{-1}AX_Z = Q^{-1}AQX_W = BX_W$$

where

$$B = Q^{-1}AQ \qquad (12.2)$$

Two matrices A and B such that there exists a non-singular matrix Q for which $B = Q^{-1}AQ$ are called **similar**. We have proved

VIII. If $Y_Z = AX_Z$ is a linear transformation of $V_n(F)$ relative to a given basis (Z-basis) and $Y_W = BX_W$ is the same linear transformation relative to another basis (W-basis), then A and B are similar.

Note. Since $Q = P^{-1}$, **(12.2)** might have been written as $B = PAP^{-1}$. A study of similar matrices will be made later. There we shall agree to write $B = R^{-1}AR$ instead of $B = SAS^{-1}$ but for no compelling reason.

Example 2. Let $Y = AX = \begin{bmatrix} 1 & 1 & 3 \\ 1 & 2 & 1 \\ 1 & 3 & 2 \end{bmatrix} X$ be a linear transformation relative to the E-basis and let $W_1 = [1,2,1]'$, $W_2 = [1,-1,2]'$, $W_3 = [1,-1,-1]'$ be a new basis. (*a*) Given the vector $X = [3,0,2]'$, find the coordinates of its image relative to the W-basis. (*b*) Find the linear transformation $Y_W = BX_W$ corresponding to $Y = AX$. (*c*) Use the result of (*b*) to find the image Y_W of $X_W = [1,3,3]'$.

Write $W = [W_1, W_2, W_3] = \begin{bmatrix} 1 & 1 & 1 \\ 2 & -1 & -1 \\ 1 & 2 & -1 \end{bmatrix}$; then $W^{-1} = \frac{1}{9}\begin{bmatrix} 3 & 3 & 0 \\ 1 & -2 & 3 \\ 5 & -1 & -3 \end{bmatrix}$.

(*a*) Relative to the W-basis, the vector $X = [3,0,2]'$ has coordinates $X_W = W^{-1}X = [1,1,1]'$. The image of X is $Y = AX = [9,5,7]'$ which, relative to the W-basis is $Y_W = W^{-1}Y = [14/3, 20/9, 19/9]'$.

(*b*) $Y_W = W^{-1}Y = W^{-1}AX = (W^{-1}AW)X_W = BX_W = \frac{1}{9}\begin{bmatrix} 36 & 21 & -15 \\ 21 & 10 & -11 \\ -3 & 23 & -1 \end{bmatrix} X_W$

(*c*) $Y_W = \frac{1}{9}\begin{bmatrix} 36 & 21 & -15 \\ 21 & 10 & -11 \\ -3 & 23 & -1 \end{bmatrix}\begin{bmatrix} 1 \\ 3 \\ 3 \end{bmatrix} = \begin{bmatrix} 6 \\ 2 \\ 7 \end{bmatrix} = [6,2,7]'$.

See Problem 5.

SOLVED PROBLEMS

1. (*a*) Set up the linear transformation $Y = AX$ which carries E_1 into $Y_1 = [1,2,3]'$, E_2 into $[3,1,2]'$, and E_3 into $Y_3 = [2,1,3]'$.
   (*b*) Find the images of $X_1 = [1,1,1]'$, $X_2 = [3,-1,4]'$, and $X_3 = [4,0,5]'$.
   (*c*) Show that X_1 and X_2 are linearly independent as also are their images.
   (*d*) Show that X_1, X_2, and X_3 are linearly dependent as also are their images.

(*a*) By Theorem I, $A = [Y_1, Y_2, Y_3]$; the equation of the linear transformation is $Y = AX = \begin{bmatrix} 1 & 3 & 2 \\ 2 & 1 & 1 \\ 3 & 2 & 3 \end{bmatrix} X$.

(*b*) The image of $X_1 = [1,1,1]'$ is $Y_1 = \begin{bmatrix} 1 & 3 & 2 \\ 2 & 1 & 1 \\ 3 & 2 & 3 \end{bmatrix}\begin{bmatrix} 1 \\ 1 \\ 1 \end{bmatrix} = [6,4,8]'$. The image of X_2 is $Y_2 = [8,9,19]'$ and the image of X_3 is $Y_3 = [14,13,27]'$.

(*c*) The rank of $[X_1, X_2] = \begin{bmatrix} 1 & 3 \\ 1 & -1 \\ 1 & 4 \end{bmatrix}$ is 2 as also is that of $[Y_1, Y_2] = \begin{bmatrix} 6 & 8 \\ 4 & 9 \\ 8 & 19 \end{bmatrix}$. Thus, X_1 and X_2 are linearly independent as also are their images.

(*d*) We may compare the ranks of $[X_1, X_2, X_3]$ and $[Y_1, Y_2, Y_3]$; however, $X_3 = X_1 + X_2$ and $Y_3 = Y_1 + Y_2$ so that both sets are linearly dependent.

2. Prove: A linear transformation **(12.1)** is non-singular if and only if A is non-singular.

Suppose A is non-singular and the transforms of $X_1 \neq X_2$ are $Y = AX_1 = AX_2$. Then $A(X_1 - X_2) = 0$ and the system of homogeneous linear equations $AX = 0$ has the non-trivial solution $X = X_1 - X_2$. This is possible if and only if $|A| = 0$, a contradiction of the hypothesis that A is non-singular.

3. Prove: A non-singular linear transformation carries linearly independent vectors into linearly independent vectors.

Assume the contrary, that is, suppose that the images $Y_i = AX_i$, $(i = 1,2,\ldots,p)$ of the linearly independent vectors $X_1, X_2, \ldots, X_p$ are linearly dependent. Then there exist scalars $s_1, s_2, \ldots, s_p$, not all zero, such that

$$\sum_{i=1}^{p} s_i Y_i \quad = \quad s_1 Y_1 + s_2 Y_2 + \cdots + s_p Y_p \quad = \quad 0$$

or

$$\sum_{i=1}^{p} s_i(AX_i) \quad = \quad A(s_1 X_1 + s_2 X_2 + \cdots + s_p X_p) \quad = \quad 0$$

Since A is non-singular, $s_1 X_1 + s_2 X_2 + \cdots + s_p X_p = 0$. But this is contrary to the hypothesis that the X_i are linearly independent. Hence, the Y_i are linearly independent.

4. A certain linear transformation $Y = AX$ carries $X_1 = [1,0,1]'$ into $[2,3,-1]'$, $X_2 = [1,-1,1]'$ into $[3,0,-2]'$, and $X_3 = [1,2,-1]'$ into $[-2,7,-1]'$. Find the images of E_1, E_2, E_3 and write the equation of the transformation.

Let $aX_1 + bX_2 + cX_3 = E_1$; then $\begin{cases} a + b + \ c = 1 \\ \quad -b + 2c = 0 \\ a + b - \ c = 0 \end{cases}$ and $a = -\frac{1}{2}$, $b = 1$, $c = \frac{1}{2}$. Thus, $E_1 = -\frac{1}{2}X_1 + X_2 + \frac{1}{2}X_3$

and its image is $Y_1 = -\frac{1}{2}[2,3,-1]' + [3,0,-2]' + \frac{1}{2}[-2,7,-1]' = [1,2,-2]'$. Similarly, the image of E_2 is $Y_2 = [-1,3,1]'$ and the image of E_3 is $Y_3 = [1,1,1]'$. The equation of the transformation is

$$Y \quad = \quad [Y_1, Y_2, Y_3]X \quad = \quad \begin{bmatrix} 1 & -1 & 1 \\ 2 & 3 & 1 \\ -2 & 1 & 1 \end{bmatrix} X$$

5. If $Y_Z = AX_Z = \begin{bmatrix} 1 & 1 & 2 \\ 2 & 2 & 1 \\ 3 & 1 & 2 \end{bmatrix} X_Z$ is a linear transformation relative to the Z-basis of Problem 12, Chapter 11, find the same transformation $Y_W = BX_W$ relative to the W-basis of that problem.

From Problem 12, Chapter 11, $X_W = PX_Z = \frac{1}{3}\begin{bmatrix} -1 & 4 & 1 \\ 2 & 1 & 1 \\ 0 & -3 & 0 \end{bmatrix} X_Z$. Then

$$X_Z \quad = \quad P^{-1}X_W \quad = \quad \begin{bmatrix} -1 & 1 & -1 \\ 0 & 0 & -1 \\ 2 & 1 & 3 \end{bmatrix} X_W \quad = \quad QX_W$$

and

$$Y_W \quad = \quad PY_Z \quad = \quad Q^{-1}AX_Z \quad = \quad Q^{-1}AQX_W \quad = \quad \frac{1}{3}\begin{bmatrix} -2 & 14 & -6 \\ 7 & 14 & 9 \\ 0 & -9 & 3 \end{bmatrix} X_W$$

SUPPLEMENTARY PROBLEMS

6. In Problem 1 show: (a) the transformation is non-singular, (b) $X = A^{-1}Y$ carries the column vectors of A into the elementary vectors.

7. Using the transformation of Problem 1, find (a) the image of $X = [1,1,2]'$, (b) the vector X whose image is $[-2,-5,-5]'$. *Ans.* (a) $[8,5,11]'$, (b) $[-3,-1,2]'$

8. Study the effect of the transformation $Y = IX$, also $Y = kIX$.

9. Set up the linear transformation which carries E_1 into $[1,2,3]'$, E_2 into $[3,1,2]'$, and E_3 into $[2,-1,-1]'$. Show that the transformation is singular and carries the linearly independent vectors $[1,1,1]'$ and $[2,0,2]'$ into the same image vector.

10. Suppose (12.1) is non-singular and show that if $X_1, X_2, \ldots, X_n$ are linearly dependent so also are their images $Y_1, Y_2, \ldots, Y_n$.

11. Use Theorem III to show that under a non-singular transformation the dimension of a vector space is unchanged. *Hint.* Consider the images of a basis of $V_n^k(F)$.

12. Given the linear transformation $Y = \begin{bmatrix} 1 & 1 & 0 \\ 2 & 3 & 1 \\ -2 & 3 & 5 \end{bmatrix} X$, show (a) it is singular, (b) the images of the linearly independent vectors $X_1 = [1,1,1]'$, $X_2 = [2,1,2]'$, and $X_3 = [1,2,3]'$ are linearly dependent, (c) the image of $V_3(R)$ is a $V_3^2(R)$.

13. Given the linear transformation $Y = \begin{bmatrix} 1 & 1 & 3 \\ 1 & 2 & 4 \\ 1 & 1 & 3 \end{bmatrix} X$, show (a) it is singular, (b) the image of every vector of the $V_3^2(R)$ spanned by $[1,1,1]'$ and $[3,2,0]'$ lies in the $V_3^1(R)$ spanned by $[5,7,5]'$.

14. Prove Theorem VII.
Hint. Let X_i and Y_i, $(i = 1,2,\ldots,n)$ be the given sets of vectors. Let $Z = AX$ carry the set X_i into E_i and $Y = BZ$ carry the E_i into Y_i.

15. Prove: Similar matrices have equal determinants.

16. Let $Y = AX = \begin{bmatrix} 1 & 2 & 3 \\ 3 & 2 & 1 \\ 1 & 1 & 1 \end{bmatrix} X$ be a linear transformation relative to the E-basis and let a new basis, say $Z_1 = [1,1,0]'$, $Z_2 = [1,0,1]'$, $Z_3 = [1,1,1]'$ be chosen. Let $X = [1,2,3]'$ relative to the E-basis. Show that
(a) $Y = [14,10,6]'$ is the image of X under the transformation,
(b) X, when referred to the new basis, has coordinates $X_Z = [-2,-1,4]'$ and Y has coordinates $Y_Z = [8,4,2]'$.
(c) $X_Z = PX$ and $Y_Z = PY$, where $P = \begin{bmatrix} 1 & 0 & -1 \\ 1 & -1 & 0 \\ -1 & 1 & 1 \end{bmatrix} = [Z_1, Z_2, Z_3]^{-1}$
(d) $Y_Z = Q^{-1}AQX_Z$, where $Q = P^{-1}$.

17. Given the linear transformation $Y_W = \begin{bmatrix} 1 & 1 & 0 \\ 0 & 1 & 1 \\ 1 & 0 & 1 \end{bmatrix} X_W$, relative to the W-basis: $W_1 = [0,-1,2]'$, $W_2 = [4,1,0]'$,

$W_3 = [-2,0,-4]'$. Find the representation relative to the Z-basis: $Z_1 = [1,-1,1]'$, $Z_2 = [1,0,-1]'$, $Z_3 = [1,2,1]'$.

Ans. $Y_Z = \begin{bmatrix} -1 & 0 & 3 \\ 2 & 2 & -5 \\ -1 & 0 & 2 \end{bmatrix} X_Z$

18. If, in the linear transformation $Y = AX$, A is singular, then the null space of A is the vector space each of whose vectors transforms into the zero vector. Determine the null space of the transformation of

(*a*) Problem 12, (*b*) Problem 13, (*c*) $Y = \begin{bmatrix} 1 & 2 & 3 \\ 2 & 4 & 6 \\ 3 & 6 & 9 \end{bmatrix} X$.

Ans. (*a*) $V_3^1(R)$ spanned by $[1,-1,1]'$
(*b*) $V_3^1(R)$ spanned by $[2,1,-1]'$
(*c*) $V_3^2(R)$ spanned by $[2,-1,0]'$ and $[3,0,-1]'$

19. If $Y = AX$ carries every vector of a vector space V_n^h into a vector of that same space, V_n^h is called an **invariant space of the transformation**. Show that in the real space $V_3(R)$ under the linear transformation

(*a*) $Y = \begin{bmatrix} 1 & 0 & -1 \\ 1 & 2 & 1 \\ 2 & 2 & 3 \end{bmatrix} X$, the V_3^1 spanned by $[1,-1,0]'$, the V_3^1 spanned by $[2,-1,-2]'$, and the V_3^1 spanned by $[1,-1,-2]'$ are invariant vector spaces.

(*b*) $Y = \begin{bmatrix} 2 & 2 & 1 \\ 1 & 3 & 1 \\ 1 & 2 & 2 \end{bmatrix} X$, the V_3^1 spanned by $[1,1,1]'$ and the V_3^2 spanned by $[1,0,-1]'$ and $[2,-1,0]'$ are invariant spaces. (Note that every vector of the V_3^2 is carried into itself.)

(*c*) $Y = \begin{bmatrix} 0 & 1 & 0 & 0 \\ 0 & 0 & 1 & 0 \\ 0 & 0 & 0 & 1 \\ -1 & 4 & -6 & 4 \end{bmatrix} X$, the V_4^1 spanned by $[1,1,1,1]'$ is an invariant vector space.

20. Consider the linear transformation $Y = PX$: $y_i = x_{j_i}$, $(i = 1, 2, \ldots, n)$ in which $j_1, j_2, \ldots, j_n$ is a permutation of $1, 2, \ldots, n$.

(*a*) Describe the **permutation matrix** P.

(*b*) Prove: There are $n!$ permutation matrices of order n.

(*c*) Prove: If P_1 and P_2 are permutation matrices so also are $P_3 = P_1P_2$ and $P_4 = P_2P_1$.

(*d*) Prove: If P is a permutation matrix so also are P' and $PP' = I$.

(*e*) Show that each permutation matrix P can be expressed as a product of a number of the elementary column matrices $K_{12}, K_{23}, \ldots, K_{n-1,n}$.

(*f*) Write $P = [E_{i_1}, E_{i_2}, \ldots, E_{i_n}]$ where $i_1, i_2, \ldots, i_n$ is a permutation of $1, 2, \ldots, n$ and E_{i_j} are the elementary n-vectors. Find a rule (other than $P^{-1} = P'$) for writing P^{-1}. For example, when $n = 4$ and $P = [E_3, E_1, E_4, E_2]$, then $P^{-1} = [E_2, E_4, E_1, E_3]$; when $P = [E_4, E_2, E_1, E_3]$, then $P^{-1} = [E_3, E_2, E_4, E_1]$.

Chapter 13

Vectors Over the Real Field

INNER PRODUCT. In this chapter all vectors are real and $V_n(R)$ is the space of all real n-vectors. If $X = [x_1, x_2, \ldots, x_n]'$ and $Y = [y_1, y_2, \ldots, y_n]'$ are two vectors of $V_n(R)$, their **inner product** is defined to be the scalar

$$X \cdot Y \quad = \quad x_1y_1 + x_2y_2 + \cdots + x_ny_n \tag{13.1}$$

Example 1. For the vectors $X_1 = [1,1,1]'$, $X_2 = [2,1,2]'$, $X_3 = [1,-2,1]'$:

(*a*) $X_1 \cdot X_2 = 1\cdot 2 + 1\cdot 1 + 1\cdot 2 = 5$

(*b*) $X_1 \cdot X_3 = 1\cdot 1 + 1(-2) + 1\cdot 1 = 0$

(*c*) $X_1 \cdot X_1 = 1\cdot 1 + 1\cdot 1 + 1\cdot 1 = 3$

(*d*) $X_1 \cdot 2X_2 = 1\cdot 4 + 1\cdot 2 + 1\cdot 4 = 10 = 2(X_1 \cdot X_2)$

Note. The inner product is frequently defined as

$$X \cdot Y \quad = \quad X'Y \quad = \quad Y'X \tag{13.1$'$}$$

The use of $X'Y$ and $Y'X$ is helpful; however, $X'Y$ and $Y'X$ are 1x1 matrices while $X{\cdot}Y$ is the element of the matrix. With this understanding, **(13.1$'$)** will be used here. Some authors write $X|Y$ for $X{\cdot}Y$. In vector analysis, the inner product is called the **dot product**.

The following rules for inner products are immediate

(13.2)

(*a*) $X_1 \cdot X_2 = X_2 \cdot X_1, \qquad X_1 \cdot kX_2 = k(X_1 \cdot X_2)$

(*b*) $X_1 \cdot (X_2 + X_3) = (X_2 + X_3) \cdot X_1 = X_1 \cdot X_2 + X_1 \cdot X_3$

(*c*) $(X_1 + X_2) \cdot (X_3 + X_4) = X_1 \cdot X_3 + X_1 \cdot X_4 + X_2 \cdot X_3 + X_2 \cdot X_4$

ORTHOGONAL VECTORS. Two vectors X and Y of $V_n(R)$ are said to be **orthogonal** if their inner product is 0. The vectors X_1 and X_3 of Example 1 are orthogonal.

THE LENGTH OF A VECTOR X of $V_n(R)$, denoted by $\|X\|$, is defined as the square root of the inner product of X and X; thus,

$$\|X\| \quad = \quad \sqrt{X{\cdot}X} \quad = \quad \sqrt{x_1^2 + x_2^2 + \cdots + x_n^2} \tag{13.3}$$

Example 2. From Example 1(*c*), $\|X_1\| = \sqrt{3}$.

See Problems 1-2.

Using **(13.1)** and **(13.3)**, it may be shown that

(13.4) $$X \cdot Y = \tfrac{1}{2}\{||X+Y||^2 - ||X||^2 - ||Y||^2\}$$

A vector X whose length is $||X|| = 1$ is called a **unit** vector. The elementary vectors E_i are examples of unit vectors.

THE SCHWARZ INEQUALITY. If X and Y are vectors of $V_n(R)$, then

(13.5) $$|X \cdot Y| \leq ||X|| \cdot ||Y||$$

that is, the numerical value of the inner product of two real vectors is at most the product of their lengths.

See Problem 3.

THE TRIANGLE INEQUALITY. If X and Y are vectors of $V_n(R)$, then

(13.6) $$||X+Y|| \leq ||X|| + ||Y||$$

ORTHOGONAL VECTORS AND SPACES. If $X_1, X_2, \ldots, X_m$ are $m \leq n$ mutually orthogonal non-zero n-vectors and if $c_1X_1 + c_2X_2 + \cdots + c_mX_m = 0$, then for $i = 1,2,\ldots,m$, $(c_1X_1 + c_2X_2 + \cdots + c_mX_m) \cdot X_i = 0$. Since this requires $c_i = 0$ for $i = 1,2,\ldots,m$, we have

I. Any set of $m \leq n$ mutually orthogonal non-zero n-vectors is a linearly independent set and spans a vector space $V_n^m(R)$.

A vector Y is said to be orthogonal to a vector space $V_n^m(R)$ if it is orthogonal to every vector of the space.

II. If a vector Y is orthogonal to each of the n-vectors $X_1, X_2, \ldots, X_m$, it is orthogonal to the space spanned by them.

See Problem 4.

III. If $V_n^h(R)$ is a subspace of $V_n^k(R)$, $k > h$, there exists at least one vector X of $V_n^k(R)$ which is orthogonal to $V_n^h(R)$.

See Problem 5.

Since mutually orthogonal vectors are linearly independent, a vector space $V_n^m(R)$, $m > 0$, can contain no more than m mutually orthogonal vectors. Suppose we have found $r < m$ mutually orthogonal vectors of a $V_n^m(R)$. They span a $V_n^r(R)$, a subspace of $V_n^m(R)$, and by Theorem **III**, there exists at least one vector of $V_n^m(R)$ which is orthogonal to the $V_n^r(R)$. We now have $r+1$ mutually orthogonal vectors of $V_n^m(R)$ and by repeating the argument, we show

IV. Every vector space $V_n^m(R)$, $m > 0$, contains m but not more than m mutually orthogonal vectors.

Two vector spaces are said to be orthogonal if every vector of one is orthogonal to every vector of the other space. For example, the space spanned by $X_1 = [1,0,0,1]'$ and $X_2 = [0,1,1,0]'$ is orthogonal to the space spanned by $X_3 = [1,0,0,-1]'$ and $X_4 = [0,1,-1,0]'$ since $(aX_1 + bX_2) \cdot (cX_3 + dX_4) = 0$ for all a, b, c, d.

V. The set of all vectors orthogonal to every vector of a given $V_n^k(R)$ is a unique vector space $V_n^{n-k}(R)$.

See Problem 6.

We may associate with any vector $X \neq 0$ a unique unit vector U obtained by dividing the components of X by $\|X\|$. This operation is called **normalization**. Thus, to normalize the vector $X = [2, 4, 4]'$, divide each component by $\|X\| = \sqrt{4+16+16} = 6$ and obtain the unit vector $[1/3, 2/3, 2/3]'$.

A basis of $V_n^m(R)$ which consists of mutually orthogonal vectors is called an **orthogonal basis** of the space; if the mutually orthogonal vectors are also unit vectors, the basis is called a normal orthogonal or **orthonormal** basis. The elementary vectors are an orthonormal basis of $V_n(R)$.

See Problem 7.

THE GRAM-SCHMIDT ORTHOGONALIZATION PROCESS. Suppose $X_1, X_2, \ldots, X_m$ are a basis of $V_n^m(R)$. Define

$$Y_1 = X_1$$

$$Y_2 = X_2 - \frac{Y_1 \cdot X_2}{Y_1 \cdot Y_1} Y_1$$

$$Y_3 = X_3 - \frac{Y_2 \cdot X_3}{Y_2 \cdot Y_2} Y_2 - \frac{Y_1 \cdot X_3}{Y_1 \cdot Y_1} Y_1$$

$$\cdots\cdots\cdots\cdots\cdots\cdots$$

$$Y_m = X_m - \frac{Y_{m-1} \cdot X_m}{Y_{m-1} \cdot Y_{m-1}} Y_{m-1} - \cdots - \frac{Y_1 \cdot X_m}{Y_1 \cdot Y_1} Y_1$$

Then the unit vectors $G_i = \dfrac{Y_i}{\|Y_i\|}$, $(i = 1, 2, \ldots, m)$ are mutually orthogonal and are an orthonormal basis of $V_n^m(R)$.

Example 3. Construct, using the Gram-Schmidt process, an orthogonal basis of $V_3(R)$, given a basis $X_1 = [1,1,1]'$, $X_2 = [1,-2,1]'$, $X_3 = [1,2,3]'$.

(i) $Y_1 = X_1 = [1,1,1]'$

(ii) $Y_2 = X_2 - \dfrac{Y_1 \cdot X_2}{Y_1 \cdot Y_1} Y_1 = [1,-2,1]' - \dfrac{0}{3} Y_1 = [1,-2,1]'$

(iii) $Y_3 = X_3 - \dfrac{Y_2 \cdot X_3}{Y_2 \cdot Y_2} Y_2 - \dfrac{Y_1 \cdot X_3}{Y_1 \cdot Y_1} Y_1 = [1,2,3]' - \dfrac{0}{6} Y_2 - \dfrac{6}{3}[1,1,1]' = [-1,0,1]'$

The vectors $G_1 = \dfrac{Y_1}{\|Y_1\|} = [1/\sqrt{3}, 1/\sqrt{3}, 1/\sqrt{3}]'$,

$G_2 = \dfrac{Y_2}{\|Y_2\|} = [1/\sqrt{6}, -2/\sqrt{6}, 1/\sqrt{6}]'$ and $G_3 = \dfrac{Y_3}{\|Y_3\|} = [-1/\sqrt{2}, 0, 1/\sqrt{2}]'$

are an orthonormal basis of $V_3(R)$. Each vector G_i is a unit vector and each product $G_i \cdot G_j = 0$. Note that $Y_2 = X_2$ here because X_1 and X_2 are orthogonal vectors.

See Problems 8-9.

Let $X_1, X_2, \ldots, X_m$ be a basis of a $V_n^m(R)$ and suppose that $X_1, X_2, \ldots, X_s, (1 \leq s < m)$, are mutually orthogonal. Then, by the Gram-Schmidt process, we may obtain an orthogonal basis $Y_1, Y_2, \ldots, Y_m$ of the space of which, it is easy to show, $Y_i = X_i$, $(i = 1,2,\ldots,s)$. Thus,

VI. If $X_1, X_2, \ldots, X_s, (1 \leq s < m)$, are mutually orthogonal unit vectors of a $V_n^m(R)$, there exist unit vectors $X_{s+1}, X_{s+2}, \ldots, X_m$ in the space such that the set $X_1, X_2, \ldots, X_m$ is an orthonormal basis.

THE GRAMIAN. Let $X_1, X_2, \ldots, X_p$ be a set of real n-vectors and define the **Gramian** matrix

$$G = \begin{bmatrix} X_1 \cdot X_1 & X_1 \cdot X_2 & \ldots & X_1 \cdot X_p \\ X_2 \cdot X_1 & X_2 \cdot X_2 & \ldots & X_2 \cdot X_p \\ \cdots & \cdots & \cdots & \cdots \\ X_p \cdot X_1 & X_p \cdot X_2 & \ldots & X_p \cdot X_p \end{bmatrix} = \begin{bmatrix} X_1'X_1 & X_1'X_2 & \ldots & X_1'X_p \\ X_2'X_1 & X_2'X_2 & \ldots & X_2'X_p \\ \cdots & \cdots & \cdots & \cdots \\ X_p'X_1 & X_p'X_2 & \ldots & X_p'X_p \end{bmatrix} \tag{13.8}$$

Clearly, the vectors are mutually orthogonal if and only if G is diagonal.

In Problem 14, Chapter 17, we shall prove

VII. For a set of real n-vectors $X_1, X_2, \ldots, X_p$, $|G| \geq 0$. The equality holds if and only if the vectors are linearly dependent.

ORTHOGONAL MATRICES. A square matrix A is called **orthogonal** if

$$AA' = A'A = I \tag{13.9}$$

that is, if

$$A' = A^{-1} \tag{13.9'}$$

From **(13.9)** it is clear that the column vectors (row vectors) of an orthogonal matrix A are mutually orthogonal unit vectors.

Example 4. By Example 3, $A = \begin{bmatrix} 1/\sqrt{3} & 1/\sqrt{6} & -1/\sqrt{2} \\ 1/\sqrt{3} & -2/\sqrt{6} & 0 \\ 1/\sqrt{3} & 1/\sqrt{6} & 1/\sqrt{2} \end{bmatrix}$ is orthogonal.

There follow readily

VIII. If the real n-square matrix A is orthogonal, its column vectors (row vectors) are an orthonormal basis of $V_n(R)$, and conversely.

IX. The inverse and the transpose of an orthogonal matrix are orthogonal.

X. The product of two or more orthogonal matrices is orthogonal.

XI. The determinant of an orthogonal matrix is ± 1.

ORTHOGONAL TRANSFORMATIONS. Let

$$Y = AX \tag{13.10}$$

be a linear transformation in $V_n(R)$ and let the images of the n-vectors X_1 and X_2 be denoted by Y_1 and Y_2 respectively. From **(13.4)** we have

$$X_1 \cdot X_2 = \tfrac{1}{2}\{\,\|X_1 + X_2\|^2 - \|X_1\|^2 - \|X_2\|^2\,\}$$

and

$$Y_1 \cdot Y_2 = \tfrac{1}{2}\{\,\|Y_1 + Y_2\|^2 - \|Y_1\|^2 - \|Y_2\|^2\,\}$$

Comparing right and left members, we see that if **(13.10)** preserves lengths it preserves inner products, and conversely. Thus,

XII. A linear transformation preserves lengths if and only if it preserves inner products.

A linear transformation $Y = AX$ is called **orthogonal** if its matrix A is orthogonal. In Problem 10, we prove

XIII. A linear transformation preserves lengths if and only if its matrix is orthogonal.

Example 5. The linear transformation $Y = AX = \begin{bmatrix} 1/\sqrt{3} & 1/\sqrt{6} & -1/\sqrt{2} \\ 1/\sqrt{3} & -2/\sqrt{6} & 0 \\ 1/\sqrt{3} & 1/\sqrt{6} & 1/\sqrt{2} \end{bmatrix} X$ is orthogonal. The image of $X = [a, b, c]'$ is

$$Y = \left[\frac{a}{\sqrt{3}} + \frac{b}{\sqrt{6}} - \frac{c}{\sqrt{2}},\ \frac{a}{\sqrt{3}} - \frac{2b}{\sqrt{6}},\ \frac{a}{\sqrt{3}} + \frac{b}{\sqrt{6}} + \frac{c}{\sqrt{2}}\right]'$$

and both vectors are of length $\sqrt{a^2 + b^2 + c^2}$.

XIV. If **(13.10)** is a transformation of coordinates from the E-basis to another, the Z-basis, then the Z-basis is orthonormal if and only if A is orthogonal.

SOLVED PROBLEMS

1. Given the vectors $X_1 = [1,2,3]'$ and $X_2 = [2,-3,4]'$, find:
(*a*) their inner product, (*b*) the length of each.

(*a*) $X_1 \cdot X_2 = X_1'X_2 = [1,2,3]\begin{bmatrix} 2 \\ -3 \\ 4 \end{bmatrix} = 1(2) + 2(-3) + 3(4) = 8$

(*b*) $\|X_1\|^2 = X_1 \cdot X_1 = X_1'X_1 = [1,2,3]\begin{bmatrix} 1 \\ 2 \\ 3 \end{bmatrix} = 14$ and $\|X_1\| = \sqrt{14}$

$\|X_2\|^2 = 2(2) + (-3)(-3) + 4(4) = 29$ and $\|X_2\| = \sqrt{29}$

2. (*a*) Show that $X = [1/3, -2/3, -2/3]'$ and $Y = [2/3, -1/3, 2/3]'$ are orthogonal.
(*b*) Find a vector Z orthogonal to both X and Y.

(*a*) $X\cdot Y \;=\; X'Y \;=\; [1/3, -2/3, -2/3]\begin{bmatrix}2/3\\-1/3\\2/3\end{bmatrix} \;=\; 0$ and the vectors are orthogonal.

(*b*) Write $[X,Y,0] = \begin{bmatrix}1/3 & 2/3 & 0\\ -2/3 & -1/3 & 0\\ -2/3 & 2/3 & 0\end{bmatrix}$ and compute the cofactors $-2/3, -2/3, 1/3$ of the elements of the column of zeros. Then by (**3.11**) $Z = [-2/3, -2/3, 1/3]'$ is orthogonal to both X and Y.

3. Prove the Schwarz Inequality: If X and Y are vectors of $V_n(R)$, then $|X\cdot Y| \leq \|X\|\cdot\|Y\|$.

Clearly, the theorem is true if X or Y is the zero vector. Assume then that X and Y are non-zero vectors. If a is any real number,

$$\begin{aligned}\|aX+Y\|^2 &= (aX+Y)\cdot(aX+Y)\\ &= [ax_1+y_1, ax_2+y_2, \ldots, ax_n+y_n]\cdot[ax_1+y_1, ax_2+y_2, \ldots, ax_n+y_n]'\\ &= (a^2x_1^2 + 2ax_1y_1 + y_1^2) + (a^2x_2^2 + 2ax_2y_2 + y_2^2) + \cdots + (a^2x_n^2 + 2ax_ny_n + y_n^2)\\ &= a^2\|X\|^2 + 2aX\cdot Y + \|Y\|^2 \geq 0\end{aligned}$$

Now a quadratic polynomial in a is greater than or equal to zero for **all** real values of a if and only if its discriminant is less than or equal to zero. Thus,

$$4(X\cdot Y)^2 - 4\|X\|^2\cdot\|Y\|^2 \leq 0$$

and

$$|X\cdot Y| \leq \|X\|\cdot\|Y\|$$

4. Prove: If a vector Y is orthogonal to each of the n-vectors $X_1, X_2, \ldots, X_m$, it is orthogonal to the space spanned by these vectors.

Any vector of the space spanned by the X's can be written as $a_1X_1 + a_2X_2 + \cdots + a_mX_m$. Then

$$(a_1X_1 + a_2X_2 + \cdots + a_mX_m)\cdot Y \;=\; a_1X_1\cdot Y + a_2X_2\cdot Y + \cdots + a_mX_m\cdot Y \;=\; 0$$

since $X_i\cdot Y = 0$, $(i = 1,2,\ldots,m)$. Thus, Y is orthogonal to every vector of the space and by definition is orthogonal to the space. In particular, if Y is orthogonal to every vector of a basis of a vector space, it is orthogonal to that space.

5. Prove: If a $V_n^h(R)$ is a subspace of a $V_n^k(R)$, $k > h$, then there exists at least one vector X of $V_n^k(R)$ which is orthogonal to the $V_n^h(R)$.

Let $X_1, X_2, \ldots, X_h$ be a basis of the $V_n^h(R)$, let X_{h+1} be a vector in the $V_n^k(R)$ but not in the $V_n^h(R)$, and consider the vector

(i) $$X \;=\; a_1X_1 + a_2X_2 + \cdots + a_hX_h + a_{h+1}X_{h+1}$$

The condition that X be orthogonal to each of $X_1, X_2, \ldots, X_h$ consists of h homogeneous linear equations

$$\begin{aligned} a_1X_1\cdot X_1 + a_2X_2\cdot X_1 + \cdots + a_hX_h\cdot X_1 + a_{h+1}X_{h+1}\cdot X_1 &= 0 \\ a_1X_1\cdot X_2 + a_2X_2\cdot X_2 + \cdots + a_hX_h\cdot X_2 + a_{h+1}X_{h+1}\cdot X_2 &= 0 \\ \cdots\cdots\cdots\cdots\cdots\cdots\cdots\cdots\cdots\cdots \\ a_1X_1\cdot X_h + a_2X_2\cdot X_h + \cdots + a_hX_h\cdot X_h + a_{h+1}X_{h+1}\cdot X_h &= 0 \end{aligned}$$

in the $h+1$ unknowns $a_1, a_2, \ldots, a_{h+1}$. By Theorem **IV**, Chapter 10, a non-trivial solution exists. When these values are substituted in **(i)**, we have a non-zero (why?) vector X orthogonal to the basis vectors of the $V_n^h(R)$ and hence to that space.

6. Prove: The set of all vectors orthogonal to every vector of a given $V_n^k(R)$ is a unique vector space $V_n^{n-k}(R)$.

Let $X_1, X_2, \ldots, X_k$ be a basis of the $V_n^k(R)$. The n-vectors X orthogonal to each of the X_i satisfy the system of homogeneous equations

(i) $$X_1\cdot X = 0,\ X_2\cdot X = 0,\ \ldots,\ X_k\cdot X = 0$$

Since the X_i are linearly independent, the coefficient matrix of the system **(i)** is of rank k; hence, there are n-k linearly independent solutions (vectors) which span a $V_n^{n-k}(R)$. (See Theorem **VI**, Chapter 10.)

Uniqueness follows from the fact that the intersection space of the $V_n^k(R)$ and the $V_n^{n-k}(R)$ is the zero-space so that the sum space is $V_n(R)$.

7. Find an orthonormal basis of $V_3(R)$, given $X = [1/\sqrt{6}, 2/\sqrt{6}, 1/\sqrt{6}]'$.

Note that X is a unit vector. Choose $Y = [1/\sqrt{2}, 0, -1/\sqrt{2}]'$ another unit vector such that $X\cdot Y = 0$. Then, as in Problem 2(*a*), obtain $Z = [1/\sqrt{3}, -1/\sqrt{3}, 1/\sqrt{3}]'$ to complete the set.

8. Derive the Gram-Schmidt equations **(13.7)**.

Let $X_1, X_2, \ldots, X_m$ be a given basis of $V_n^m(R)$ and denote by $Y_1, Y_2, \ldots, Y_m$ the set of mutually orthogonal vectors to be found.

(*a*) Take $Y_1 = X_1$.

(*b*) Take $Y_2 = X_2 + aY_1$. Since Y_1 and Y_2 are to be mutually orthogonal,

$$Y_1\cdot Y_2 = Y_1\cdot X_2 + Y_1\cdot aY_1 = Y_1\cdot X_2 + aY_1\cdot Y_1 = 0$$

and $a = -\dfrac{Y_1\cdot X_2}{Y_1\cdot Y_1}$. Thus, $Y_2 = X_2 - \dfrac{Y_1\cdot X_2}{Y_1\cdot Y_1}Y_1$.

(*c*) Take $Y_3 = X_3 + aY_2 + bY_1$. Since Y_1, Y_2, Y_3 are to be mutually orthogonal,

$$Y_1\cdot Y_3 = Y_1\cdot X_3 + aY_1\cdot Y_2 + bY_1\cdot Y_1 = Y_1\cdot X_3 + bY_1\cdot Y_1 = 0$$

and

$$Y_2\cdot Y_3 = Y_2\cdot X_3 + aY_2\cdot Y_2 + bY_2\cdot Y_1 = Y_2\cdot X_3 + aY_2\cdot Y_2 = 0$$

Then $a = -\dfrac{Y_2\cdot X_3}{Y_2\cdot Y_2}$, $b = -\dfrac{Y_1\cdot X_3}{Y_1\cdot Y_1}$, and $Y_3 = X_3 - \dfrac{Y_2\cdot X_3}{Y_2\cdot Y_2}Y_2 - \dfrac{Y_1\cdot X_3}{Y_1\cdot Y_1}Y_1$.

(*d*) Continue the process until Y_m is obtained.

9. Construct an orthonormal basis of V_3, given the basis $X_1 = [2,1,3]'$, $X_2 = [1,2,3]'$, and $X_3 = [1,1,1]'$.

Take $Y_1 = X_1 = [2,1,3]'$. Then

$$Y_2 = X_2 - \frac{Y_1 \cdot X_2}{Y_1 \cdot Y_1} Y_1 = [1,2,3]' - \frac{13}{14}[2,1,3]' = [-6/7, 15/14, 3/14]'$$

$$Y_3 = X_3 - \frac{Y_2 \cdot X_3}{Y_2 \cdot Y_2} Y_2 - \frac{Y_1 \cdot X_3}{Y_1 \cdot Y_1} Y_1$$

$$= [1,1,1]' - \frac{2}{9}\left[-\frac{6}{7}, \frac{15}{14}, \frac{3}{14}\right]' - \frac{3}{7}[2,1,3]' = \left[\frac{1}{3}, \frac{1}{3}, -\frac{1}{3}\right]'$$

Normalizing the Y's, we have $[2/\sqrt{14}, 1/\sqrt{14}, 3/\sqrt{14}]'$, $[-4/\sqrt{42}, 5/\sqrt{42}, 1/\sqrt{42}]'$, $[1/\sqrt{3}, 1/\sqrt{3}, -1/\sqrt{3}]'$ as the required orthonormal basis.

10. Prove: A linear transformation preserves lengths if and only if its matrix is orthogonal.

Let Y_1, Y_2 be the respective images of X_1, X_2 under the linear transformation $Y = AX$.

Suppose A is orthogonal so that $A'A = I$. Then

(i) $$Y_1 \cdot Y_2 = Y_1' Y_2 = (X_1'A')(AX_2) = X_1'X_2 = X_1 \cdot X_2$$

and, by Theorem **XII** lengths are preserved.

Conversely, suppose lengths (also inner products) are preserved. Then

$$Y_1 \cdot Y_2 = X_1'(A'A)X_2 = X_1'X_2, \qquad A'A = I$$

and A is orthogonal.

SUPPLEMENTARY PROBLEMS

11. Given the vectors $X_1 = [1,2,1]'$, $X_2 = [2,1,2]'$, $X_3 = [2,1,-4]'$, find:
(a) the inner product of each pair,
(b) the length of each vector,
(c) a vector orthogonal to the vectors X_1, X_2; X_1, X_3.
Ans. (a) 6, 0, −3 (b) $\sqrt{6}$, 3, $\sqrt{21}$ (c) $[1,0,-1]'$, $[3,-2,1]'$

12. Using arbitrary vectors of $V_3(R)$, verify **(13.2)**.

13. Prove **(13.4)**.

14. Let $X = [1,2,3,4]'$ and $Y = [2,1,-1,1]'$ be a basis of a $V_4^2(R)$ and $Z = [4,2,3,1]'$ lie in a $V_4^3(R)$ containing X and Y.
(a) Show that Z is not in the $V_4^2(R)$.
(b) Write $W = aX + bY + cZ$ and find a vector W of the $V_4^3(R)$ orthogonal to both X and Y.

15. (a) Prove: A vector of $V_n(R)$ is orthogonal to itself if and only if it is the zero vector.
(b) Prove: If X_1, X_2, X_3 are a set of linearly dependent non-zero n-vectors and if $X_1 \cdot X_2 = X_1 \cdot X_3 = 0$, then X_2 and X_3 are linearly dependent.

16. Prove: A vector X is orthogonal to every vector of a $V_n^m(R)$ if and only if it is orthogonal to every vector of a basis of the space.

17. Prove: If two spaces $V_n^h(R)$ and $V_n^k(R)$ are orthogonal, their intersection space is $V_n^0(R)$.

18. Prove: The Triangle Inequality.
Hint. Show that $\|X+Y\|^2 \leq (\|X\| + \|Y\|)^2$, using the Schwarz Inequality.

19. Prove: $\|X+Y\| = \|X\| + \|Y\|$ if and only if X and Y are linearly dependent.

20. Normalize the vectors of Problem 11.
Ans. $[1/\sqrt{6}, 2/\sqrt{6}, 1/\sqrt{6}]'$, $[2/3, 1/3, 2/3]'$, $[2/\sqrt{21}, 1/\sqrt{21}, -4/\sqrt{21}]'$

21. Show that the vectors X, Y, Z of Problem 2 are an orthonormal basis of $V_3(R)$.

22. (*a*) Show that if $X_1, X_2, \ldots, X_m$ are linearly independent so also are the unit vectors obtained by normalizing them.
(*b*) Show that if the vectors of (*a*) are mutually orthogonal non-zero vectors, so also are the unit vectors obtained by normalizing them.

23. Prove: (*a*) If A is orthogonal and $|A| = 1$, each element of A is equal to its cofactor in $|A|$.
(*b*) If A is orthogonal and $|A| = -1$, each element of A is equal to the negative of its cofactor in $|A|$.

24. Prove Theorems **VIII**, **IX**, **X**, **XI**.

25. Prove: If A and B commute and C is orthogonal, then $C'AC$ and $C'BC$ commute.

26. Prove that AA' (or $A'A$), where A is n-square, is a diagonal matrix if and only if the rows (or columns) of A are orthogonal.

27. Prove: If X and Y are n-vectors, then $XY' + YX'$ is symmetric.

28. Prove: If X and Y are n-vectors and A is n-square, then $X \cdot (AY) = (A'X) \cdot Y$.

29. Prove: If $X_1, X_2, \ldots, X_n$ be an orthonormal basis and if $X = \sum_{i=1}^{n} c_i X_i$, then (*a*) $X \cdot X_i = c_i$, $(i = 1, 2, \ldots, n)$; (*b*) $X \cdot X = c_1^2 + c_2^2 + \cdots + c_n^2$.

30. Find an orthonormal basis of $V_3(R)$, given (*a*) $X_1 = [3/\sqrt{17}, -2/\sqrt{17}, 2/\sqrt{17}]'$; (*b*) $[3, 0, 2]'$
Ans. (*a*) X_1, $[0, 1/\sqrt{2}, 1/\sqrt{2}]'$, $[-4/\sqrt{34}, -3/\sqrt{34}, 3/\sqrt{34}]'$
(*b*) $[3/\sqrt{13}, 0, 2/\sqrt{13}]'$, $[2/\sqrt{13}, 0, -3/\sqrt{13}]'$, $[0, 1, 0]'$

31. Construct orthonormal bases of $V_3(R)$ by the Gram-Schmidt process, using the given vectors in order:
(*a*) $[1, -1, 0]'$, $[2, -1, -2]'$, $[1, -1, -2]'$
(*b*) $[1, 0, 1]'$, $[1, 3, 1]'$, $[3, 2, 1]'$
(*c*) $[2, -1, 0]'$, $[4, -1, 0]'$, $[4, 0, -1]'$
Ans. (*a*) $[\frac{1}{2}\sqrt{2}, -\frac{1}{2}\sqrt{2}, 0]'$, $[\sqrt{2}/6, \sqrt{2}/6, -2\sqrt{2}/3]'$, $[-2/3, -2/3, -1/3]'$
(*b*) $[\frac{1}{2}\sqrt{2}, 0, \frac{1}{2}\sqrt{2}]'$, $[0, 1, 0]'$, $[\frac{1}{2}\sqrt{2}, 0, -\frac{1}{2}\sqrt{2}]'$
(*c*) $[2\sqrt{5}/5, -\sqrt{5}/5, 0]'$, $[\sqrt{5}/5, 2\sqrt{5}/5, 0]'$, $[0, 0, -1]'$

32. Obtain an orthonormal basis of $V_3(R)$, given $X_1 = [1, 1, -1]'$ and $X_2 = [2, 1, 0]'$.
Hint. Take $Y_1 = X_1$, obtain Y_2 by the Gram-Schmidt process, and Y_3 by the method of Problem 2(*b*).
Ans. $[\sqrt{3}/3, \sqrt{3}/3, -\sqrt{3}/3]'$, $[\frac{1}{2}\sqrt{2}, 0, \frac{1}{2}\sqrt{2}]'$, $[\sqrt{6}/6, -\sqrt{6}/3, -\sqrt{6}/6]'$

33. Obtain an orthonormal basis of $V_3(R)$, given $X_1 = [7,-1,-1]'$.

34. Show in two ways that the vectors $[1,2,3,4]'$, $[1,-1,-2,-3]'$, and $[5,4,5,6]'$ are linearly dependent.

35. Prove: If A is skew-symmetric, and $I+A$ is non-singular, then $B = (I-A)(I+A)^{-1}$ is orthogonal.

36. Use Problem 35 to obtain the orthogonal matrix B, given

$(a)\ A = \begin{bmatrix} 0 & 5 \\ -5 & 0 \end{bmatrix}$, $\qquad (b)\ A = \begin{bmatrix} 0 & 1 & 2 \\ -1 & 0 & 3 \\ -2 & -3 & 0 \end{bmatrix}$

Ans. $(a)\ \dfrac{1}{13}\begin{bmatrix} -12 & -5 \\ 5 & -12 \end{bmatrix}$, $\qquad (b)\ \dfrac{1}{15}\begin{bmatrix} 5 & -14 & 2 \\ -10 & -5 & -10 \\ 10 & 2 & -11 \end{bmatrix}$

37. Prove: If A is an orthogonal matrix and if $B = AP$, where P is non-singular, then PB^{-1} is orthogonal.

38. In a transformation of coordinates from the E-basis to an orthonormal Z-basis with matrix P, $Y = AX$ becomes $Y_1 = P^{-1}APX_1$ or $Y_1 = BX_1$ (see Chapter 12). Show that if A is orthogonal so also is B, and conversely, to prove Theorem **XIV**.

39. Prove: If A is orthogonal and $I+A$ is non-singular then $B = (I-A)(I+A)^{-1}$ is skew-symmetric.

40. Let $X = [x_1, x_2, x_3]'$ and $Y = [y_1, y_2, y_3]'$ be two vectors of $V_3(R)$ and define the **vector product**, $X \times Y$, of X and Y as $Z = X \times Y = [z_1, z_2, z_3]'$ where $z_1 = \begin{vmatrix} x_2 & y_2 \\ x_3 & y_3 \end{vmatrix}$, $z_2 = \begin{vmatrix} x_3 & y_3 \\ x_1 & y_1 \end{vmatrix}$, $z_3 = \begin{vmatrix} x_1 & y_1 \\ x_2 & y_2 \end{vmatrix}$. After identifying the z_i as cofactors of the elements of the third column of $[X_1, Y_1, 0]$, establish:

(a) The vector product of two linearly dependent vectors is the zero vector.

(b) The vector product of two linearly independent vectors is orthogonal to each of the two vectors.

(c) $X \times Y = -(Y \times X)$

(d) $(kX) \times Y = k(X \times Y) = X \times (kY)$, k a scalar.

41. If W, X, Y, Z are four vectors of $V_3(R)$, establish:

(a) $X \times (Y+Z) = X \times Y + X \times Z$

(b) $X \cdot (Y \times Z) = Y \cdot (Z \times X) = Z \cdot (X \times Y) = |XYZ|$

(c) $(W \times X) \cdot (Y \times Z) = \begin{vmatrix} W \cdot Y & W \cdot Z \\ X \cdot Y & X \cdot Z \end{vmatrix}$

(d) $(X \times Y) \cdot (X \times Y) = \begin{vmatrix} X \cdot X & X \cdot Y \\ Y \cdot X & Y \cdot Y \end{vmatrix}$

Chapter 14

Vectors Over the Complex Field

COMPLEX NUMBERS. If x and y are real numbers and i is defined by the relation $i^2 = -1$, $z = x+iy$ is called a **complex number**. The real number x is called the **real part** and the real number y is called the **imaginary part** of $x+iy$.

Two complex numbers are **equal** if and only if the real and imaginary parts of one are equal respectively to the real and imaginary parts of the other.

A complex number $x+iy = 0$ if and only if $x = y = 0$.

The **conjugate** of the complex number $z = x+iy$ is given by $\bar{z} = \overline{x+iy} = x-iy$. The sum (product) of any complex number and its conjugate is a real number.

The **absolute value** $|z|$ of the complex number $z = x+iy$ is given by $|z| = \sqrt{z\cdot\bar{z}} = \sqrt{x^2+y^2}$. It follows immediately that for any complex number $z = x+iy$,

(14.1) $$|z| \geq |x| \quad \text{and} \quad |z| \geq |y|$$

VECTORS. Let X be an n-vector over the complex field C. The totality of such vectors constitutes the vector space $V_n(C)$. Since $V_n(R)$ is a subfield, it is to be expected that each theorem concerning vectors of $V_n(C)$ will reduce to a theorem of Chapter 13 when only real vectors are considered.

If $X = [x_1, x_2, \ldots, x_n]'$ and $Y = [y_1, y_2, \ldots, y_n]'$ are two vectors of $V_n(C)$, their **inner product** is defined as

(14.2) $$X\cdot Y \;=\; \bar{X}'Y \;=\; \bar{x}_1 y_1 + \bar{x}_2 y_2 + \cdots + \bar{x}_n y_n$$

The following laws governing inner products are readily verified:

(14.3)

(*a*) $X\cdot Y = \overline{Y\cdot X}$

(*b*) $(cX)\cdot Y = \bar{c}(X\cdot Y)$

(*c*) $X\cdot(cY) = c(X\cdot Y)$

(*d*) $X\cdot(Y+Z) = X\cdot Y + X\cdot Z$

(*e*) $(Y+Z)\cdot X = Y\cdot X + Z\cdot X$

(*f*) $X\cdot Y + Y\cdot X = 2R(X\cdot Y)$ where $R(X\cdot Y)$ is the real part of $X\cdot Y$.

(*g*) $X\cdot Y - Y\cdot X = 2C(X\cdot Y)$ where $C(X\cdot Y)$ is the imaginary part of $X\cdot Y$.

See Problem 1.

The **length** of a vector X is given by $\|X\| = \sqrt{X\cdot X} = \sqrt{x_1\bar{x}_1 + x_2\bar{x}_2 + \cdots + x_n\bar{x}_n}$.

Two vectors X and Y are **orthogonal** if $X\cdot Y = Y\cdot X = 0$.

For vectors of $V_n(C)$, the Triangle Inequality

(14.4) $$\|X+Y\| \leq \|X\| + \|Y\|$$

and the Schwarz Inequality (see Problem 2)

(14.5) $$|X\cdot Y| \leq \|X\|\cdot\|Y\|$$

hold. Moreover, we have (see Theorems **I-IV** of Chapter 13)

I. Any set of m mutually orthogonal non-zero n-vectors over C is linearly independent and, hence, spans a vector space $V_n^m(C)$.

II. If a vector Y is orthogonal to each of the n-vectors $X_1, X_2, \ldots, X_m$, then it is orthogonal to the space spanned by these vectors.

III. If $V_n^h(C)$ is a subspace of $V_n^k(C)$, $k>h$, then there exists at least one vector X in $V_n^k(C)$ which is orthogonal to $V_n^h(C)$.

IV. Every vector space $V_n^m(C)$, $m>0$, contains m but not more than m mutually orthogonal vectors.

A basis of $V_n^m(C)$ which consists of mutually orthogonal vectors is called an **orthogonal basis**. If the mutually orthogonal vectors are also unit vectors, the basis is called a **normal** or **orthonormal** basis.

THE GRAM-SCHMIDT PROCESS. Let $X_1, X_2, \ldots, X_m$ be a basis for $V_n^m(C)$. Define

$$Y_1 = X_1$$

$$Y_2 = X_2 - \frac{Y_1 \cdot X_2}{Y_1 \cdot Y_1} Y_1$$

(14.6)
$$Y_3 = X_3 - \frac{Y_2 \cdot X_3}{Y_2 \cdot Y_2} Y_2 - \frac{Y_1 \cdot X_3}{Y_1 \cdot Y_1} Y_1$$

$$\cdots\cdots\cdots\cdots\cdots\cdots$$

$$Y_m = X_m - \frac{Y_{m-1} \cdot X_m}{Y_{m-1} \cdot Y_{m-1}} Y_{m-1} - \cdots - \frac{Y_1 \cdot X_m}{Y_1 \cdot Y_1} Y_1$$

The unit vectors $G_i = \dfrac{Y_i}{\|Y_i\|}$ $(i = 1, 2, \ldots, m)$ are an orthonormal basis for $V_n^m(C)$.

V. If $X_1, X_2, \ldots, X_s$, $(1 \leq s < m)$, are mutually orthogonal unit vectors of $V_n^m(C)$, there exist unit vectors (obtained by the Gram-Schmidt Process) $X_{s+1}, X_{s+2}, \ldots, X_m$ in the space such that the set $X_1, X_2, \ldots, X_m$ is an orthonormal basis.

THE GRAMIAN. Let $X_1, X_2, \ldots, X_p$ be a set of n-vectors with complex elements and define the **Gramian** matrix.

(14.7)
$$G = \begin{bmatrix} X_1 \cdot X_1 & X_1 \cdot X_2 & \cdots & X_1 \cdot X_p \\ X_2 \cdot X_1 & X_2 \cdot X_2 & \cdots & X_2 \cdot X_p \\ \cdots & \cdots & \cdots & \cdots \\ X_p \cdot X_1 & X_p \cdot X_2 & \cdots & X_p \cdot X_p \end{bmatrix} = \begin{bmatrix} \bar{X}_1' X_1 & \bar{X}_1' X_2 & \cdots & \bar{X}_1' X_p \\ \bar{X}_2' X_1 & \bar{X}_2' X_2 & \cdots & \bar{X}_2' X_p \\ \cdots & \cdots & \cdots & \cdots \\ \bar{X}_p' X_1 & \bar{X}_p' X_2 & \cdots & \bar{X}_p' X_p \end{bmatrix}$$

Clearly, the vectors are mutually orthogonal if and only if G is diagonal.

Following Problem 14, Chapter 17, we may prove

VI. For a set of n-vectors $X_1, X_2, \ldots, X_p$ with complex elements, $|G| \geq 0$. The equality holds if and only if the vectors are linearly dependent.

UNITARY MATRICES. An n-square matrix A is called **unitary** if $(\bar{A})'A = A(\bar{A})' = I$, that is if $(\bar{A})' = A^{-1}$. The column vectors (row vectors) of a unitary matrix are mutually orthogonal unit vectors.

Paralleling the theorems on orthogonal matrices of Chapter 13, we have

VII. The column vectors (row vectors) of an n-square unitary matrix are an orthonormal basis of $V_n(C)$, and conversely.

VIII. The inverse and the transpose of a unitary matrix are unitary.

IX. The product of two or more unitary matrices is unitary.

X. The determinant of a unitary matrix has absolute value 1.

UNITARY TRANSFORMATIONS. The linear transformation

$$Y = AX \tag{14.8}$$

where A is unitary, is called a **unitary transformation**.

XI. A linear transformation preserves lengths (and hence, inner products) if and only if its matrix is unitary.

XII. If $Y = AX$ is a transformation of coordinates from the E-basis to another the Z-basis, then the Z-basis is orthonormal if and only if A is unitary.

SOLVED PROBLEMS

1. Given $X = [1+i, -i, 1]'$ and $Y = [2+3i, 1-2i, i]'$,
   (a) find $X \cdot Y$ and $Y \cdot X$
   (b) verify $X \cdot Y = \overline{Y \cdot X}$
   (c) verify $X \cdot Y + Y \cdot X = 2R(X \cdot Y)$
   (d) verify $X \cdot Y - Y \cdot X = 2C(X \cdot Y)$

(a) $X \cdot Y = \bar{X}'Y = [1-i, i, 1]\begin{bmatrix} 2+3i \\ 1-2i \\ i \end{bmatrix} = (1-i)(2+3i) + i(1-2i) + 1(i) = 7+3i$

$Y \cdot X = \bar{Y}'X = [2-3i, 1+2i, -i]\begin{bmatrix} 1+i \\ -i \\ 1 \end{bmatrix} = 7-3i$

(b) From (a): $\overline{Y \cdot X}$, the conjugate of $Y \cdot X$, is $7+3i = X \cdot Y$.

(c) $X \cdot Y + Y \cdot X = (7+3i) + (7-3i) = 14 = 2(7) = 2R(X \cdot Y)$

(d) $X \cdot Y - Y \cdot X = (7+3i) - (7-3i) = 6i = 2(3i) = 2C(X \cdot Y)$

2. Prove the Schwarz Inequality: $|X \cdot Y| \leq \|X\| \cdot \|Y\|$.

As in the case of real vectors, the inequality is true if $X = 0$ or $Y = 0$. When X and Y are non-zero vectors and a is real, then

$$\|aX+Y\|^2 = (aX+Y)\cdot(aX+Y) = a^2 X \cdot X + a(X \cdot Y + Y \cdot X) + Y \cdot Y = a^2\|X\|^2 + 2aR(X \cdot Y) + \|Y\|^2 \geq 0.$$

Since the quadratic function in a is non-negative if and only if its discriminant is non-positive,

$$R(X\cdot Y)^2 - \|X\|^2\|Y\|^2 \leq 0 \qquad \text{and} \qquad R(X\cdot Y) \leq \|X\|\cdot\|Y\|$$

If $X\cdot Y = 0$, then $|X\cdot Y| = R(X\cdot Y) \leq \|X\|\cdot\|Y\|$. If $X\cdot Y \neq 0$, define $c = \dfrac{X\cdot Y}{|X\cdot Y|}$.

Then $R(cX\cdot Y) \leq \|cX\|\cdot\|Y\| = |c|\,\|X\|\cdot\|Y\| = \|X\|\cdot\|Y\|$ while, by **(14.3(*b*))**, $R(cX\cdot Y) = R[\bar{c}(X\cdot Y)] = |X\cdot Y|$. Thus, $|X\cdot Y| \leq \|X\|\cdot\|Y\|$ for all X and Y.

3. Prove: $B = (\bar{A})'A$ is Hermitian for any square matrix A.

$$(\bar{B})' = \{\overline{(\bar{A})'A}\}' = (\overline{A'\bar{A}}) = (\bar{A})'A = B \quad \text{and } B \text{ is Hermitian.}$$

4. If $A = B+iC$ is Hermitian, show that $(\bar{A})'A$ is real if and only if B and C anti-commute.

Since $B+iC$ is Hermitian, $(\overline{B+iC})' = B+iC$; thus,

$$(\bar{A})'A = (\overline{B+iC})'(B+iC) = (B+iC)(B+iC) = B^2 + i(BC+CB) - C^2$$

This is real if and only if $BC+CB = 0$ or $BC = -CB$; thus, if and only if B and C anti-commute.

5. Prove: If A is skew-Hermitian, then $\pm iA$ is Hermitian.

Consider $B = -iA$. Since A is skew-Hermitian, $(\bar{A})' = -A$. Then

$$(\bar{B})' = (\overline{-iA})' = i(\bar{A})' = i(-A) = -iA = B$$

and B is Hermitian. The reader will consider the case $B = iA$.

SUPPLEMENTARY PROBLEMS

6. Given the vectors $X_1 = [i, 2i, 1]'$, $X_2 = [1, 1+i, 0]'$, and $X_3 = [i, 1-i, 2]'$,

(*a*) find $X_1\cdot X_2$ and $X_2\cdot X_3$,

(*b*) find the length of each vector X_i,

(*c*) show that $[1-i, -1, 1-i]'$ is orthogonal to both X_1 and X_2,

(*d*) find a vector orthogonal to both X_1 and X_3.

Ans. (*a*) $2-3i, -i$ (*b*) $\sqrt{6}, \sqrt{3}, \sqrt{7}$ (*d*) $[-1-5i, i, 3-i]$

7. Show that $[1+i, i, 1]'$, $[i, 1-i, 0]'$, and $[1-i, 1, 3i]'$ are both linearly independent and mutually orthogonal.

8. Prove the relations **(14.3)**.

9. Prove the Triangle Inequality.

10. Prove Theorems **I-IV**.

11. Derive the relations **(14.6)**.

12. Using the relations (**14.6**) and the given vectors in order, construct an orthonormal basis for $V_3(C)$ when the vectors are
(*a*) $[0,1,-1]'$, $[1+i,1,1]'$, $[1-i,1,1]'$
(*b*) $[1+i,i,1]'$, $[2,1-2i,2+i]'$, $[1-i,0,-i]'$.

Ans. (*a*) $[0,\frac{1}{2}\sqrt{2},-\frac{1}{2}\sqrt{2}]'$, $[\frac{1}{2}(1+i),\frac{1}{2},\frac{1}{2}]'$, $[-\frac{1}{\sqrt{2}}i,\frac{1}{2\sqrt{2}}(1+i),\frac{1}{2\sqrt{2}}(1+i)]'$

(*b*) $[\frac{1}{2}(1+i),\frac{1}{2}i,\frac{1}{2}]'$, $[\frac{1}{2\sqrt{3}},\frac{1-5i}{4\sqrt{3}},\frac{3+3i}{4\sqrt{3}}]'$, $[\frac{7-i}{2\sqrt{30}},\frac{-5}{2\sqrt{30}},\frac{-6+3i}{2\sqrt{30}}]'$

13. Prove: If A is a matrix over the complex field, then $A+\bar{A}$ has only real elements and $A-\bar{A}$ has only pure imaginary elements.

14. Prove Theorem **V**.

15. If A is n-square, show
(*a*) $\bar{A}'A$ is diagonal if and only if the columns of A are mutually orthogonal vectors.
(*b*) $\bar{A}'A = I$ if and only if the columns of A are mutually orthogonal unit vectors.

16. Prove: If X and Y are n-vectors and A is n-square, then $X\cdot AY = \bar{A}'X\cdot Y$.

17. Prove Theorems **VII-X**.

18. Prove: If A is skew-Hermitian such that $I+A$ is non-singular, then $B = (I-A)(I+A)^{-1}$ is unitary.

19. Use Problem 18 to form a unitary matrix, given (*a*) $\begin{bmatrix} 0 & 1+i \\ -1+i & i \end{bmatrix}$, (*b*) $\begin{bmatrix} 0 & i & 1+i \\ i & 0 & i \\ -1+i & i & 0 \end{bmatrix}$.

Ans. (*a*) $\frac{1}{5}\begin{bmatrix} -1+2i & -4-2i \\ 2-4i & -2-i \end{bmatrix}$, (*b*) $\frac{1}{29}\begin{bmatrix} -9+8i & -10-4i & -16-18i \\ -2-24i & 1+12i & -10-4i \\ 4-10i & -2-24i & -9+8i \end{bmatrix}$

20. Prove: If A and B are unitary and of the same order, then AB and BA are unitary.

21. Follow the proof in Problem 10, Chapter 13, to prove Theorem **XI**.

22. Prove: If A is unitary and Hermitian, then A is involutory.

23. Show that $\begin{bmatrix} \frac{1}{2}(1+i) & i/\sqrt{3} & \frac{3+i}{2\sqrt{15}} \\ -\frac{1}{2} & 1/\sqrt{3} & \frac{4+3i}{2\sqrt{15}} \\ \frac{1}{2} & -i/\sqrt{3} & \frac{5i}{2\sqrt{15}} \end{bmatrix}$ is unitary.

24. Prove: If A is unitary and if $B = AP$ where P is non-singular, then PB^{-1} is unitary.

25. Prove: If A is unitary and $I+A$ is non-singular, then $B = (I-A)(I+A)^{-1}$ is skew-Hermitian.

Chapter 15

Congruence

CONGRUENT MATRICES. Two n-square matrices A and B over F are called congruent, $\overset{C}{\sim}$, over F if there exists a non-singular matrix P over F such that

$$(15.1) \qquad B \;=\; P'AP$$

Clearly, congruence is a special case of equivalence so that congruent matrices have the same rank.

When P is expressed as a product of elementary column matrices, P' is the product in reverse order of the same elementary row matrices; that is, A and B are congruent provided A can be reduced to B by a sequence of **pairs** of elementary transformations, each pair consisting of an elementary row transformation followed by the same elementary column transformation.

SYMMETRIC MATRICES. In Problem 1, we prove

I. Every symmetric matrix A over F of rank r is congruent over F to a diagonal matrix whose first r diagonal elements are non-zero while all other elements are zero.

Example 1. Find a non-singular matrix P with rational elements such that $D = P'AP$ is diagonal, given

$$A \;=\; \begin{bmatrix} 1 & 2 & 3 & 2 \\ 2 & 3 & 5 & 8 \\ 3 & 5 & 8 & 10 \\ 2 & 8 & 10 & -8 \end{bmatrix}$$

In reducing A to D, we use $[A\ I]$ and calculate en route the matrix P'. First we use $H_{21}(-2)$ and $K_{21}(-2)$, then $H_{31}(-3)$ and $K_{31}(-3)$, then $H_{41}(-2)$ and $K_{41}(-2)$ to obtain zeros in the first row and in the first column. Considerable time is saved, however, if the three row transformations are made first and then the three column transformations. If A is not then transformed into a symmetric matrix, an error has been made. We have

$$[A\ H] \;=\; \left[\begin{array}{cccc|cccc} 1 & 2 & 3 & 2 & 1 & 0 & 0 & 0 \\ 2 & 3 & 5 & 8 & 0 & 1 & 0 & 0 \\ 3 & 5 & 8 & 10 & 0 & 0 & 1 & 0 \\ 2 & 8 & 10 & -8 & 0 & 0 & 0 & 1 \end{array}\right] \overset{C}{\sim} \left[\begin{array}{cccc|cccc} 1 & 0 & 0 & 0 & 1 & 0 & 0 & 0 \\ 0 & -1 & -1 & 4 & -2 & 1 & 0 & 0 \\ 0 & -1 & -1 & 4 & -3 & 0 & 1 & 0 \\ 0 & 4 & 4 & -12 & -2 & 0 & 0 & 1 \end{array}\right]$$

$$\overset{C}{\sim} \left[\begin{array}{cccc|cccc} 1 & 0 & 0 & 0 & 1 & 0 & 0 & 0 \\ 0 & -1 & 0 & 0 & -2 & 1 & 0 & 0 \\ 0 & 0 & 0 & 0 & -1 & -1 & 1 & 0 \\ 0 & 0 & 0 & 4 & -10 & 4 & 0 & 1 \end{array}\right] \overset{C}{\sim} \left[\begin{array}{cccc|cccc} 1 & 0 & 0 & 0 & 1 & 0 & 0 & 0 \\ 0 & -1 & 0 & 0 & -2 & 1 & 0 & 0 \\ 0 & 0 & 4 & 0 & -10 & 4 & 0 & 1 \\ 0 & 0 & 0 & 0 & -1 & -1 & 1 & 0 \end{array}\right]$$

$$=\; [D\ P']$$

Then
$$P = \begin{bmatrix} 1 & -2 & -10 & -1 \\ 0 & 1 & 4 & -1 \\ 0 & 0 & 0 & 1 \\ 0 & 0 & 1 & 0 \end{bmatrix}$$

The matrix D to which A has been reduced is not unique. The additional transformations $H_3(\tfrac{1}{2})$ and $K_3(\tfrac{1}{2})$, for example, will replace D by the diagonal matrix $\begin{bmatrix} 1 & 0 & 0 & 0 \\ 0 & -1 & 0 & 0 \\ 0 & 0 & 1 & 0 \\ 0 & 0 & 0 & 0 \end{bmatrix}$ while the transformations $H_2(3)$ and $K_2(3)$ replace D by $\begin{bmatrix} 1 & 0 & 0 & 0 \\ 0 & -9 & 0 & 0 \\ 0 & 0 & 4 & 0 \\ 0 & 0 & 0 & 0 \end{bmatrix}$. There is, however, no pair of rational or real transformations which will replace D by a diagonal matrix having only non-negative elements in the diagonal.

REAL SYMMETRIC MATRICES. Let the real symmetric matrix A be reduced by real elementary transformations to a congruent diagonal matrix D, that is, let $P'AP = D$. While the non-zero diagonal elements of D depend both on A and P, it will be shown in Chapter 17 that the number of positive non-zero diagonal elements depends solely on A.

By a sequence of row and the same column transformations of type 1 the diagonal elements of D may be rearranged so that the positive elements precede the negative elements. Then a sequence of real row and the same column transformations of type 2 may be used to reduce the diagonal matrix to one in which the non-zero diagonal elements are either +1 or −1. We have

II. A real symmetric matrix of rank r is congruent over the real field to a canonical matrix

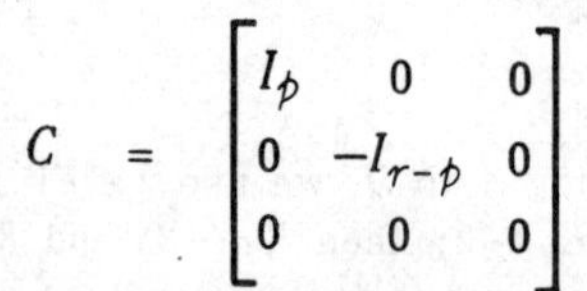

$$C = \begin{bmatrix} I_p & 0 & 0 \\ 0 & -I_{r-p} & 0 \\ 0 & 0 & 0 \end{bmatrix} \tag{15.2}$$

The integer p of **(15.2)** is called the **index** of the matrix and $s = p-(r-p)$ is called the **signature**.

Example 2. Applying the transformations H_{23}, K_{23}, and $H_2(\tfrac{1}{2}), K_2(\tfrac{1}{2})$ to the result of Example 1, we have

$$[A \mid I] \underset{C}{\sim} \left[\begin{array}{cccc|cccc} 1 & 0 & 0 & 0 & 1 & 0 & 0 & 0 \\ 0 & -1 & 0 & 0 & -2 & 1 & 0 & 0 \\ 0 & 0 & 4 & 0 & -10 & 4 & 0 & 1 \\ 0 & 0 & 0 & 0 & -1 & -1 & 1 & 0 \end{array}\right] \underset{C}{\sim} \left[\begin{array}{cccc|cccc} 1 & 0 & 0 & 0 & 1 & 0 & 0 & 0 \\ 0 & 1 & 0 & 0 & -5 & 2 & 0 & \tfrac{1}{2} \\ 0 & 0 & -1 & 0 & -2 & 1 & 0 & 0 \\ 0 & 0 & 0 & 0 & -1 & -1 & 1 & 0 \end{array}\right] = [C \mid Q']$$

and $Q'AQ = C$. Thus, A is of rank $r = 3$, index $p = 2$, and signature $s = 1$.

III. Two n-square real symmetric matrices are congruent over the real field if and only if they have the same rank and the same index or the same rank and the same signature.

In the real field the set of all n-square matrices of the type **(15.2)** is a canonical set over congruence for real n-square symmetric matrices.

IN THE COMPLEX FIELD, we have

IV. Every n-square complex symmetric matrix of rank r is congruent over the field of complex numbers to a canonical matrix

$$(15.3) \qquad C = \begin{bmatrix} I_r & 0 \\ 0 & 0 \end{bmatrix}$$

Example 3. Applying the transformations $H_3(i)$ and $K_3(i)$ to the result of Example 2, we have

$$[A \mid I] \overset{C}{\sim} \left[\begin{array}{cccc|cccc} 1 & 0 & 0 & 0 & 1 & 0 & 0 & 0 \\ 0 & 1 & 0 & 0 & -5 & 2 & 0 & \frac{1}{2} \\ 0 & 0 & -1 & 0 & -2 & 1 & 0 & 0 \\ 0 & 0 & 0 & 0 & -1 & -1 & 1 & 0 \end{array}\right] \overset{C}{\sim} \left[\begin{array}{cccc|cccc} 1 & 0 & 0 & 0 & 1 & 0 & 0 & 0 \\ 0 & 1 & 0 & 0 & -5 & 2 & 0 & \frac{1}{2} \\ 0 & 0 & 1 & 0 & -2i & i & 0 & 0 \\ 0 & 0 & 0 & 0 & -1 & -1 & 1 & 0 \end{array}\right] = [D \mid R']$$

and $R'AR = D = \begin{bmatrix} I_3 & 0 \\ 0 & 0 \end{bmatrix}$.

See Problems 2-3.

V. Two n-square complex symmetric matrices are congruent over the field of complex numbers if and only if they have the same rank.

SKEW-SYMMETRIC MATRICES. If A is skew-symmetric, then

$$(P'AP)' = P'A'P = P'(-A)P = -P'AP$$

Thus,

VI. Every matrix $B = P'AP$ congruent to a skew-symmetric matrix A is also skew-symmetric.

In Problem 4, we prove

VII. Every n-square skew-symmetric matrix A over F is congruent over F to a canonical matrix

$$(15.4) \qquad B = \text{diag}\,(D_1, D_2, \ldots, D_t, 0, \ldots, 0)$$

where $D_i = \begin{bmatrix} 0 & 1 \\ -1 & 0 \end{bmatrix}$, $(i = 1, 2, \ldots, t)$. The rank of A is $r = 2t$.

See Problem 5.

There follows

VIII. Two n-square skew-symmetric matrices over F are congruent over F if and only if they have the same rank.

The set of all matrices of the type **(15.4)** is a canonical set over congruence for n-square skew-symmetric matrices.

HERMITIAN MATRICES. Two n-square Hermitian matrices A and B are called **Hermitely congruent**, $[\overset{HC}{\sim}]$, or **conjunctive** if there exists a non-singular matrix P such that

$$(15.5) \qquad B = \overline{P}'AP$$

Thus,

IX. Two n-square Hermitian matrices are conjunctive if and only if one can be obtained from the other by a sequence of pairs of elementary transformations, each pair consisting of a column transformation and the corresponding conjugate row transformation.

X. An Hermitian matrix A of rank r is conjunctive to a canonical matrix

$$C = \begin{bmatrix} I_p & 0 & 0 \\ 0 & -I_{r-p} & 0 \\ 0 & 0 & 0 \end{bmatrix} \tag{15.6}$$

The integer p of **(15.6)** is called the **index** of A and $s = p-(r-p)$ is called the **signature**.

XI. Two n-square Hermitian matrices are conjunctive if and only if they have the same rank and index or the same rank and the same signature.

The reduction of an Hermitian matrix to the canonical form **(15.6)** follows the procedures of Problem 1 with attention to the proper pairs of elementary transformations. The extreme troublesome case is covered in Problem 7.

See Problems 6-7.

SKEW-HERMITIAN MATRICES. If A is skew-Hermitian, then

$$(\overline{\bar{P}'AP})' = (\bar{P}'\bar{A}'\bar{\bar{P}}) = -\bar{P}'AP$$

Thus,

XII. Every matrix $B = \bar{P}'AP$ conjunctive to a skew-Hermitian matrix A is also skew-Hermitian.

By Problem 5, Chapter 14, $H = -iA$ is Hermitian if A is skew-Hermitian. By Theorem **X** there exists a non-singular matrix P such that

$$\bar{P}'HP = C = \begin{bmatrix} I_p & 0 & 0 \\ 0 & -I_{r-p} & 0 \\ 0 & 0 & 0 \end{bmatrix}$$

Then $i\bar{P}'HP = i\bar{P}'(-iA)P = \bar{P}'AP = iC$ and

$$B = \bar{P}'AP = \begin{bmatrix} iI_p & 0 & 0 \\ 0 & -iI_{r-p} & 0 \\ 0 & 0 & 0 \end{bmatrix} \tag{15.7}$$

Thus,

XIII. Every n-square skew-Hermitian matrix A is conjunctive to a matrix **(15.7)** in which r is the rank of A and p is the index of $-iA$.

XIV. Two n-square skew-Hermitian matrices A and B are conjunctive if and only if they have the same rank while $-iA$ and $-iB$ have the same index.

See Problem 8.

SOLVED PROBLEMS

1. Prove: Every symmetric matrix over F of rank r can be reduced to a diagonal matrix having exactly r non-zero elements in the diagonal.

Suppose the symmetric matrix $A = [a_{ij}]$ is not diagonal. If $a_{11} \neq 0$, a sequence of pairs of elementary transformations of type 3, each consisting of a row transformation and the same column transformation, will reduce A to

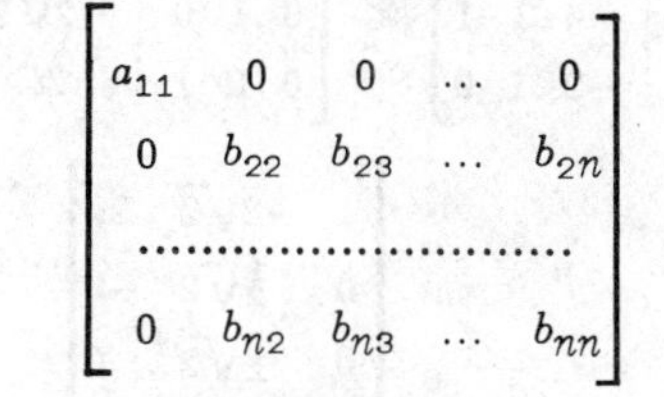

Now the continued reduction is routine so long as $b_{22}, c_{33}, \ldots$ are different from zero. Suppose then that along in the reduction, we have obtained the matrix

$$\begin{bmatrix} a_{11} & 0 & \cdots & & & & & 0 \\ 0 & b_{22} & \cdots & & & & & 0 \\ \cdots & \cdots & \cdots & \cdots & \cdots & \cdots & \cdots & \cdots \\ 0 & 0 & \cdots & h_{ss} & 0 & \cdots & & 0 \\ 0 & 0 & \cdots & 0 & 0 & k_{s+1,\,s+2} & \cdots & k_{s+1,\,n} \\ 0 & 0 & \cdots & 0 & k_{s+2,\,s+1} & k_{s+2,\,s+2} & \cdots & k_{s+2,\,n} \\ \cdots & \cdots & \cdots & \cdots & \cdots & \cdots & \cdots & \cdots \\ 0 & 0 & \cdots & 0 & k_{n,\,s+1} & k_{n,\,s+2} & \cdots & k_{nn} \end{bmatrix}$$

in which the diagonal element $k_{s+1,\,s+1} = 0$. If every $k_{ij} = 0$, we have proved the theorem with $s = r$. If, however, some k_{ij}, say $k_{s+u,\,s+v} \neq 0$, we move it into the $(s+1, s+1)$ position by the proper row and column transformation of type 1 when $u = v$; otherwise, we add the $(s+u)$th row to the $(s+v)$th row and after the corresponding column transformation have a diagonal element different from zero. (When $a_{11} = 0$, we proceed as in the case $k_{s+1,\,s+1} = 0$ above.)

Since we are led to a sequence of equivalent matrices, A is ultimately reduced to a diagonal matrix whose first r diagonal elements are non-zero while all other elements are zero.

2. Reduce the symmetric matrix $A = \begin{bmatrix} 1 & 2 & 2 \\ 2 & 3 & 5 \\ 2 & 5 & 5 \end{bmatrix}$ to canonical form **(15.2)** and to canonical form **(15.3)**.

In each obtain the matrix P which effects the reduction.

$$[A \mid I] = \left[\begin{array}{ccc|ccc} 1 & 2 & 2 & 1 & 0 & 0 \\ 2 & 3 & 5 & 0 & 1 & 0 \\ 2 & 5 & 5 & 0 & 0 & 1 \end{array}\right] \underset{\sim}{C} \left[\begin{array}{ccc|ccc} 1 & 0 & 0 & 1 & 0 & 0 \\ 0 & -1 & 1 & -2 & 1 & 0 \\ 0 & 1 & 1 & -2 & 0 & 1 \end{array}\right] \underset{\sim}{C} \left[\begin{array}{ccc|ccc} 1 & 0 & 0 & 1 & 0 & 0 \\ 0 & -1 & 0 & -2 & 1 & 0 \\ 0 & 0 & 2 & -4 & 1 & 1 \end{array}\right] \underset{\sim}{C} \left[\begin{array}{ccc|ccc} 1 & 0 & 0 & 1 & 0 & 0 \\ 0 & 2 & 0 & -4 & 1 & 1 \\ 0 & 0 & -1 & -2 & 1 & 0 \end{array}\right]$$

$$= [D \mid P_1']$$

To obtain **(15.2)**, we have

$$[D \mid P_1'] = \left[\begin{array}{ccc|ccc} 1 & 0 & 0 & 1 & 0 & 0 \\ 0 & 2 & 0 & -4 & 1 & 1 \\ 0 & 0 & -1 & -2 & 1 & 0 \end{array}\right] \underset{\sim}{C} \left[\begin{array}{ccc|ccc} 1 & 0 & 0 & 1 & 0 & 0 \\ 0 & 1 & 0 & -2\sqrt{2} & \frac{1}{2}\sqrt{2} & \frac{1}{2}\sqrt{2} \\ 0 & 0 & -1 & -2 & 1 & 0 \end{array}\right] = [C \mid P']$$

and
$$P = \begin{bmatrix} 1 & -2\sqrt{2} & -2 \\ 0 & \frac{1}{2}\sqrt{2} & 1 \\ 0 & \frac{1}{2}\sqrt{2} & 0 \end{bmatrix}$$

To obtain **(15.3)**, we have

$$[D \mid P_1'] = \left[\begin{array}{ccc|ccc} 1 & 0 & 0 & 1 & 0 & 0 \\ 0 & 2 & 0 & -4 & 1 & 1 \\ 0 & 0 & -1 & -2 & 1 & 0 \end{array}\right] \overset{C}{\sim} \left[\begin{array}{ccc|ccc} 1 & 0 & 0 & 1 & 0 & 0 \\ 0 & 1 & 0 & -2\sqrt{2} & \frac{1}{2}\sqrt{2} & \frac{1}{2}\sqrt{2} \\ 0 & 0 & 1 & 2i & -i & 0 \end{array}\right] = [C \mid P']$$

and
$$P = \begin{bmatrix} 1 & -2\sqrt{2} & 2i \\ 0 & \frac{1}{2}\sqrt{2} & -i \\ 0 & \frac{1}{2}\sqrt{2} & 0 \end{bmatrix}$$

3. Find a non-singular matrix P such that $P'AP$ is in canonical form **(15.3)**, given

$$A = \begin{bmatrix} 1 & i & 1+i \\ i & 0 & 2-i \\ 1+i & 2-i & 10+2i \end{bmatrix}$$

$$[A \mid I] = \left[\begin{array}{ccc|ccc} 1 & i & 1+i & 1 & 0 & 0 \\ i & 0 & 2-i & 0 & 1 & 0 \\ 1+i & 2-i & 10+2i & 0 & 0 & 1 \end{array}\right] \overset{C}{\sim} \left[\begin{array}{ccc|ccc} 1 & 0 & 0 & 1 & 0 & 0 \\ 0 & 1 & 3-2i & -i & 1 & 0 \\ 0 & 3-2i & 10 & -1-i & 0 & 1 \end{array}\right]$$

$$\overset{C}{\sim} \left[\begin{array}{ccc|ccc} 1 & 0 & 0 & 1 & 0 & 0 \\ 0 & 1 & 0 & -i & 1 & 0 \\ 0 & 0 & 5+12i & 1+2i & -3+2i & 1 \end{array}\right] \overset{C}{\sim} \left[\begin{array}{ccc|ccc} 1 & 0 & 0 & 1 & 0 & 0 \\ 0 & 1 & 0 & -i & 1 & 0 \\ 0 & 0 & 1 & \frac{7+4i}{13} & \frac{-5+12i}{13} & \frac{3-2i}{13} \end{array}\right]$$

$$= [C \mid P']$$

Here,
$$P = \begin{bmatrix} 1 & -i & \frac{7+4i}{13} \\ 0 & 1 & \frac{-5+12i}{13} \\ 0 & 0 & \frac{3-2i}{13} \end{bmatrix}$$

4. Prove: Every n-square skew-symmetric matrix A over F of rank $2t$ is congruent over F to a matrix

$$B = \operatorname{diag}(D_1, D_2, \ldots, D_t, 0, \ldots, 0)$$

where $D_i = \begin{bmatrix} 0 & 1 \\ -1 & 0 \end{bmatrix}$, $(i = 1, 2, \ldots, t)$.

If $A = 0$, then $B = A$. If $A \neq 0$, then some $a_{ij} = -a_{ji} \neq 0$. Interchange the ith and first rows and the jth and second rows; then interchange the ith and first columns and the jth and second columns to replace A by the skew-symmetric matrix $\left[\begin{array}{cc|c} 0 & a_{ij} & E_2 \\ -a_{ij} & 0 & \\ \hline E_3 & & E_4 \end{array}\right]$. Next multiply the first row and the first column by $1/a_{ij}$

to obtain $\begin{bmatrix} \begin{matrix} 0 & 1 \\ -1 & 0 \end{matrix} & F_2 \\ F_3 & E_4 \end{bmatrix}$ and from it, by elementary row and column transformations of type 3, obtain

$$\begin{bmatrix} \begin{matrix} 0 & 1 \\ -1 & 0 \end{matrix} & 0 \\ 0 & F_4 \end{bmatrix} = \begin{bmatrix} D_1 & 0 \\ 0 & F_4 \end{bmatrix}$$

If $F_4 = 0$, the reduction is complete; otherwise, the process is repeated on F_4, ... until B is obtained.

5. Find a non-singular matrix P such that $P'AP$ is in canonical form **(15.4)**, given

$$A = \begin{bmatrix} 0 & 0 & 2 & 4 \\ 0 & 0 & 1 & -3 \\ -2 & -1 & 0 & -2 \\ -4 & 3 & 2 & 0 \end{bmatrix}$$

Using $a_{13} \neq 0$, we need only interchange the third and second rows followed by the interchange of the third and second columns of $\left[\begin{array}{cccc|cccc} 0 & 0 & 2 & 4 & 1 & 0 & 0 & 0 \\ 0 & 0 & 1 & -3 & 0 & 1 & 0 & 0 \\ -2 & -1 & 0 & -2 & 0 & 0 & 1 & 0 \\ -4 & 3 & 2 & 0 & 0 & 0 & 0 & 1 \end{array}\right]$ to obtain $\left[\begin{array}{cccc|cccc} 0 & 2 & 0 & 4 & 1 & 0 & 0 & 0 \\ -2 & 0 & -1 & -2 & 0 & 0 & 1 & 0 \\ 0 & 1 & 0 & -3 & 0 & 1 & 0 & 0 \\ -4 & 2 & 3 & 0 & 0 & 0 & 0 & 1 \end{array}\right]$

Next, multiply the first row and first column by $\frac{1}{2}$; then proceed to clear the first two rows and the first two columns of non-zero elements. We have, in turn,

$$\left[\begin{array}{cccc|cccc} 0 & 1 & 0 & 2 & \frac{1}{2} & 0 & 0 & 0 \\ -1 & 0 & -1 & -2 & 0 & 0 & 1 & 0 \\ 0 & 1 & 0 & -3 & 0 & 1 & 0 & 0 \\ -2 & 2 & 3 & 0 & 0 & 0 & 0 & 1 \end{array}\right] \quad \text{and} \quad \left[\begin{array}{cccc|cccc} 0 & 1 & 0 & 0 & \frac{1}{2} & 0 & 0 & 0 \\ -1 & 0 & 0 & 0 & 0 & 0 & 1 & 0 \\ 0 & 0 & 0 & -5 & -\frac{1}{2} & 1 & 0 & 0 \\ 0 & 0 & 5 & 0 & -1 & 0 & -2 & 1 \end{array}\right]$$

Finally, multiply the third row and third column by $-1/5$ to obtain

$$\left[\begin{array}{cccc|cccc} 0 & 1 & 0 & 0 & 1/2 & 0 & 0 & 0 \\ -1 & 0 & 0 & 0 & 0 & 0 & 1 & 0 \\ 0 & 0 & 0 & 1 & 1/10 & -1/5 & 0 & 0 \\ 0 & 0 & -1 & 0 & -1 & 0 & -2 & 1 \end{array}\right] = \begin{bmatrix} D_1 & 0 \\ 0 & D_2 \end{bmatrix} P'$$

Thus when $P = \begin{bmatrix} 1/2 & 0 & 1/10 & -1 \\ 0 & 0 & -1/5 & 0 \\ 0 & 1 & 0 & -2 \\ 0 & 0 & 0 & 1 \end{bmatrix}$, $P'AP = \text{diag}(D_1, D_2)$.

6. Find a non-singular matrix P such that $\overline{P}'AP$ is in canonical form **(15.6)**, given

$$A = \begin{bmatrix} 1 & 1-i & -3+2i \\ 1+i & 2 & -i \\ -3-2i & i & 0 \end{bmatrix}$$

$$[A \mid I] = \begin{bmatrix} 1 & 1-i & -3+2i & | & 1 & 0 & 0 \\ 1+i & 2 & -i & | & 0 & 1 & 0 \\ -3-2i & i & 0 & | & 0 & 0 & 1 \end{bmatrix} \underset{\sim}{C} \begin{bmatrix} 1 & 0 & 0 & | & 1 & 0 & 0 \\ 0 & 0 & 5 & | & -1-i & 1 & 0 \\ 0 & 5 & -13 & | & 3+2i & 0 & 1 \end{bmatrix}$$

$$\underset{\sim}{C} \begin{bmatrix} 1 & 0 & 0 & | & 1 & 0 & 0 \\ 0 & \frac{25}{13} & 0 & | & \frac{2-3i}{13} & 1 & \frac{5}{13} \\ 0 & 0 & -13 & | & 3+2i & 0 & 1 \end{bmatrix} \underset{\sim}{C} \begin{bmatrix} 1 & 0 & 0 & | & 1 & 0 & 0 \\ 0 & 1 & 0 & | & \frac{2-3i}{5\sqrt{13}} & \frac{13}{5\sqrt{13}} & \frac{1}{\sqrt{13}} \\ 0 & 0 & -1 & | & \frac{3+2i}{\sqrt{13}} & 0 & \frac{1}{\sqrt{13}} \end{bmatrix}$$

$$= [C \mid \overline{P}']$$

and

$$P = \begin{bmatrix} 1 & \frac{2+3i}{5\sqrt{13}} & \frac{3-2i}{\sqrt{13}} \\ 0 & \frac{13}{5\sqrt{13}} & 0 \\ 0 & \frac{1}{\sqrt{13}} & \frac{1}{\sqrt{13}} \end{bmatrix}$$

7. Find a non-singular matrix P such that $\overline{P}'AP$ is in canonical form **(15.6)**, given

$$A = \begin{bmatrix} 1 & 1+2i & 2-3i \\ 1-2i & 5 & -4-2i \\ 2+3i & -4+2i & 13 \end{bmatrix}$$

$$[A \mid I] = \begin{bmatrix} 1 & 1+2i & 2-3i & | & 1 & 0 & 0 \\ 1-2i & 5 & -4-2i & | & 0 & 1 & 0 \\ 2+3i & -4+2i & 13 & | & 0 & 0 & 1 \end{bmatrix} \underset{\sim}{HC} \begin{bmatrix} 1 & 0 & 0 & | & 1 & 0 & 0 \\ 0 & 0 & 5i & | & -1+2i & 1 & 0 \\ 0 & -5i & 0 & | & -2-3i & 0 & 1 \end{bmatrix}$$

$$\underset{\sim}{HC} \begin{bmatrix} 1 & 0 & 0 & | & 1 & 0 & 0 \\ 0 & 10 & 5i & | & 2 & 1 & i \\ 0 & -5i & 0 & | & -2-3i & 0 & 1 \end{bmatrix} \underset{\sim}{HC} \begin{bmatrix} 1 & 0 & 0 & | & 1 & 0 & 0 \\ 0 & 10 & 0 & | & 2 & 1 & i \\ 0 & 0 & -5/2 & | & -2-2i & \frac{1}{2}i & \frac{1}{2} \end{bmatrix}$$

$$\underset{\sim}{HC} \begin{bmatrix} 1 & 0 & 0 & | & 1 & 0 & 0 \\ 0 & 1 & 0 & | & \frac{2}{\sqrt{10}} & \frac{1}{\sqrt{10}} & \frac{i}{\sqrt{10}} \\ 0 & 0 & -1 & | & \frac{-4-4i}{\sqrt{10}} & \frac{i}{\sqrt{10}} & \frac{1}{\sqrt{10}} \end{bmatrix}$$

$$= [C \mid \overline{P}']$$

and

$$P = \begin{bmatrix} 1 & \frac{2}{\sqrt{10}} & \frac{-4+4i}{\sqrt{10}} \\ 0 & \frac{1}{\sqrt{10}} & \frac{-i}{\sqrt{10}} \\ 0 & \frac{-i}{\sqrt{10}} & \frac{1}{\sqrt{10}} \end{bmatrix}$$

. Find a non-singular matrix P such that $\bar{P}'AP$ is in canonical form **(15.7)**, given

$$A = \begin{bmatrix} i & -1 & -1+i \\ 1 & 0 & 1+2i \\ 1+i & -1+2i & 2i \end{bmatrix}$$

Consider the Hermitian matrix $H = -iA = \begin{bmatrix} 1 & i & 1+i \\ -i & 0 & 2-i \\ 1-i & 2+i & 2 \end{bmatrix}$.

The non-singular matrix $P = \begin{bmatrix} 1 & -1-2i & -i \\ 0 & 1 & 1 \\ 0 & 1 & 0 \end{bmatrix}$ is such that $\bar{P}'HP = \text{diag}[1, 1, -1]$.

Then $\bar{P}'AP = \text{diag}[i, i, -i]$.

SUPPLEMENTARY PROBLEMS

9. Find a non-singular matrix P such that $P'AP$ is in canonical form **(15.2)**, given

(a) $A = \begin{bmatrix} 1 & -2 \\ -2 & 3 \end{bmatrix}$, (b) $A = \begin{bmatrix} 1 & 2 & 0 \\ 2 & 3 & -1 \\ 0 & -1 & -2 \end{bmatrix}$, (c) $A = \begin{bmatrix} 0 & 1 & 2 \\ 1 & 0 & 4 \\ 2 & 4 & 0 \end{bmatrix}$, (d) $A = \begin{bmatrix} 1 & -1 & 0 \\ -1 & 2 & 1 \\ 0 & 1 & 1 \end{bmatrix}$

Ans. (a) $P = \begin{bmatrix} 1 & 2 \\ 0 & 1 \end{bmatrix}$, (b) $P = \begin{bmatrix} 1 & -2 & 2 \\ 0 & 1 & -1 \\ 0 & 0 & 1 \end{bmatrix}$, (c) $P = \begin{bmatrix} \frac{1}{2}\sqrt{2} & -\frac{1}{2}\sqrt{2} & -1 \\ \frac{1}{2}\sqrt{2} & \frac{1}{2}\sqrt{2} & -\frac{1}{2} \\ 0 & 0 & \frac{1}{4} \end{bmatrix}$, (d) $P = \begin{bmatrix} 1 & 1 & -1 \\ 0 & 1 & -1 \\ 0 & 0 & 1 \end{bmatrix}$

0. Find a non-singular matrix P such that $P'AP$ is in canonical form **(15.3)**, given

(a) $A = \begin{bmatrix} 1 & 1+2i \\ 1+2i & 1+4i \end{bmatrix}$, (b) $A = \begin{bmatrix} 2i & 1+i & 2-4i \\ 1+i & 1+i & -1-2i \\ 2-4i & -1-2i & -3-5i \end{bmatrix}$

Ans. (a) $P = \begin{bmatrix} 1 & \frac{1}{2}(-1-2i) \\ 0 & \frac{1}{2} \end{bmatrix}$, (b) $P = \begin{bmatrix} \frac{1}{2}(1-i) & i/\sqrt{2} & (1+i)/2 \\ 0 & (1-i)/\sqrt{2} & (-3-2i)/13 \\ 0 & 0 & (3+2i)/13 \end{bmatrix}$

1. Find a non-singular matrix P such that $P'AP$ is in canonical form **(15.4)**, given

(a) $A = \begin{bmatrix} 0 & 1 & 2 \\ -1 & 0 & -3 \\ -2 & 3 & 0 \end{bmatrix}$ (b) $A = \begin{bmatrix} 0 & 0 & 2 \\ 0 & 0 & 3 \\ -2 & -3 & 0 \end{bmatrix}$ (c) $A = \begin{bmatrix} 0 & 0 & 0 & 0 \\ 0 & 0 & 1 & -2 \\ 0 & -1 & 0 & 3 \\ 0 & 2 & -3 & 0 \end{bmatrix}$ (d) $A = \begin{bmatrix} 0 & 1 & 2 & -2 \\ -1 & 0 & -1 & 1 \\ -2 & 1 & 0 & 3 \\ 2 & -1 & -3 & 0 \end{bmatrix}$

Ans.

(a) $P = \begin{bmatrix} 1 & 0 & -3 \\ 0 & 1 & -2 \\ 0 & 0 & 1 \end{bmatrix}$ (b) $P = \begin{bmatrix} \frac{1}{2} & 0 & -3/2 \\ 0 & 0 & 1 \\ 0 & 1 & 0 \end{bmatrix}$ (c) $P = \begin{bmatrix} 0 & 0 & 1 & 0 \\ 1 & 0 & 0 & 3 \\ 0 & 1 & 0 & 2 \\ 0 & 0 & 0 & 1 \end{bmatrix}$ (d) $P = \begin{bmatrix} 1 & 0 & -1/3 & 1 \\ 0 & 1 & -2/3 & 2 \\ 0 & 0 & 1/3 & 0 \\ 0 & 0 & 0 & 1 \end{bmatrix}$

12. Find a non-singular matrix P such that $\overline{P}'AP$ is in canonical form **(15.6)**, given

(a) $A = \begin{bmatrix} 1 & 1-3i \\ 1+3i & 10 \end{bmatrix}$ (b) $A = \begin{bmatrix} 1 & 1+i & 2 \\ 1-i & 3 & -i \\ 2 & i & 4 \end{bmatrix}$, (c) $A = \begin{bmatrix} 1 & 1+i & 3-2i \\ 1-i & 3 & 3-4i \\ 3+2i & 3+4i & 18 \end{bmatrix}$

Ans. (a) $P = \begin{bmatrix} 1 & -1+3i \\ 0 & 1 \end{bmatrix}$ (b) $P = \begin{bmatrix} 1 & -1-i & (-5-i)/\sqrt{5} \\ 0 & 1 & (2-i)/\sqrt{5} \\ 0 & 0 & 1/\sqrt{5} \end{bmatrix}$ (c) $P = \begin{bmatrix} 1 & -1-i & (-2+5i) \\ 0 & 1 & (-2-i) \\ 0 & 0 & 1 \end{bmatrix}$

13. Find a non-singular matrix P such that $\overline{P}'AP$ is in canonical form **(15.7)**, given

(a) $A = \begin{bmatrix} i & 1+i \\ -1+i & i \end{bmatrix}$ (c) $A = \begin{bmatrix} i & -1-i & -1 \\ 1-i & 0 & 1-i \\ 1 & -1-i & -i \end{bmatrix}$

(b) $A = \begin{bmatrix} i & -1 & 1+i \\ 1 & 2i & i \\ -1+i & i & 6i \end{bmatrix}$ (d) $A = \begin{bmatrix} 0 & 1 & 2+i \\ -1 & 0 & 1-2i \\ -2+i & -1-2i & 0 \end{bmatrix}$

Ans. (a) $P = \begin{bmatrix} 1 & -1+i \\ 0 & 1 \end{bmatrix}$ (c) $P = \begin{bmatrix} 1 & (1-i)/\sqrt{2} & -1 \\ 0 & 1/\sqrt{2} & -1 \\ 0 & 0 & 1 \end{bmatrix}$

(b) $P = \begin{bmatrix} 1 & -i & -2+3i \\ 0 & 1 & -2-i \\ 0 & 0 & 1 \end{bmatrix}$ (d) $P = \begin{bmatrix} 1/\sqrt{2} & (1-2i)/\sqrt{10} & -1/\sqrt{2} \\ i/\sqrt{2} & (-2-i)/\sqrt{10} & i/\sqrt{2} \\ 0 & 1/\sqrt{10} & 0 \end{bmatrix}$

14. If $D = \begin{bmatrix} 0 & 1 \\ -1 & 0 \end{bmatrix}$ show that a 2-square matrix C satisfies $C'DC = D$ if and only if $|C| = 1$.

15. Let A be a non-singular n-square real symmetric matrix of index p. Show that $|A| > 0$ if and only if $n-p$ is even.

16. Prove: A non-singular symmetric matrix A is congruent to its inverse.
Hint. Take $P = BB'$ where $B'AB = I$ and show that $P'AP = A^{-1}$.

17. Rewrite the discussion of symmetric matrices including the proof of Theorem I to obtain **(15.6)** for Hermitian matrices.

18. Prove: If $A \overset{C}{\sim} B$ then A is symmetric (skew-symmetric) if and only if B is symmetric (skew-symmetric).

19. Let S be a non-singular symmetric matrix and T be a skew-symmetric matrix such that $(S+T)(S-T)$ is non-singular. Show that $P'SP = S$ when

$$P = (S+T)^{-1}(S-T)$$

Hint. $P'SP = [(S-T)^{-1}(S+T)S^{-1}(S-T)(S+T)^{-1}]^{-1}$.

20. Let S be a non-singular symmetric matrix and let T be such that $(S+T)(S-T)$ is non-singular. Show that if $P'SP = S$ when $P = (S+T)^{-1}(S-T)$ and $I+P$ is non-singular, then T is skew-symmetric.
Hint. $T = S(I-P)(I+P)^{-1} = S(I+P)^{-1}(I-P)$.

21. Show that congruence of n-square matrices is an equivalence relation.

Chapter 16

Bilinear Forms

AN EXPRESSION which is linear and homogeneous in each of the sets of variables $(x_1, x_2, \ldots, x_m)$ and $(y_1, y_2, \ldots, y_n)$ is called a **bilinear form** in these variables. For example,

$$x_1y_1 + 2x_1y_2 - 13x_1y_3 - 4x_2y_1 + 15x_2y_2 - x_2y_3$$

is a bilinear form in the variables (x_1, x_2) and (y_1, y_2, y_3).

The most general bilinear form in the variables $(x_1, x_2, \ldots, x_m)$ and $(y_1, y_2, \ldots, y_n)$ may be written as

$$\begin{aligned} f(x,y) \quad = \quad & a_{11}x_1y_1 + a_{12}x_1y_2 + \cdots + a_{1n}x_1y_n \\ & + a_{21}x_2y_1 + a_{22}x_2y_2 + \cdots + a_{2n}x_2y_n \\ & + \cdots\cdots\cdots\cdots\cdots\cdots\cdots \\ & + a_{m1}x_my_1 + a_{m2}x_my_2 + \cdots + a_{mn}x_my_n \end{aligned}$$

or, more briefly, as

$$\textbf{(16.1)} \qquad f(x,y) \quad = \quad \sum_{i=1}^{m}\sum_{j=1}^{n} a_{ij}x_iy_j$$

$$= \quad \begin{bmatrix} x_1, x_2, \ldots, x_m \end{bmatrix} \begin{bmatrix} a_{11} & a_{12} & \cdots & a_{1n} \\ a_{21} & a_{22} & \cdots & a_{2n} \\ \cdots & \cdots & \cdots & \cdots \\ a_{m1} & a_{m2} & \cdots & a_{mn} \end{bmatrix} \begin{bmatrix} y_1 \\ y_2 \\ \vdots \\ y_n \end{bmatrix}$$

$$= \quad X'AY$$

where $X = [x_1, x_2, \ldots, x_m]'$, $A = [a_{ij}]$, and $Y = [y_1, y_2, \ldots, y_n]'$.

The matrix A of the coefficients is called the **matrix** of the bilinear form and the rank of A is called the **rank** of the form.

See Problem 1.

Example 1. The bilinear form

$$x_1y_1 + x_1y_3 + x_2y_1 + x_2y_2 + x_3y_3 \quad = \quad \begin{bmatrix} x_1, x_2, x_3 \end{bmatrix} \begin{bmatrix} 1 & 0 & 1 \\ 1 & 1 & 0 \\ 0 & 0 & 1 \end{bmatrix} \begin{bmatrix} y_1 \\ y_2 \\ y_3 \end{bmatrix}$$

$$= \quad X'AY$$

CANONICAL FORMS. Let the m x's of (16.1) be replaced by new variables u's by means of the linear transformation

$$(16.2)\qquad x_i = \sum_{j=1}^{n} b_{ij}u_j, \quad (i = 1,2,\ldots,m) \text{ or } X = BU$$

and the n y's be replaced by new variables v's by means of the linear transformation

$$(16.3)\qquad y_i = \sum_{j=1}^{n} c_{ij}v_j, \quad (i = 1,2,\ldots,n) \text{ or } Y = CV$$

We have $X'AY = (BU)'A(CV) = U'(B'AC)V$. Now applying the linear transformations $U = IX$, $V = IY$ we obtain a new bilinear form in the original variables $X'(B'AC)Y = X'DY$.

Two bilinear forms are called equivalent if and only if there exist non-singular transformations which carry one form into the other.

I. Two bilinear forms with $m\times n$ matrices A and B over F are equivalent over F if and only if they have the same rank.

If the rank of (16.1) is r, there exist (see Chapter 5) non-singular matrices P and Q such that

$$PAQ = \begin{bmatrix} I_r & 0 \\ 0 & 0 \end{bmatrix}$$

Taking $B = P'$ in (16.2) and $C = Q$ in (16.3), the bilinear form is reduced to

$$(16.4)\qquad U'(PAQ)V = U'\begin{bmatrix} I_r & 0 \\ 0 & 0 \end{bmatrix}V = u_1v_1 + u_2v_2 + \cdots + u_rv_r$$

Thus,

II. Any bilinear form over F of rank r can be reduced by non-singular linear transformations over F to the canonical form $u_1v_1 + u_2v_2 + \cdots + u_rv_r$.

Example 2. For the matrix of the bilinear form $X'AY = X'\begin{bmatrix} 1 & 0 & 1 \\ 1 & 1 & 0 \\ 0 & 0 & 1 \end{bmatrix}Y$ of Example 1,

$$\begin{matrix} I_3 \\ \\ A\;\; I_3 \end{matrix} = \begin{matrix} 1 & 0 & 0 & & & \\ 0 & 1 & 0 & & & \\ 0 & 0 & 1 & & & \\ 1 & 0 & 1 & 1 & 0 & 0 \\ 1 & 1 & 0 & 0 & 1 & 0 \\ 0 & 0 & 1 & 0 & 0 & 1 \end{matrix} \rightarrow \begin{matrix} 1 & 0 & 0 & & & \\ 0 & 1 & 0 & & & \\ 0 & 0 & 1 & & & \\ 1 & 0 & 1 & 1 & 0 & 0 \\ 0 & 1 & -1 & -1 & 1 & 0 \\ 0 & 0 & 1 & 0 & 0 & 1 \end{matrix} \rightarrow \begin{matrix} 1 & 0 & -1 & & & \\ 0 & 1 & 0 & & & \\ 0 & 0 & 1 & & & \\ 1 & 0 & 0 & 1 & 0 & 0 \\ 0 & 1 & -1 & -1 & 1 & 0 \\ 0 & 0 & 1 & 0 & 0 & 1 \end{matrix} \rightarrow \begin{matrix} 1 & 0 & -1 & & & \\ 0 & 1 & 1 & & & \\ 0 & 0 & 1 & & & \\ 1 & 0 & 0 & 1 & 0 & 0 \\ 0 & 1 & 0 & -1 & 1 & 0 \\ 0 & 0 & 1 & 0 & 0 & 1 \end{matrix}$$

$$= \begin{matrix} Q \\ \\ I_3\;\; P' \end{matrix}$$

Thus, $X = PU = \begin{bmatrix} 1 & -1 & 0 \\ 0 & 1 & 0 \\ 0 & 0 & 1 \end{bmatrix}U$ and $Y = QV = \begin{bmatrix} 1 & 0 & -1 \\ 0 & 1 & 1 \\ 0 & 0 & 1 \end{bmatrix}V$ reduce $X'AY$ to

$$U'\begin{bmatrix} 1 & 0 & 0 \\ -1 & 1 & 0 \\ 0 & 0 & 1 \end{bmatrix}\begin{bmatrix} 1 & 0 & 1 \\ 1 & 1 & 0 \\ 0 & 0 & 1 \end{bmatrix}\begin{bmatrix} 1 & 0 & -1 \\ 0 & 1 & 1 \\ 0 & 0 & 1 \end{bmatrix}V = U'I_3V = u_1v_1 + u_2v_2 + u_3v_3$$

The equations of the transformations are

$$\begin{cases} x_1 = u_1 - u_2 \\ x_2 = \quad\; u_2 \\ x_3 = \quad\quad u_3 \end{cases} \quad \text{and} \quad \begin{cases} y_1 = v_1 \quad\; - v_3 \\ y_2 = \quad v_2 + v_3 \\ y_3 = \quad\quad\; v_3 \end{cases}$$

See Problem 2.

TYPES OF BILINEAR FORMS. A bilinear form $\sum_{i=1}^{n}\sum_{j=1}^{n} a_{ij}\,x_i y_j = X'AY$ is called

symmetric		**symmetric**
alternate	according as A is	**skew-symmetric**
Hermitian		**Hermitian**
alternate Hermitian		**skew-Hermitian**

COGREDIENT TRANSFORMATIONS. Consider a bilinear form $X'AY$ in the two sets of n variables $(x_1, x_2, \ldots, x_n)$ and $(y_1, y_2, \ldots, y_n)$. When the x's and y's are subjected to the same transformation $X = CU$ and $Y = CV$ the variables are said to be transformed **cogrediently**. We have

III. Under cogredient transformations $X = CU$ and $Y = CV$, the bilinear form $X'AY$, where A is n-square, is carried into the bilinear form $U'(C'AC)V$.

If A is symmetric, so also is $C'AC$; hence,

IV. A symmetric bilinear form remains symmetric under cogredient transformations of the variables.

V. Two bilinear forms over F are equivalent under cogredient transformations of the variables if and only if their matrices are congruent over F.

From Theorem **I**, Chapter 15, we have

VI. A symmetric bilinear form of rank r can be reduced by non-singular cogredient transformations of the variables to

$$a_1x_1y_1 + a_2x_2y_2 + \cdots + a_rx_ry_r \tag{16.5}$$

From Theorems **II** and **IV**, Chapter 15, follows

VII. A real symmetric bilinear form of rank r can be reduced by non-singular cogredient transformations of the variables in the real field to

$$x_1y_1 + x_2y_2 + \cdots + x_py_p - x_{p+1}y_{p+1} - \cdots - x_ry_r \tag{16.6}$$

and in the complex field to

$$x_1y_1 + x_2y_2 + \cdots + x_ry_r \tag{16.7}$$

See Problem 3.

CONTRAGREDIENT TRANSFORMATIONS. Let the bilinear form be that of the section above. When the x's are subjected to the transformation $X = (C^{-1})'U$ and the y's are subjected to the transformation $Y = CV$, the variables are said to be transformed **contragrediently**. We have

VIII. Under contragredient transformations $X = (C^{-1})'U$ and $Y = CV$, the bilinear form $X'AY$, where A is n-square, is carried into the bilinear form $U'(C^{-1}AC)V$.

IX. The bilinear form $X'IY = x_1y_1 + x_2y_2 + \cdots + x_ny_n$ is transformed into itself if and only if the two sets of variables are transformed contragrediently.

FACTORABLE BILINEAR FORMS. In Problem 4, we prove

X. A non-zero bilinear form is factorable if and only if its rank is one.

SOLVED PROBLEMS

1. $x_1y_1 + 2x_1y_2 - 13x_1y_3 - 4x_2y_1 + 15x_2y_2 - x_2y_3 = \begin{bmatrix} x_1, x_2 \end{bmatrix}\begin{bmatrix} 1 & 2 & -13 \\ -4 & 15 & -1 \end{bmatrix}\begin{bmatrix} y_1 \\ y_2 \\ y_3 \end{bmatrix} = X'\begin{bmatrix} 1 & 2 & -13 \\ -4 & 15 & -1 \end{bmatrix}Y.$

2. Reduce $x_1y_1 + 2x_1y_2 + 3x_1y_3 - 2x_1y_4 + 2x_2y_1 - 2x_2y_2 + x_2y_3 + 3x_2y_4 + 3x_3y_1 + 4x_3y_3 + x_3y_4$ to canonical form.

The matrix of the form is $A = \begin{bmatrix} 1 & 2 & 3 & -2 \\ 2 & -2 & 1 & 3 \\ 3 & 0 & 4 & 1 \end{bmatrix}$ By Problem 6, Chapter 5, the non-singular matrices

$P = \begin{bmatrix} 1 & 0 & 0 \\ -2 & 1 & 0 \\ -1 & -1 & 1 \end{bmatrix}$ and $Q = \begin{bmatrix} 1 & 1/3 & -4/3 & -1/3 \\ 0 & -1/6 & -5/6 & 7/6 \\ 0 & 0 & 1 & 0 \\ 0 & 0 & 0 & 1 \end{bmatrix}$ are such that $PAQ = \begin{bmatrix} I_2 & 0 \\ 0 & 0 \end{bmatrix}$. Thus, the linear transformations

$$X = P'U \quad\text{or}\quad \begin{cases} x_1 = u_1 - 2u_2 - u_3 \\ x_2 = u_2 - u_3 \\ x_3 = u_3 \end{cases} \quad\text{and}\quad Y = QV \quad\text{or}\quad \begin{cases} y_1 = v_1 + \frac{1}{3}v_2 - \frac{4}{3}v_3 - \frac{1}{3}v_4 \\ y_2 = - \frac{1}{6}v_2 - \frac{5}{6}v_3 + \frac{7}{6}v_4 \\ y_3 = v_3 \\ y_4 = v_4 \end{cases}$$

reduce $X'AY$ to $u_1v_1 + u_2v_2$.

3. Reduce the symmetric bilinear form $X'AY = X'\begin{bmatrix} 1 & 2 & 3 & 2 \\ 2 & 3 & 5 & 8 \\ 3 & 5 & 8 & 10 \\ 2 & 8 & 10 & -8 \end{bmatrix}Y$ by means of cogredient transformations to (**16.5**) in the rational field, (*b*) (**16.6**) in the real field, and (*c*) (**16.7**) in the complex field.

(*a*) From Example 1, Chapter 15, the linear transformations $X = \begin{bmatrix} 1 & -2 & -10 & -1 \\ 0 & 1 & 4 & -1 \\ 0 & 0 & 0 & 1 \\ 0 & 0 & 1 & 0 \end{bmatrix}U$ $Y = \begin{bmatrix} 1 & -2 & -10 & -1 \\ 0 & 1 & 4 & -1 \\ 0 & 0 & 0 & 1 \\ 0 & 0 & 1 & 0 \end{bmatrix}V$

reduce $X'AY$ to $u_1v_1 - u_2v_2 + 4u_3v_3$.

(*b*) From Example 2, Chapter 15, the linear transformations $X = \begin{bmatrix} 1 & -5 & -2 & -1 \\ 0 & 2 & 1 & -1 \\ 0 & 0 & 0 & 1 \\ 0 & \frac{1}{2} & 0 & 0 \end{bmatrix} U$, $Y = \begin{bmatrix} 1 & -5 & -2 & -1 \\ 0 & 2 & 1 & -1 \\ 0 & 0 & 0 & 1 \\ 0 & \frac{1}{2} & 0 & 0 \end{bmatrix} V$

reduce $X'AY$ to $u_1v_1 + u_2v_2 - u_3v_3$.

(*c*) From the result of Example 2, Chapter 15, we may obtain

$$\left[\begin{array}{cccc|cccc} 1 & 0 & 0 & 0 & 1 & 0 & 0 & 0 \\ 0 & 1 & 0 & 0 & -5 & 2 & 0 & \frac{1}{2} \\ 0 & 0 & -1 & 0 & -2 & 1 & 0 & 0 \\ 0 & 0 & 0 & 0 & -1 & -1 & 1 & 0 \end{array}\right] \overset{C}{\sim} \left[\begin{array}{cccc|cccc} 1 & 0 & 0 & 0 & 1 & 0 & 0 & 0 \\ 0 & 1 & 0 & 0 & -5 & 2 & 0 & \frac{1}{2} \\ 0 & 0 & 1 & 0 & -2i & i & 0 & 0 \\ 0 & 0 & 0 & 0 & -1 & -1 & 1 & 0 \end{array}\right]$$

Thus, the linear transformations $X = \begin{bmatrix} 1 & -5 & -2i & -1 \\ 0 & 2 & i & -1 \\ 0 & 0 & 0 & 1 \\ 0 & \frac{1}{2} & 0 & 0 \end{bmatrix} U$, $Y = \begin{bmatrix} 1 & -5 & -2i & -1 \\ 0 & 2 & i & -1 \\ 0 & 0 & 0 & 1 \\ 0 & \frac{1}{2} & 0 & 0 \end{bmatrix} V$ reduce $X'AY$ to

$u_1v_1 + u_2v_2 + u_3v_3$.

4. Prove: A non-zero bilinear form $f(x,y)$ is factorable if and only if its rank is 1.

Suppose the form is factorable so that

$$\sum\sum a_{ij}x_iy_j \quad = \quad (\sum b_ix_i)(\sum c_jy_j) \quad = \quad \sum\sum b_i c_j x_i y_j$$

and, hence, $a_{ij} = b_ic_j$. Clearly, any second order minor of $A = [a_{ij}]$, as

$$\begin{vmatrix} a_{ij} & a_{is} \\ a_{kj} & a_{ks} \end{vmatrix} \quad = \quad \begin{vmatrix} a_ib_j & a_ib_s \\ a_kb_j & a_kb_s \end{vmatrix} \quad = \quad b_jb_s\begin{vmatrix} a_i & a_i \\ a_k & a_k \end{vmatrix}$$

vanishes. Thus the rank of A is 1.

Conversely, suppose that the bilinear form is of rank 1. Then by Theorem I there exist non-singular linear transformations which reduce the form to $U'(B'AC)V = u_1v_1$. Now the inverses of the transformations

$$u_i \quad = \quad \sum_j r_{ij}x_j \qquad \text{and} \qquad v_i \quad = \quad \sum_j s_{ij}y_j$$

carry u_1v_1 into $(\sum_j r_{ij}x_j)(\sum_j s_{ij}y_j) = f(x,y)$ and so $f(x,y)$ is factorable.

SUPPLEMENTARY PROBLEMS

5. Obtain linear transformations which reduce each of the following bilinear forms to canonical form **(16.4)**

(*a*) $x_1y_1 - 2x_1y_3 + 3x_2y_1 + x_2y_2 - 3x_2y_3 - x_3y_1 - x_3y_2 - x_3y_3$

(*b*) $X'\begin{bmatrix} 2 & -5 & 1 & 0 \\ -4 & -11 & 2 & 7 \\ 5 & -5 & 1 & 0 \end{bmatrix}Y$, (*c*) $X'\begin{bmatrix} 3 & 1 & 0 \\ -2 & 8 & -2 \\ 1 & 1 & -2 \\ 3 & 5 & 0 \end{bmatrix}Y$, (*d*) $X'\begin{bmatrix} 7 & 4 & 5 & 8 \\ 4 & 7 & 3 & 5 \\ 5 & 3 & 12 & 6 \\ 8 & 5 & 6 & 10 \end{bmatrix}Y$

6. Obtain cogredient transformations which reduce

(*a*) $X'\begin{bmatrix} 1 & 2 & 3 \\ 2 & 5 & 4 \\ 3 & 4 & 14 \end{bmatrix}Y$ and (*b*) $X'\begin{bmatrix} 1 & 3 & 1 \\ 3 & 10 & 2 \\ 1 & 2 & 5 \end{bmatrix}Y$ to canonical form **(16.6)**.

Ans. (*a*) $C = \begin{bmatrix} 1 & -2 & -7 \\ 0 & 1 & 2 \\ 0 & 0 & 1 \end{bmatrix}$ (*b*) $C = \begin{bmatrix} 1 & -3 & -4\sqrt{3}/3 \\ 0 & 1 & \sqrt{3}/3 \\ 0 & 0 & \sqrt{3}/3 \end{bmatrix}$

7. If B_1, B_2, C_1, C_2 are non-singular n-square matrices such that $B_1A_1C_1 = B_2A_2C_2 = \begin{bmatrix} I_r & 0 \\ 0 & 0 \end{bmatrix}$, find the transformation which carries $X'A_1Y$ into $U'A_2V$.

Ans. $X = (B_2^{-1}B_1)'U$, $Y = C_1C_2^{-1}V$

8. Interpret Problem 23, Chapter 5, in terms of a pair of bilinear forms.

9. Write the transformation contragredient to $X = \begin{bmatrix} 1 & 1 & 0 \\ 0 & 1 & 1 \\ 1 & 0 & 1 \end{bmatrix}U$. *Ans.* $Y = \begin{bmatrix} \frac{1}{2} & \frac{1}{2} & -\frac{1}{2} \\ -\frac{1}{2} & \frac{1}{2} & \frac{1}{2} \\ \frac{1}{2} & -\frac{1}{2} & \frac{1}{2} \end{bmatrix}V$

10. Prove that an orthogonal transformation is contragredient to itself, that is $X = PU$, $Y = PV$.

11. Prove: Theorem **IX**.

12. If $X'AY$ is a real non-singular bilinear form then $X'A^{-1}Y$ is called its **reciprocal bilinear form**. Show that when reciprocal bilinear forms are transformed cogrediently by the same orthogonal transformation, reciprocal bilinear forms result.

13. Use Problem 4, Chapter 15, to show that there exist cogredient transformations $X = PU$, $Y = PV$ which reduce an alternate bilinear form of rank $r = 2t$ to the canonical form

$$u_1v_2 - u_2v_1 + u_3v_4 - u_4v_3 + \cdots + u_{2t-1}v_{2t} - u_{2t}v_{2t-1}$$

14. Determine canonical forms for Hermitian and alternate Hermitian bilinear forms.
Hint. See **(15.6)** and **(15.7)**.

Chapter 17

Quadratic Forms

A HOMOGENEOUS POLYNOMIAL of the type

(17.1)
$$q \;=\; X'AX \;=\; \sum_{i=1}^{n}\sum_{j=1}^{n} a_{ij}x_i x_j$$

whose coefficients a_{ij} are elements of F is called a quadratic form over F in the variables $x_1, x_2, \ldots, x_n$.

Example 1. $q = x_1^2 + 2x_2^2 - 7x_3^2 - 4x_1x_2 + 8x_1x_3$ is a quadratic form in the variables x_1, x_2, x_3. The matrix of the form may be written in various ways according as the cross-product terms $-4x_1x_2$ and $8x_1x_3$ are separated to form the terms $a_{12}x_1x_2, a_{21}x_2x_1$ and $a_{13}x_1x_3, a_{31}x_3x_1$. We shall agree that **the matrix of a quadratic form be symmetric** and shall always separate the cross-product terms so that $a_{ij} = a_{ji}$. Thus,

$$q \;=\; x_1^2 + 2x_2^2 - 7x_3^2 - 4x_1x_2 + 8x_1x_3$$
$$= \; X'\begin{bmatrix} 1 & -2 & 4 \\ -2 & 2 & 0 \\ 4 & 0 & -7 \end{bmatrix}X$$

The symmetric matrix $A = [a_{ij}]$ is called the **matrix** of the quadratic form and the rank of A is called the **rank** of the form. If the rank is $r<n$, the quadratic form is called **singular**; otherwise, **non-singular**.

TRANSFORMATIONS. The linear transformation over F, $X = BY$, carries the quadratic form **(17.1)** with symmetric matrix A over F into the quadratic form

(17.2)
$$(BY)'A(BY) \;=\; Y'(B'AB)Y$$

with symmetric matrix $B'AB$.

Two quadratic forms in the same variables $x_1, x_2, \ldots, x_n$ are called **equivalent** if and only if there exists a non-singular linear transformation $X = BY$ which, together with $Y = IX$, carries one of the forms into the other. Since $B'AB$ is congruent to A, we have

I. The rank of a quadratic form is invariant under a non-singular transformation of the variables.

II. Two quadratic forms over F are equivalent over F if and only if their matrices are congruent over F.

From Problem 1, Chapter 15, it follows that a quadratic form of rank r can be reduced to the form

(17.3)
$$h_1y_1^2 + h_2y_2^2 + \cdots + h_ry_r^2, \qquad h_i \neq 0$$

in which only terms in the squares of the variables occur, by a non-singular linear transformation $X = BY$. We recall that the matrix B is the product of elementary column matrices while B' is the product in reverse order of the same elementary row matrices.

Example 2. Reduce $q = X'\begin{bmatrix} 1 & -2 & 4 \\ -2 & 2 & 0 \\ 4 & 0 & -7 \end{bmatrix}X$ of Example 1 to the form **(17.3)**.

We have
$$[A \; I] = \left[\begin{array}{rrr|rrr} 1 & -2 & 4 & 1 & 0 & 0 \\ -2 & 2 & 0 & 0 & 1 & 0 \\ 4 & 0 & -7 & 0 & 0 & 1 \end{array}\right] \underset{\sim}{C} \left[\begin{array}{rrr|rrr} 1 & 0 & 0 & 1 & 0 & 0 \\ 0 & -2 & 8 & 2 & 1 & 0 \\ 0 & 8 & -23 & -4 & 0 & 1 \end{array}\right]$$
$$\underset{\sim}{C} \left[\begin{array}{rrr|rrr} 1 & 0 & 0 & 1 & 0 & 0 \\ 0 & -2 & 0 & 2 & 1 & 0 \\ 0 & 0 & 9 & 4 & 4 & 1 \end{array}\right] = [D \; B']$$

Thus, $X = BY = \begin{bmatrix} 1 & 2 & 4 \\ 0 & 1 & 4 \\ 0 & 0 & 1 \end{bmatrix}Y$ reduces q to $q' = y_1^2 - 2y_2^2 + 9y_3^2$.

See Problems 1-2.

LAGRANGE'S REDUCTION. The reduction of a quadratic form to the form **(17.3)** can be carried out by a procedure, known as Lagrange's Reduction, which consists essentially of repeated completing of the square.

Example 3.
$$\begin{aligned} q &= x_1^2 + 2x_2^2 - 7x_3^2 - 4x_1x_2 + 8x_1x_3 \\ &= \{x_1^2 - 4x_1(x_2 - 2x_3)\} + 2x_2^2 - 7x_3^2 \\ &= \{x_1^2 - 4x_1(x_2 - 2x_3) + 4(x_2 - 2x_3)^2\} + 2x_2^2 - 7x_3^2 - 4(x_2 - 2x_3)^2 \\ &= (x_1 - 2x_2 + 4x_3)^2 - 2(x_2^2 - 8x_2x_3) - 23x_3^2 \\ &= (x_1 - 2x_2 + 4x_3)^2 - 2(x_2^2 - 8x_2x_3 + 16x_3^2) + 9x_3^2 \\ &= (x_1 - 2x_2 + 4x_3)^2 - 2(x_2 - 4x_3)^2 + 9x_3^2 \end{aligned}$$

Thus,
$$\begin{cases} y_1 = x_1 - 2x_2 + 4x_3 \\ y_2 = x_2 - 4x_3 \\ y_3 = x_3 \end{cases} \quad \text{or} \quad \begin{cases} x_1 = y_1 + 2y_2 + 4y_3 \\ x_2 = y_2 + 4y_3 \\ x_3 = y_3 \end{cases}$$

reduces q to $y_1^2 - 2y_2^2 + 9y_3^2$.

See Problem 3.

REAL QUADRATIC FORMS. Let the real quadratic form $q = X'AX$ be reduced by a real non-singular transformation to the form **(17.3)**. If one or more of the h_i are negative, there exists a non-singular transformation $X = CZ$, where C is obtained from B by a sequence of row and column transformations of type 1, which carries q into

(17.4)
$$s_1z_1^2 + s_2z_2^2 + \cdots + s_pz_p^2 - s_{p+1}z_{p+1}^2 - \cdots - s_rz_r^2$$

in which the terms with positive coefficients precede those with negative coefficients.

Now the non-singular transformation

$$w_i = \sqrt{s_i}\, z_i, \qquad (i = 1, 2, \ldots, r)$$

$$w_j = z_j \quad, \qquad (j = r+1, r+2, \ldots, n)$$

or

$$Z = \operatorname{diag}\left(\frac{1}{\sqrt{s_1}}, \frac{1}{\sqrt{s_2}}, \ldots, \frac{1}{\sqrt{s_r}}, 1, 1, \ldots, 1\right) W$$

carries **(17.4)** into the canonical form

(17.5) $$w_1^2 + w_2^2 + \cdots + w_p^2 - w_{p+1}^2 - \cdots - w_r^2$$

Thus, since the product of non-singular transformations is a non-singular transformation, we have

III. Every real quadratic form can be reduced by a real non-singular transformation to the canonical form **(17.5)** where p, the number of positive terms, is called the **index** and r is the rank of the given quadratic form.

Example 4. In Example 2, the quadratic form $q = x_1^2 + 2x_2^2 - 7x_3^2 - 4x_1x_2 + 8x_1x_3$ was reduced to $q' = y_1^2 - 2y_2^2 + 9y_3^2$. The non-singular transformation $y_1 = z_1$, $y_2 = z_3$, $y_3 = z_2$ carries q' into $q'' = z_1^2 + 9z_2^2 - 2z_3^2$ and the non-singular transformation $z_1 = w_1$, $z_2 = w_2/3$, $z_3 = w_3/\sqrt{2}$ reduces q'' to $q''' = w_1^2 + w_2^2 - w_3^2$.

Combining the transformations, we have that the non-singular linear transformation

$$\begin{aligned} x_1 &= w_1 + \tfrac{4}{3}w_2 + \sqrt{2}\, w_3 \\ x_2 &= \tfrac{4}{3}w_2 + \tfrac{1}{2}\sqrt{2}\, w_3 \\ x_3 &= \tfrac{1}{3}w_2 \end{aligned} \qquad \text{or} \qquad X = \begin{bmatrix} 1 & 4/3 & \sqrt{2} \\ 0 & 4/3 & \tfrac{1}{2}\sqrt{2} \\ 0 & 1/3 & 0 \end{bmatrix} W$$

reduces q to $q''' = w_1^2 + w_2^2 - w_3^2$. The quadratic form is of rank 3 and index 2.

SYLVESTER'S LAW OF INERTIA. In Problem 5, we prove the law of inertia:

IV. If a real quadratic form is reduced by two real non-singular transformations to canonical forms **(17.5)**, they have the same rank and the same index.

Thus, the index of a real symmetric matrix depends upon the matrix and not upon the elementary transformations which produce **(15.2)**.

The difference between the number of positive and negative terms, $p - (r - p)$, in **(17.5)** is called the **signature** of the quadratic form. As a consequence of Theorem **IV**, we have

V. Two real quadratic forms each in n variables are equivalent over the real field if and only if they have the same rank and the same index or the same rank and the same signature.

COMPLEX QUADRATIC FORMS. Let the complex quadratic form $X'AX$ be reduced by a non-singular transformation to the form **(17.3)**. It is clear that the non-singular transformation

$$z_i = \sqrt{h_i}\, y_i, \qquad (i = 1, 2, \ldots, r)$$

$$z_j = y_j \quad, \qquad (j = r+1, r+2, \ldots, n)$$

or

$$Y = \operatorname{diag}\left(\frac{1}{\sqrt{h_1}}, \frac{1}{\sqrt{h_2}}, \ldots, \frac{1}{\sqrt{h_r}}, 1, 1, \ldots, 1\right) Z$$

carries **(17.3)** into

(17.6)
$$z_1^2 + z_2^2 + \cdots + z_r^2$$

Thus,

VI. Every quadratic form over the complex field of rank r can be reduced by a non-singular transformation over the complex field to the canonical form **(17.6)**.

VII. Two complex quadratic forms each in n variables are equivalent over the complex field if and only if they have the same rank.

DEFINITE AND SEMI-DEFINITE FORMS. A real non-singular quadratic form $q = X'AX$, $|A| \neq 0$, in n variables is called **positive definite** if its rank and index are equal. Thus, in the real field a positive definite quadratic form can be reduced to $y_1^2 + y_2^2 + \cdots + y_n^2$ and for any non-trivial set of values of the x's, $q > 0$.

A real singular quadratic form $q = X'AX$, $|A| = 0$, is called **positive semi-definite** if its rank and index are equal, i.e., $r = p < n$. Thus, in the real field a positive semi-definite quadratic form can be reduced to $y_1^2 + y_2^2 + \cdots + y_r^2$, $r < n$, and for any non-trivial set of values of the x's, $q \geq 0$.

A real non-singular quadratic form $q = X'AX$ is called **negative definite** if its index $p = 0$, i.e., $r = n$, $p = 0$. Thus, in the real field a negative definite form can be reduced to $-y_1^2 - y_2^2 - \cdots - y_n^2$ and for any non-trivial set of values of the x's, $q < 0$.

A real singular quadratic form $q = X'AX$ is called **negative semi-definite** if its index $p = 0$, i.e., $r < n$, $p = 0$. Thus, in the real field a negative semi-definite form can be reduced to $-y_1^2 - y_2^2 - \cdots y_r^2$ and for any non-trivial set of values of the x's, $q \leq 0$.

Clearly, if q is negative definite (semi-definite), then $-q$ is positive definite (semi-definite).

For positive definite quadratic forms, we have

VIII. If $q = X'AX$ is positive definite, then $|A| > 0$.

PRINCIPAL MINORS. A minor of a matrix A is called **principal** if it is obtained by deleting certain rows and the same numbered columns of A. Thus, the diagonal elements of a principal minor of A are diagonal elements of A.

In Problem 6, we prove

IX. Every symmetric matrix of rank r has at least one principal minor of order r different from zero.

DEFINITE AND SEMI-DEFINITE MATRICES. The matrix A of a real quadratic form $q = X'AX$ is called definite or semi-definite according as the quadratic form is definite or semi-definite. We have

X. A real symmetric matrix A is positive definite if and only if there exists a non-singular matrix C such that $A = C'C$.

XI. A real symmetric matrix A of rank r is positive semi-definite if and only if there exists a matrix C of rank r such that $A = C'C$.

See Problem 7.

XII. If A is positive definite, every principal minor of A is positive.

See Problem 8.

XIII. If A is positive semi-definite, every principal minor of A is non-negative.

REGULAR QUADRATIC FORMS. For a symmetric matrix $A = [a_{ij}]$ over F, we define the **leading principal minors** as

$$(17.7) \qquad p_0 = 1, \quad p_1 = a_{11}, \quad p_2 = \begin{vmatrix} a_{11} & a_{12} \\ a_{21} & a_{22} \end{vmatrix}, \quad p_3 = \begin{vmatrix} a_{11} & a_{12} & a_{13} \\ a_{21} & a_{22} & a_{23} \\ a_{31} & a_{32} & a_{33} \end{vmatrix}, \quad \ldots, \quad p_n = |A|$$

In Problem 9, we prove

XIV. Any n-square non-singular symmetric matrix A can be rearranged by interchanges of certain rows and the interchanges of corresponding columns so that not both p_{n-1} and p_{n-2} are zero.

XV. If A is a symmetric matrix and if $p_{n-2}p_n \neq 0$ but $p_{n-1} = 0$, then p_{n-2} and p_n have opposite signs.

Example 5. For the quadratic form $X'AX = X'\begin{bmatrix} 1 & 1 & 2 & 1 \\ 1 & 1 & 2 & 2 \\ 2 & 2 & 3 & 4 \\ 1 & 2 & 4 & 1 \end{bmatrix}X$, $p_0 = 1$, $p_1 = 1$, $p_2 = 0$, $p_3 = 0$, $p_4 = |A| = 1$.

Here $\alpha_{33} \neq 0$; the transformation $X = K_{34}\tilde{X}$ yields

$$(\text{i}) \qquad \tilde{X}'\begin{bmatrix} 1 & 1 & 1 & 2 \\ 1 & 1 & 2 & 2 \\ 1 & 2 & 1 & 4 \\ 2 & 2 & 4 & 3 \end{bmatrix}\tilde{X}$$

for which $p_0 = 1$, $p_1 = 1$, $p_2 = 0$, $p_3 = -1$, $p_4 = 1$. Thus, for (**i**) not both p_2 and p_3 are zero.

A symmetric matrix A of rank r is said to be **regularly arranged** if no two consecutive p's in the sequence $p_0, p_1, \ldots, p_r$ are zero. When A is regularly arranged the quadratic form $X'AX$ is said to be **regular**. In Example 5, the given form is not regular; the quadratic form (**i**) in the same example is regular.

Let A be a symmetric matrix of rank r. By Theorem **IX**, A contains at least one non-vanishing r-square principal minor M whose elements can be brought into the upper left corner of A. Then $p_r \neq 0$ while $p_{r+1} = p_{r+2} = \ldots = p_n = 0$. By Theorem **XIV**, the first r rows and the first r columns may be rearranged so that at least one of p_{r-1} and p_{r-2} is different from zero. If $p_{r-1} \neq 0$ and $p_{r-2} = 0$, we apply the above procedure to the matrix of p_{r-1}; if $p_{r-2} \neq 0$, we apply the procedure to the matrix of p_{r-2}; and so on, until M is regularly arranged, Thus,

XVI. Any symmetric matrix (quadratic form) of rank r can be regularly arranged.

See Problem 10.

XVII. A real quadratic form $X'AX$ is positive definite if and only if its rank is n and all leading principal minors are positive.

XVIII. A real quadratic form $X'AX$ of rank r is positive semi-definite if and only if each of the principal minors $p_0, p_1, \ldots, p_r$ is positive.

KRONECKER'S METHOD OF REDUCTION. The method of Kronecker for reducing a quadratic form into one in which only the squared terms of the variables appear is based on

XIX. If $q = X'AX$ is a quadratic form over F in n variables of rank r, then by a non-singular linear transformation over F it can be brought to $q' = \tilde{X}'B\tilde{X}$ in which a non-singular r-rowed minor C of A occupies the upper left corner of B. Moreover, there exists a non-singular linear transformation over F which reduces q to $q'' = \tilde{X}'C\tilde{X}$, a non-singular quadratic form in r variables.

XX. If $q = X'AX$ is a non-singular quadratic form over F in n variables and if $p_{n-1} = \alpha_{nn} \neq 0$, the non-singular transformation

$$\begin{cases} x_i = y_i + \alpha_{in} y_n, & (i = 1, 2, \ldots, n-1) \\ x_n = \alpha_{nn} y_n \end{cases}$$

or

$$X = BY = \begin{bmatrix} 1 & 0 & 0 & \ldots & 0 & \alpha_{1n} \\ 0 & 1 & 0 & \ldots & 0 & \alpha_{2n} \\ \ldots & \ldots & \ldots & \ldots & \ldots & \ldots \\ 0 & 0 & 0 & \ldots & 1 & \alpha_{n-1,\,n} \\ 0 & 0 & 0 & \ldots & 0 & \alpha_{nn} \end{bmatrix} Y$$

carries q into $\sum_{i=1}^{n-1}\sum_{j=1}^{n-1} a_{ij}y_iy_j + p_{n-1}p_n y_n^2$ in which one squared term in the variables has been isolated.

Example 6. For the quadratic form $X'AX = X'\begin{bmatrix} 1 & -2 & 4 \\ -2 & 2 & 0 \\ 4 & 0 & -7 \end{bmatrix}X$, $p_2 = \alpha_{33} = \begin{vmatrix} 1 & -2 \\ -2 & 2 \end{vmatrix} = -2 \neq 0$. The non-singular transformation

$$\begin{cases} x_1 = y_1 + \alpha_{13}y_3 = y_1 \qquad - 8y_3 \\ x_2 = y_2 + \alpha_{23}y_3 = \qquad y_2 - 8y_3 \\ x_3 = \qquad \alpha_{33}y_3 = \qquad\quad - 2y_3 \end{cases} \quad \text{or} \quad X = \begin{bmatrix} 1 & 0 & -8 \\ 0 & 1 & -8 \\ 0 & 0 & -2 \end{bmatrix} Y$$

reduces $X'AX$ to

$$Y'\begin{bmatrix} 1 & 0 & 0 \\ 0 & 1 & 0 \\ -8 & -8 & -2 \end{bmatrix}\begin{bmatrix} 1 & -2 & 4 \\ -2 & 2 & 0 \\ 4 & 0 & -7 \end{bmatrix}\begin{bmatrix} 1 & 0 & -8 \\ 0 & 1 & -8 \\ 0 & 0 & -2 \end{bmatrix}Y = Y'\begin{bmatrix} 1 & -2 & 0 \\ -2 & 2 & 0 \\ 0 & 0 & 36 \end{bmatrix}Y$$

in which the variable y_3 appears only in squared form.

XXI. If $q = X'AX$ is a non-singular quadratic form over F and if $\alpha_{n-1,\,n-1} = \alpha_{nn} = 0$ but $\alpha_{n,n-1} \neq 0$, the non-singular transformation over F

$$\begin{cases} x_i = y_i + \alpha_{i,\,n-1}y_{n-1} + \alpha_{in}y_n, & (i = 1, 2, \ldots, n-2) \\ x_{n-1} = \alpha_{n-1,n}\,y_n, \quad x_n = \alpha_{n,\,n-1}\,y_{n-1} \end{cases}$$

or

$$X = BY = \begin{bmatrix} 1 & 0 & 0 & \ldots & 0 & \alpha_{1,\,n-1} & \alpha_{1n} \\ 0 & 1 & 0 & \ldots & 0 & \alpha_{2,\,n-1} & \alpha_{2n} \\ \ldots & \ldots & \ldots & \ldots & \ldots & \ldots & \ldots \\ 0 & 0 & 0 & \ldots & 1 & \alpha_{n-2,\,n-1} & \alpha_{n-2,\,n} \\ 0 & 0 & 0 & \ldots & 0 & 0 & \alpha_{n-1,\,n} \\ 0 & 0 & 0 & \ldots & 0 & \alpha_{n,\,n-1} & 0 \end{bmatrix} Y$$

carries q into $\sum_{i=1}^{n-2}\sum_{j=1}^{n-2} a_{ij}y_i y_j + 2\alpha_{n,n-1}p_n y_{n-1}y_n$.

The further transformation

$$\begin{cases} y_i = z_i, & (i = 1, 2, \ldots, n-2) \\ y_{n-1} = z_{n-1} - z_n \\ y_n = z_{n-1} + z_n \end{cases}$$

yields $\sum_{i=1}^{n-2}\sum_{j=1}^{n-2} a_{ij}z_i z_j + 2\alpha_{n,n-1}p_n(z_{n-1}^2 - z_n^2)$ in which two squared terms with opposite signs are isolated.

Example 7. For the quadratic form

$$X'AX = X'\begin{bmatrix} 1 & 2 & 1 \\ 2 & 4 & 3 \\ 1 & 3 & 1 \end{bmatrix}X$$

$\alpha_{22} = \alpha_{33} = 0$ but $\alpha_{32} = -1 \neq 0$. The non-singular transformation

$$\begin{cases} x_1 = y_1 + \alpha_{12}y_2 + \alpha_{13}y_3 \\ x_2 = \alpha_{23}y_3 \\ x_3 = \alpha_{32}y_2 \end{cases} \quad \text{or} \quad X = \begin{bmatrix} 1 & 1 & 2 \\ 0 & 0 & -1 \\ 0 & -1 & 0 \end{bmatrix}Y$$

reduces $X'AX$ to

$$Y'\begin{bmatrix} 1 & 0 & 0 \\ 1 & 0 & -1 \\ 2 & -1 & 0 \end{bmatrix}\begin{bmatrix} 1 & 2 & 1 \\ 2 & 4 & 3 \\ 1 & 3 & 1 \end{bmatrix}\begin{bmatrix} 1 & 1 & 2 \\ 0 & 0 & -1 \\ 0 & -1 & 0 \end{bmatrix}Y = Y'\begin{bmatrix} 1 & 0 & 0 \\ 0 & 0 & 1 \\ 0 & 1 & 0 \end{bmatrix}Y = Y'BY = y_1^2 + 2y_2y_3$$

The transformation

$$\begin{cases} y_1 = z_1 \\ y_2 = z_2 - z_3 \\ y_3 = z_2 + z_3 \end{cases} \quad \text{or} \quad Y = \begin{bmatrix} 1 & 0 & 0 \\ 0 & 1 & -1 \\ 0 & 1 & 1 \end{bmatrix}Z$$

carries $Y'BY$ into

$$Z'\begin{bmatrix} 1 & 0 & 0 \\ 0 & 1 & 1 \\ 0 & -1 & 1 \end{bmatrix}\begin{bmatrix} 1 & 0 & 0 \\ 0 & 0 & 1 \\ 0 & 1 & 0 \end{bmatrix}\begin{bmatrix} 1 & 0 & 0 \\ 0 & 1 & -1 \\ 0 & 1 & 1 \end{bmatrix}Z = Z'\begin{bmatrix} 1 & 0 & 0 \\ 0 & 2 & 0 \\ 0 & 0 & -2 \end{bmatrix}Z = z_1^2 + 2z_2^2 - 2z_3^2$$

Consider now a quadratic form in n variables of rank r. By Theorem **XIX**, q can be reduced to $q_1 = X'AX$ where A has a non-singular r-square minor in the upper left-hand corner and zeros elsewhere. By Theorem **XVI**, A may be regularly arranged.

If $p_{r-1} \neq 0$, Theorem **XX** can be used to isolate one squared term

$$(17.8) \qquad p_{r-1}p_r y_r^2$$

If $p_{r-1} = 0$ but $\alpha_{r-1,r-1} \neq 0$, interchanges of the last two rows and the last two columns yield a matrix in which the new $p_{r-1} = \alpha_{r-1,r-1} \neq 0$. Since $p_{r-2} \neq 0$, Theorem **XX** can be used twice to isolate two squared terms

$$(17.9) \qquad p_{r-2}\alpha_{r-1,r-1}y_{r-1}^2 + \alpha_{r-1,r-1}p_r y_r^2$$

which have opposite signs since p_{r-2} and p_r have opposite signs by Theorem **XV**.

If $p_{r-1}=0$ and $\alpha_{r-1,\,r-1}=0$, then (see Problem 9) $\alpha_{r,\,r-1}\neq 0$ and Theorem **XXI** can be used to isolate two squared terms

(17.10)
$$2\alpha_{r,\,r-1}\,p_r(y_{r-1}^2 - y_r^2)$$

having opposite signs.

This process may be repeated until the given quadratic form is reduced to another containing only squared terms of the variables.

In **(17.8)** the isolated term will be positive or negative according as the sequence p_{r-1}, p_r presents a permanence or a variation of sign. In **(17.9)** and **(17.10)** it is seen that the sequences $p_{r-2}, \alpha_{r-1,\,r-1}, p_r$ and $p_{r-2}, \alpha_{r,\,r-1}, p_r$ present one permanence and one variation of sign regardless of the sign of $\alpha_{r-1,\,r-1}$ and $\alpha_{r,\,r-1}$. Thus,

XXII. If $q = X'AX$, a regular quadratic form of rank r, is reduced to canonical form by the method of Kronecker, the number of positive terms is exactly the number of permanences of sign and the number of negative terms is exactly the number of variations of sign in the sequence $p_0, p_1, p_2, \ldots, p_r$, where a zero in the sequence can be counted either as positive or negative but must be counted.

See Problems 11-13.

FACTORABLE QUADRATIC FORMS. Let $X'AX \neq 0$, with complex coefficients, be the given quadratic form.

Suppose $X'AX$ factors so that

(i)
$$X'AX = (a_1x_1 + a_2x_2 + \cdots + a_nx_n)(b_1x_1 + b_2x_2 + \cdots + b_nx_n)$$

If the factors are linearly independent, at least one matrix $\begin{bmatrix} a_i & a_j \\ b_i & b_j \end{bmatrix}$ is non-singular. Let the variables and their coefficients be renumbered so that $\begin{bmatrix} a_i & a_j \\ b_i & b_j \end{bmatrix}$ becomes $\begin{bmatrix} a_1 & a_2 \\ b_1 & b_2 \end{bmatrix}$.

The non-singular transformation

$$\begin{cases} y_1 = a_1x_1 + a_2x_2 + \cdots + a_nx_n \\ y_2 = b_1x_1 + b_2x_2 + \cdots + b_nx_n \\ y_3 = x_3, \;\ldots,\; y_n = x_n \end{cases}$$

transforms (i) into y_1y_2 of rank 2. Thus, (i) is of rank 2.

If the factors are linearly dependent, at least one element $a_i \neq 0$. Let the variables and their coefficients be renumbered so that a_i is a_1. The non-singular transformation

$$\begin{cases} y_1 = a_1x_1 + a_2x_2 + \cdots + a_nx_n \\ y_2 = x_2, \;\ldots,\; y_n = x_n \end{cases}$$

transforms (i) into $\dfrac{b_1}{a_1}y_1^2$ of rank 1. Thus, (i) is of rank 1.

Conversely, if $X'AX$ has rank 1 or 2 it may be reduced respectively by Theorem **VI** to y_1^2 or $y_1^2 + y_2^2$, each of which may be written in the complex field as the product of two linear factors.

We have proved

XXIII. A quadratic form $X'AX \neq 0$ with complex coefficients is the product of two linear factors if and only if its rank is $r \leq 2$.

SOLVED PROBLEMS

1. Reduce $q = X'AX = X'\begin{bmatrix} 1 & 2 & 3 & 2 \\ 2 & 3 & 5 & 8 \\ 3 & 5 & 8 & 10 \\ 2 & 8 & 10 & -8 \end{bmatrix}X$ to the form **(17.3)**.

From Example 1, Chapter 15,

$$[A \mid I] = \left[\begin{array}{cccc|cccc} 1 & 2 & 3 & 2 & 1 & 0 & 0 & 0 \\ 2 & 3 & 5 & 8 & 0 & 1 & 0 & 0 \\ 3 & 5 & 8 & 10 & 0 & 0 & 1 & 0 \\ 2 & 8 & 10 & -8 & 0 & 0 & 0 & 1 \end{array}\right] \overset{C}{\sim} \left[\begin{array}{cccc|cccc} 1 & 0 & 0 & 0 & 1 & 0 & 0 & 0 \\ 0 & -1 & 0 & 0 & -2 & 1 & 0 & 0 \\ 0 & 0 & 4 & 0 & -10 & 4 & 0 & 1 \\ 0 & 0 & 0 & 0 & -1 & -1 & 1 & 0 \end{array}\right] = [D \mid P']$$

Thus, the transformation $X = PY = \begin{bmatrix} 1 & -2 & -10 & -1 \\ 0 & 1 & 4 & -1 \\ 0 & 0 & 0 & 1 \\ 0 & 0 & 1 & 0 \end{bmatrix}Y$ reduces q to the required form $y_1^2 - y_2^2 + 4y_3^2$.

2. Reduce $q = X'AX = X'\begin{bmatrix} 1 & 2 & 2 \\ 2 & 4 & 8 \\ 2 & 8 & 4 \end{bmatrix}X$ to the form **(17.3)**.

We find

$$[A \mid I] \overset{C}{\sim} \left[\begin{array}{ccc|ccc} 1 & 0 & 0 & 1 & 0 & 0 \\ 0 & 8 & 0 & -4 & 1 & 1 \\ 0 & 0 & -2 & 0 & -\frac{1}{2} & \frac{1}{2} \end{array}\right] = [D \mid P']$$

Thus, the transformation $X = PY = \begin{bmatrix} 1 & -4 & 0 \\ 0 & 1 & -\frac{1}{2} \\ 0 & 1 & \frac{1}{2} \end{bmatrix}Y$ reduces q to $y_1^2 + 8y_2^2 - 2y_3^2$.

3. Lagrange reduction.

(*a*)
$$\begin{aligned} q &= 2x_1^2 + 5x_2^2 + 19x_3^2 - 24x_4^2 + 8x_1x_2 + 12x_1x_3 + 8x_1x_4 + 18x_2x_3 - 8x_2x_4 - 16x_3x_4 \\ &= 2\{x_1^2 + 2x_1(2x_2+3x_3+2x_4)\} + 5x_2^2 + 19x_3^2 - 24x_4^2 + 18x_2x_3 - 8x_2x_4 - 16x_3x_4 \\ &= 2\{x_1^2 + 2x_1(2x_2+3x_3+2x_4) + (2x_2+3x_3+2x_4)^2\} \\ &\qquad + 5x_2^2 + 19x_3^2 - 24x_4^2 + 18x_2x_3 - 8x_2x_4 - 16x_3x_4 - 2(2x_2+3x_3+2x_4)^2 \\ &= 2(x_1+2x_2+3x_3+2x_4)^2 - 3\{x_2^2 + 2x_2(x_3+4x_4)\} + x_3^2 - 32x_4^2 - 40x_3x_4 \\ &= 2(x_1+2x_2+3x_3+2x_4)^2 - 3(x_2+x_3+4x_4)^2 + 4(x_3-2x_4)^2 \end{aligned}$$

Thus, the transformation $\begin{cases} y_1 = x_1 + 2x_2 + 3x_3 + 2x_4 \\ y_2 = x_2 + x_3 + 4x_4 \\ y_3 = x_3 - 2x_4 \\ y_4 = x_4 \end{cases}$ reduces q to $2y_1^2 - 3y_2^2 + 4y_3^2$.

(*b*) For the quadratic form of Problem 2, we have

$$q = x_1^2 + 4x_1x_2 + 4x_1x_3 + 4x_2^2 + 16x_2x_3 + 4x_3^2 = (x_1 + 2x_2 + 2x_3)^2 + 8x_2x_3$$

Since there is no term in x_2^2 or x_3^2 but a term in x_2x_3, we use the non-singular transformation

(i) $x_1 = z_1, \quad x_2 = z_2, \quad x_3 = z_2 + z_3$

to obtain

$$q = (z_1+4z_2+2z_3)^2 + 8z_2^2 + 8z_2z_3 = (z_1+4z_2+2z_3)^2 + 8(z_2+\tfrac{1}{2}z_3)^2 - 2z_3^2 = y_1^2 + 8y_2^2 - 2y_3^2$$

Now $Y = \begin{bmatrix} 1 & 4 & 2 \\ 0 & 1 & \frac{1}{2} \\ 0 & 0 & 1 \end{bmatrix} Z$ and from (i), $Z = \begin{bmatrix} 1 & 0 & 0 \\ 0 & 1 & 0 \\ 0 & -1 & 1 \end{bmatrix} X$; hence, $Y = \begin{bmatrix} 1 & 4 & 2 \\ 0 & 1 & \frac{1}{2} \\ 0 & 0 & 1 \end{bmatrix}\begin{bmatrix} 1 & 0 & 0 \\ 0 & 1 & 0 \\ 0 & -1 & 1 \end{bmatrix} X = \begin{bmatrix} 1 & 2 & 2 \\ 0 & \frac{1}{2} & \frac{1}{2} \\ 0 & -1 & 1 \end{bmatrix} X.$

Thus, the non-singular transformation $X = \begin{bmatrix} 1 & -4 & 0 \\ 0 & 1 & -\frac{1}{2} \\ 0 & 1 & \frac{1}{2} \end{bmatrix} Y$ effects the reduction.

4. Using the result of Problem 2,

$$[A \mid I] \overset{C}{\sim} \left[\begin{array}{ccc|ccc} 1 & 0 & 0 & 1 & 0 & 0 \\ 0 & 8 & 0 & -4 & 1 & 1 \\ 0 & 0 & -2 & 0 & -\frac{1}{2} & \frac{1}{2} \end{array}\right]$$

and applying the transformations $H_2(\frac{1}{4}\sqrt{2})$, $K_2(\frac{1}{4}\sqrt{2})$ and $H_3(\frac{1}{2}\sqrt{2})$, $K_3(\frac{1}{2}\sqrt{2})$, we have

$$[A \mid I] \overset{C}{\sim} \left[\begin{array}{ccc|ccc} 1 & 0 & 0 & 1 & 0 & 0 \\ 0 & 8 & 0 & -4 & 1 & 1 \\ 0 & 0 & -2 & 0 & -\frac{1}{2} & \frac{1}{2} \end{array}\right] \overset{C}{\sim} \left[\begin{array}{ccc|ccc} 1 & 0 & 0 & 1 & 0 & 0 \\ 0 & 1 & 0 & -\sqrt{2} & \frac{1}{4}\sqrt{2} & \frac{1}{4}\sqrt{2} \\ 0 & 0 & -1 & 0 & -\frac{1}{4}\sqrt{2} & \frac{1}{4}\sqrt{2} \end{array}\right] = [C \mid Q']$$

Thus, the transformation $X = QY = \begin{bmatrix} 1 & -\sqrt{2} & 0 \\ 0 & \frac{1}{4}\sqrt{2} & -\frac{1}{4}\sqrt{2} \\ 0 & \frac{1}{4}\sqrt{2} & \frac{1}{4}\sqrt{2} \end{bmatrix} Y$ reduces $q = X' \begin{bmatrix} 1 & 2 & 2 \\ 2 & 4 & 8 \\ 2 & 8 & 4 \end{bmatrix} X$ to the canonical form $y_1^2 + y_2^2 - y_3^2$.

5. Prove: If a real quadratic form q is carried by two non-singular transformations into two distinct reduced forms

(i) $$y_1^2 + y_2^2 + \cdots + y_p^2 - y_{p+1}^2 - y_{p+2}^2 - \cdots - y_r^2$$

and

(ii) $$y_1^2 + y_2^2 + \cdots + y_q^2 - y_{q+1}^2 - y_{q+2}^2 - \cdots - y_r^2$$

then $p = q$.

Suppose $q > p$. Let $X = FY$ be the transformation which produces (i) and $X = GY$ be the transformation which produces (ii). Then

$$Y = F^{-1}X = \left.\begin{array}{l} b_{11}x_1 + b_{12}x_2 + \cdots + b_{1n}x_n \\ b_{21}x_1 + b_{22}x_2 + \cdots + b_{2n}x_n \\ \cdots\cdots\cdots\cdots\cdots\cdots\cdots\cdots \\ b_{n1}x_1 + b_{n2}x_2 + \cdots + b_{nn}x_n \end{array}\right\}$$

and

$$Y = G^{-1}X = \left.\begin{array}{l} c_{11}x_1 + c_{12}x_2 + \cdots + c_{1n}x_n \\ c_{21}x_1 + c_{22}x_2 + \cdots + c_{2n}x_n \\ \cdots\cdots\cdots\cdots\cdots\cdots\cdots\cdots \\ c_{n1}x_1 + c_{n2}x_2 + \cdots + c_{nn}x_n \end{array}\right\}$$

respectively carry (i) and (ii) back into q. Thus,

$$\textbf{(iii)} \quad (b_{11}x_1+b_{12}x_2+\cdots+b_{1n}x_n)^2 + \cdots + (b_{p1}x_1+b_{p2}x_2+\cdots+b_{pn}x_n)^2$$
$$- (b_{p+1,1}x_1+b_{p+1,2}x_2+\cdots+b_{p+1,n}x_n)^2 - \cdots - (b_{r1}x_1+b_{r2}x_2+\cdots+b_{rn}x_n)^2$$
$$= (c_{11}x_1+c_{12}x_2+\cdots+c_{1n}x_n)^2 + \cdots + (c_{q1}x_1+c_{q2}x_2+\cdots+c_{qn}x_n)^2$$
$$- (c_{q+1,1}x_1+c_{q+1,2}x_2+\cdots+c_{q+1,n}x_n)^2 - \cdots - (c_{r1}x_1+c_{r2}x_2+\cdots+c_{rn}x_n)^2$$

Consider the $r-q+p<r$ equations

$$\begin{cases} b_{11}x_1+b_{12}x_2+\cdots+b_{1n}x_n = 0 \\ b_{21}x_1+b_{22}x_2+\cdots+b_{2n}x_n = 0 \\ \cdots\cdots\cdots\cdots\cdots\cdots \\ b_{p1}x_1+b_{p2}x_2+\cdots+b_{pn}x_n = 0 \end{cases} \qquad \begin{cases} c_{q+1,1}x_1+c_{q+1,2}x_2+\cdots+c_{q+1,n}x_n = 0 \\ c_{q+2,1}x_1+c_{q+2,2}x_2+\cdots+c_{q+2,n}x_n = 0 \\ \cdots\cdots\cdots\cdots\cdots\cdots \\ c_{r1}x_1+c_{r2}x_2+\cdots+c_{rn}x_n = 0 \end{cases}$$

By Theorem **IV**, Chapter 10, they have a non-trivial solution, say $(\alpha_1, \alpha_2, \ldots, \alpha_n)$. When this solution is substituted into (iii), we have

$$-(b_{p+1,1}\alpha_1+b_{p+1,2}\alpha_2+\cdots+b_{p+1,n}\alpha_n)^2 - \cdots\cdots - (b_{r1}\alpha_1+b_{r2}\alpha_2+\cdots+b_{rn}\alpha_n)^2$$
$$= (c_{11}\alpha_1+c_{12}\alpha_2+\cdots+c_{1n}\alpha_n)^2 + \cdots\cdots + (c_{q1}\alpha_1+c_{q2}\alpha_2+\cdots+c_{qn}\alpha_n)^2$$

Clearly, this requires that each of the squared terms be zero. But then neither F nor G is non-singular, contrary to the hypothesis. Thus, $q \leq p$. A repetition of the above argument under the assumption that $q<p$ will also lead to a contradiction. Hence $q=p$.

6. Prove: Every symmetric matrix A of rank r has at least one principal minor of order r different from zero.

Since A is of rank r, it has at least one r-square minor which is different from zero. Suppose that it stands in the rows numbered $i_1, i_2, \ldots, i_r$. Let these rows be moved above to become the first r rows of the matrix and let the columns numbered $i_1, i_2, \ldots, i_r$ be moved in front to be the first r columns.

Now the first r rows are linearly independent while all other rows are linear combinations of them. By taking proper linear combinations of the first r rows and adding to the other rows, these last $n-r$ rows can be reduced to zero. Since A is symmetric, the same operations on the columns will reduce the last $n-r$ columns to zero. Hence, we now have

$$\left[\begin{array}{cccc|c} a_{i_1i_1} & a_{i_1i_2} & \cdots & a_{i_1i_r} & \\ a_{i_2i_1} & a_{i_2i_2} & \cdots & a_{i_2i_r} & 0 \\ \cdots & \cdots & \cdots & \cdots & \\ a_{i_ri_1} & a_{i_ri_2} & \cdots & a_{i_ri_r} & \\ \hline & & 0 & & 0 \end{array}\right]$$

in which a non-vanishing minor stands in the upper left-hand corner of the matrix. Clearly, this is a principal minor of A.

7. Prove: A real symmetric matrix A of rank r is positive semi-definite if and only if there exists a matrix C of rank r such that $A = C'C$.

Since A is of rank r, its canonical form is $N_1 = \begin{bmatrix} I_r & 0 \\ 0 & 0 \end{bmatrix}$. Then there exists a non-singular matrix B

such that $A = B'N_1B$. Since $N_1' = N_1 = N_1^2$, we have $A = B'N_1B = B'N_1N_1B = B'N_1'\cdot N_1B$. Set $C = N_1B$; then C is of rank r and $A = C'C$ as required.

Conversely, let C be a real n-square matrix of rank r; then $A = C'C$ is of rank $s \leq r$. Let its canonical form be

$$N_2 \quad = \quad \text{diag}(d_1, d_2, \ldots, d_s, 0, 0, \ldots, 0)$$

where each d_i is either +1 or −1. Then there exists a non-singular real matrix E such that $E'(C'C)E = N_2$. Set $CE = B = [b_{ij}]$. Since $B'B = N_2$, we have

$$b_{i1}^2 + b_{i2}^2 + \cdots + b_{in}^2 = d_i, \qquad (i = 1, 2, \ldots, s)$$

and

$$b_{j1}^2 + b_{j2}^2 + \cdots + b_{jn}^2 = 0, \qquad (j = s+1, s+2, \ldots, n)$$

Clearly, each $d_i > 0$ and A is positive semi-definite.

8. Prove: If A is positive definite, then every principal minor of A is positive.

Let $q = X'AX$. The principal minor of A obtained by deleting its ith row and column is the matrix A_i of the quadratic form q_1 obtained from q by setting $x_i = 0$. Now every value of q_i for non-trivial sets of values of its variables is also a value of q and, hence, is positive. Thus, A_i is positive definite.

This argument may be repeated for the principal minors $A_{ij}, A_{ijk}, \ldots$ obtained from A by deleting two, three, ... rows and the same columns of A.

By Theorem **VI**, $A_i > 0$, $A_{ij} > 0, \ldots$; thus, every principal minor is positive.

9. Prove: Any n-square non-singular matrix $A = [a_{ij}]$ can be rearranged by interchanges of certain rows and the interchanges of corresponding columns so that not both p_{n-1} and p_{n-2} are zero.

Clearly, the theorem is true for A of order 1 and of order 2. Moreover, it is true for A of order $n > 2$ when $p_{n-1} = \alpha_{nn} \neq 0$. Suppose $\alpha_{nn} = 0$; then either (a) some $\alpha_{ii} \neq 0$ or (b) all $\alpha_{ii} = 0$.

Suppose (a) some $\alpha_{ii} \neq 0$. After the ith row and the ith column have been moved to occupy the position of the last row and the last column, the new matrix has $p_{n-1} = \alpha_{ii} \neq 0$.

Suppose (b) all $\alpha_{ii} = 0$. Since $|A| \neq 0$, at least one $\alpha_{ni} \neq 0$. Move the ith row into the $(n-1)$st position and the ith column into the $(n-1)$st position. In the new matrix $\alpha_{n-1,n} = \alpha_{n,n-1} \neq 0$. By **(6.6)**, we have

$$\begin{vmatrix} \alpha_{n-1,n-1} & \alpha_{n-1,n} \\ \alpha_{n,n-1} & \alpha_{nn} \end{vmatrix} = \begin{vmatrix} 0 & \alpha_{n-1,n} \\ \alpha_{n-1,n} & 0 \end{vmatrix} = -\alpha_{n-1,n}^2 = p_{n-2}p_n$$

and $p_{n-2} \neq 0$.

Note that this also proves Theorem **XV**.

10. Renumber the variables so that $q = X'AX = X'\begin{bmatrix} 0 & 0 & 2 & 1 \\ 0 & 1 & 3 & 1 \\ 2 & 3 & 4 & 1 \\ 1 & 1 & 1 & 1 \end{bmatrix}X$ is regular.

Here $p_0 = 1$, $p_1 = 0$, $p_2 = 0$, $p_3 = -4$, $p_4 = -3$. Since $p_1 = p_2 = 0$, $p_3 \neq 0$, we examine the matrix $B = \begin{bmatrix} 0 & 0 & 2 \\ 0 & 1 & 3 \\ 2 & 3 & 4 \end{bmatrix}$ of p_3. The cofactor $B_{22} = \begin{vmatrix} 0 & 2 \\ 2 & 4 \end{vmatrix} \neq 0$; the interchange of the second and third rows and of the second and third columns of A yields

$$X'\begin{bmatrix} 0 & 2 & 0 & 1 \\ 2 & 4 & 3 & 1 \\ 0 & 3 & 1 & 1 \\ 1 & 1 & 1 & 1 \end{bmatrix}X$$

for which $p_0 = 1,\ p_1 = 0,\ p_2 = -4,\ p_3 = -4,\ p_4 = -3$. Here, x_2 has been renumbered as x_3 and x_3 as x_2.

11. Reduce by Kronecker's method $q = X'\begin{bmatrix} 1 & 2 & 3 & 4 \\ 2 & 1 & 5 & 6 \\ 3 & 5 & 2 & 3 \\ 4 & 6 & 3 & 4 \end{bmatrix}X$.

Here $p_0 = 1,\ p_1 = 1,\ p_2 = -3,\ p_3 = 20,\ p_4 = -5$ and q is regular. The sequence of p's presents one permanence and three variations in sign; the reduced form will have one positive and three negative terms.

Since each $p_i \neq 0$, repeated use of Theorem **XIX** yields the reduced form

$$p_0p_1y_1^2 + p_1p_2y_2^2 + p_2p_3y_3^2 + p_3p_4y_4^2 \quad = \quad y_1^2 - 3y_2^2 - 60y_3^2 - 100y_4^2$$

12. Reduce by Kronecker's method $q = X'AX = X'\begin{bmatrix} 1 & 2 & 3 & 1 \\ 2 & 4 & 6 & 3 \\ 3 & 6 & 9 & 2 \\ 1 & 3 & 2 & 5 \end{bmatrix}X$.

Here A is of rank 3 and $\alpha_{33} \neq 0$. An interchange of the last two rows and the last two columns carries A into $B = \begin{bmatrix} 1 & 2 & 1 & 3 \\ 2 & 4 & 3 & 6 \\ 1 & 3 & 5 & 2 \\ 3 & 6 & 2 & 9 \end{bmatrix}$ in which $C = \begin{vmatrix} 1 & 2 & 1 \\ 2 & 4 & 3 \\ 1 & 3 & 5 \end{vmatrix} \neq 0$. Since B is of rank 3, it can be reduced to $\begin{bmatrix} 1 & 2 & 1 & 0 \\ 2 & 4 & 3 & 0 \\ 1 & 3 & 5 & 0 \\ 0 & 0 & 0 & 0 \end{bmatrix}$.

Now q has been reduced to $\tilde{X}'C\tilde{X} = \tilde{X}'\begin{bmatrix} 1 & 2 & 1 \\ 2 & 4 & 3 \\ 1 & 3 & 5 \end{bmatrix}\tilde{X}$ for which $p_0 = 1,\ p_1 = 1,\ p_2 = 0,\ p_3 = -1$. The reduced form will contain two positive and one negative term. Since $p_2 = 0$ but $\gamma_{22} = \begin{vmatrix} 1 & 1 \\ 1 & 5 \end{vmatrix} = 4 \neq 0$, the reduced form is by **(16.8)**

$$p_0p_1y_1^2 + p_1\gamma_{22}y_2^2 + \gamma_{22}p_3y_3^2 \quad = \quad y_1^2 + 4y_2^2 - 4y_3^2$$

13. Reduce by Kronecker's method $q = X'\begin{bmatrix} 1 & -2 & 1 & 2 \\ -2 & 4 & 1 & -1 \\ 1 & 1 & 1 & 2 \\ 2 & -1 & 2 & 1 \end{bmatrix}X$.

Here $p_0 = 1,\ p_1 = 1,\ p_2 = 0,\ p_3 = -9,\ p_4 = 27$; the reduced form will have two positive and two negative terms. Consider the matrix $B = \begin{bmatrix} 1 & -2 & 1 \\ -2 & 4 & 1 \\ 1 & 1 & 1 \end{bmatrix}$ of p_3. Since $\beta_{33} = 0$ but $\beta_{32} = -3 \neq 0$ the reduced form is by **(16.8)** and **(16.9)**

$$p_0\,p_1\,y_1^2 + 2\beta_{32}p_3\,(y_2^2 - y_3^2) + p_3\,p_4\,y_4^2 \quad = \quad y_1^2 + 54y_2^2 - 54y_3^2 - 243y_4^2$$

14. Prove: For a set of real n-vectors $X_1, X_2, \ldots, X_p$,

$$|G| \quad = \quad \begin{vmatrix} X_1\cdot X_1 & X_1\cdot X_2 & \cdots & X_1\cdot X_p \\ X_2\cdot X_1 & X_2\cdot X_2 & \cdots & X_2\cdot X_p \\ \cdots & \cdots & \cdots & \cdots \\ X_p\cdot X_1 & X_p\cdot X_2 & \cdots & X_p\cdot X_p \end{vmatrix} \geq 0$$

the equality holding if and only if the set is linearly dependent.

(a) Suppose the vectors X_i are linearly independent and let $X = [x_1, x_2, \ldots, x_p]' \neq 0$. Then $Z = \sum_{i=1}^{p} X_i x_i \neq 0$ and $0 < Z\cdot Z = (\sum_{i=1}^{p} X_i x_i)\cdot(\sum_{j=1}^{p} X_j x_j) = X'(X_i' X_j)X = X'(X_i\cdot X_j)X = X'GX$.

Since this quadratic form is positive definite, $|G| > 0$.

(b) Suppose the vectors X_i are linearly dependent. Then there exist scalars $k_1, k_2, \ldots, k_p$, not all zero, such that $\xi = \sum_{i=1}^{p} k_i X_i = 0$ and, hence, such that

$$X_j\cdot\xi \quad = \quad k_1 X_j\cdot X_1 + k_2 X_j\cdot X_2 + \cdots + k_p X_j\cdot X_p \quad = \quad 0, \qquad (j = 1,2,\ldots,p)$$

Thus the system of homogeneous equations

$$X_j\cdot X_1 x_1 + X_j\cdot X_2 x_2 + \cdots + X_j\cdot X_p x_p \quad = \quad 0, \qquad (j = 1,2,\ldots,p)$$

has a non-trivial solution $x_i = k_i$, $(i = 1,2,\ldots,p)$, and $|G| = 0$.

We have proved that $|G| \geq 0$. To prove the converse of (b), we need only to assume $|G| = 0$ and reverse the steps of (b) to obtain $X_j\cdot\xi = 0$, $(j = 1,2,\ldots,p)$ where $\xi = \sum_{i=1}^{p} k_i X_i$. Thus, $\sum_{j=1}^{p} k_j X_j\cdot\xi = \xi\cdot\xi = 0$, $\xi = 0$, and the given vectors X_j are linearly dependent.

SUPPLEMENTARY PROBLEMS

15. Write the following quadratic forms in matrix notation:

(a) $x_1^2 + 4x_1x_2 + 3x_2^2$ (b) $2x_1^2 - 6x_1x_2 + x_3^2$ (c) $x_1^2 - 2x_2^2 - 3x_3^2 + 4x_1x_2 + 6x_1x_3 - 8x_2x_3$

Ans. (a) $X'\begin{bmatrix} 1 & 2 \\ 2 & 3 \end{bmatrix}X$ (b) $X'\begin{bmatrix} 2 & -3 & 0 \\ -3 & 0 & 0 \\ 0 & 0 & 1 \end{bmatrix}X$ (c) $X'\begin{bmatrix} 1 & 2 & 3 \\ 2 & -2 & -4 \\ 3 & -4 & -3 \end{bmatrix}X$

16. Write out in full the quadratic form in x_1, x_2, x_3 whose matrix is $\begin{bmatrix} 2 & -3 & 1 \\ -3 & 2 & 4 \\ 1 & 4 & -5 \end{bmatrix}$.

Ans. $2x_1^2 - 6x_1x_2 + 2x_1x_3 + 2x_2^2 + 8x_2x_3 - 5x_3^2$

17. Reduce by the method of Problem 1 and by Lagrange's Reduction:

(a) $X'\begin{bmatrix} 1 & 2 & 4 \\ 2 & 6 & -2 \\ 4 & -2 & 18 \end{bmatrix}X$ (b) $X'\begin{bmatrix} 1 & 1 & 1 & 1 \\ 1 & -1 & 3 & -3 \\ 1 & 3 & 3 & 1 \\ 1 & -3 & 1 & -3 \end{bmatrix}X$ (c) $X'\begin{bmatrix} 0 & 1 & 2 \\ 1 & 1 & -1 \\ 2 & -1 & 0 \end{bmatrix}X$ (d) $X'\begin{bmatrix} 0 & 0 & 1 \\ 0 & 1 & -2 \\ 1 & -2 & 3 \end{bmatrix}X$

Ans. (a) $y_1^2 + 2y_2^2 - 48y_3^2$ (b) $y_1^2 - 2y_2^2 + 4y_3^2$ (c) $y_1^2 - y_2^2 + 8y_3^2$ (d) $y_1^2 - y_2^2 + y_3^2$

Hint. In (c) and (d) use $x_1 = z_3$, $x_2 = z_1$, $x_3 = z_2$.

18. (*a*) Show that $X'\begin{bmatrix}1 & 4\\0 & 0\end{bmatrix}X = X'\begin{bmatrix}1 & 2\\2 & 0\end{bmatrix}X$ but the matrices have different ranks.

(*b*) Show that the symmetric matrix of a quadratic form is unique.

19. Show that over the real field $x_1^2+x_2^2-4x_3^2+4x_2x_3$ and $9x_1^2+2x_2^2+2x_3^2+6x_1x_2-6x_1x_3-8x_2x_3$ are equivalent.

20. Prove: A real symmetric matrix is positive (negative) definite if and only if it is congruent over the real field to I $(-I)$.

21. Show that $X'AX$ of Problem 12 is reduced to $\widetilde{X}'C\widetilde{X}$ by $X = R\widetilde{X}$, where $R = K_{34}K_{41}(-5)K_{42}(1)$. Then prove Theorem **XIX**.

22. (*a*) Show that if two real quadratic forms in the same variables are positive definite, so also is their sum.

(*b*) Show that if q_1 is a positive definite form in $x_1, x_2, \ldots, x_s$ and q_2 is a positive definite form in $x_{s+1}, x_{s+2}, \ldots, x_n$, then $q = q_1 + q_2$ is a positive definite form in $x_1, x_2, \ldots, x_n$.

23. Prove: If C is any real non-singular matrix, then $C'C$ is positive definite.
Hint: Consider $X'IX = Y'C'ICY$.

24. Prove: Every positive definite matrix A can be written as $A = C'C$. (Problems 23 and 24 complete the proof of Theorem **X**.) *Hint*: Consider $D'AD = I$.

25. Prove: If a real symmetric matrix A is positive definite, so also is A^p for p any positive integer.

26. Prove: If A is a real positive definite symmetric matrix and if B and C are such that $B'AB = I$ and $A = C'C$, then CB is orthogonal.

27. Prove: Every principal minor of a positive semi-definite matrix A is equal to or greater than zero.

28. Show that $ax_1^2 - 2bx_1x_2 + cx_2^2$ is positive definite if and only if $a>0$ and $|A| = ac - b^2 > 0$.

29. Verify the stated effect of the transformation in each of Theorems **XX** and **XXI**.

30. By Kronecker's reduction, after renumbering the variables when necessary, transform each of the following into a canonical form.

(*a*) $X'\begin{bmatrix}1 & -1 & 0\\-1 & 2 & -1\\0 & -1 & 2\end{bmatrix}X$ (*c*) $X'\begin{bmatrix}1 & 2 & 1 & 2\\2 & 0 & 1 & 2\\1 & 1 & 1 & 2\\2 & 2 & 2 & 1\end{bmatrix}X$ (*e*) $X'\begin{bmatrix}1 & 0 & -2\\0 & 0 & 1\\-2 & 1 & 3\end{bmatrix}X$ (*g*) $X'\begin{bmatrix}0 & 0 & 1\\0 & 1 & -2\\1 & -2 & 3\end{bmatrix}X$

(*b*) $X'\begin{bmatrix}4 & -4 & 2\\-4 & 3 & -3\\2 & -3 & 1\end{bmatrix}X$ (*d*) $X'\begin{bmatrix}1 & 2 & 3 & 1\\2 & -4 & 6 & 2\\3 & 6 & 9 & 3\\1 & 2 & 3 & 1\end{bmatrix}X$ (*f*) $X'\begin{bmatrix}1 & 2 & -1\\2 & 4 & 2\\-1 & 2 & 3\end{bmatrix}X$ (*h*) $X'\begin{bmatrix}0 & 2 & 0 & 1\\2 & 4 & 3 & 1\\0 & 3 & 1 & 1\\1 & 1 & 1 & 1\end{bmatrix}X$

Hint: In (*g*), renumber the variables to obtain (*e*) and also as in Problem 17(*d*).

Ans. (*a*) $p_0 = p_1 = p_2 = p_3 = 1$; $y_1^2 + y_2^2 + y_3^2$
(*b*) $4y_1^2 - 16y_2^2 + 16y_3^2$
(*c*) $y_1^2 - 4y_2^2 + 4y_3^2 - 3y_4^2$
(*d*) $y_1^2 - 8y_2^2$
(*e*) $p_0 = p_1 = 1$, $\alpha_{22} = -1$, $p_3 = -1$; $y_1^2 - y_2^2 + y_3^2$
(*f*) $p_0 = p_1 = 1$, $\alpha_{23} = -4$, $p_3 = -16$; $y_1^2 + 128y_2^2 - 128y_3^2$
(*g*) See (*e*).
(*h*) $4y_1^2 - 16y_2^2 + 16y_3^2 + 12y_4^2$

31. Show that $q = x_1^2 - 6x_2^2 - 6x_3^2 - 3x_4^2 - x_1x_2 - x_1x_3 + 2x_1x_4 + 13x_2x_3 - 11x_2x_4 + 9x_3x_4$ can be factored.

Chapter 18

Hermitian Forms

THE FORM DEFINED by

$$(18.1) \qquad h \;=\; \bar{X}'HX \;=\; \sum_{i=1}^{n}\sum_{j=1}^{n} h_{ij}\bar{x}_i x_j, \qquad \bar{h}_{ij} = h_{ji}$$

where H is Hermitian and the components of X are in the field of complex numbers, is called an **Hermitian form**. The rank of H is called the rank of the form. If the rank is $r<n$, the form is called **singular**; otherwise, **non-singular**.

If H and X are real, **(18.1)** is a real quadratic form; hence, we shall find that the theorems here are analogous to those of Chapter 17 and their proofs require only minor changes from those of that chapter.

Since H is Hermitian, every h_{ii} is real and every $h_{ii}\bar{x}_i x_i$ is real. Moreover, for the pair of cross-products $h_{ij}\bar{x}_i x_j$ and $h_{ji}\bar{x}_j x_i$,

$$h_{ij}\bar{x}_i x_j \;+\; h_{ji}\bar{x}_j x_i \;=\; h_{ij}\bar{x}_i x_j \;+\; \overline{h_{ij}\bar{x}_i x_j}$$

is real. Thus,

I. The values of an Hermitian form are real.

The non-singular linear transformation $X = BY$ carries the Hermitian form **(18.1)** into another Hermitian form

$$(18.2) \qquad (\overline{BY})'H(BY) \;=\; \bar{Y}'(\bar{B}'HB)Y$$

Two Hermitian forms in the same variables x_i are called **equivalent** if and only if there exists a non-singular linear transformation $X = BY$ which, together with $Y = IX$, carries one of the forms into the other. Since $\bar{B}'HB$ and H are conjunctive, we have

II. The rank of an Hermitian form is invariant under a non-singular transformation of the variables.

and

III. Two Hermitian forms are equivalent if and only if their matrices are conjunctive.

REDUCTION TO CANONICAL FORM. An Hermitian form **(18.1)** of rank r can be reduced to diagonal form

$$(18.3) \qquad k_1\bar{y}_1 y_1 \;+\; k_2\bar{y}_2 y_2 \;+\; \cdots \;+\; k_r\bar{y}_r y_r, \qquad k_i \neq 0 \text{ and real}$$

by a non-singular linear transformation $X = BY$. From **(18.2)** the matrix B is a product of elementary column matrices while $\bar{B}'$ is the product in reverse order of the conjugate elementary row matrices.

By a further linear transformation, **(18.3)** can be reduced to the canonical form [see **(15.6)**]

$$(18.4) \qquad \bar{z}_1 z_1 \;+\; \bar{z}_2 z_2 \;+\; \cdots \;+\; \bar{z}_p z_p \;-\; \bar{z}_{p+1} z_{p+1} \;-\; \cdots \;-\; \bar{z}_r z_r$$

of index p and signature $p-(r-p)$. Here, also, p depends upon the given form and not upon the transformation which reduces that form to **(18.4)**.

IV. Two Hermitian forms each in the same n variables are **equivalent** if and only if they have the same rank and the same index or the same rank and the same signature.

DEFINITE AND SEMI-DEFINITE FORMS. A non-singular Hermitian form $h = \bar{X}'HX$ in n variables is called **positive definite** if its rank and index are equal to n. Thus, a positive definite Hermitian form can be reduced to $\bar{y}_1 y_1 + \bar{y}_2 y_2 + \cdots + \bar{y}_n y_n$ and for any non-trivial set of values of the x's, $h>0$.

A singular Hermitian form $h = \bar{X}'HX$ is called **positive semi-definite** if its rank and index are equal, i.e., $r = p < n$. Thus, a positive semi-definite Hermitian form can be reduced to $\bar{y}_1 y_1 + \bar{y}_2 y_2 + \cdots + \bar{y}_r y_r$, $r<n$, and for any non-trivial set of values of the x's, $h \geq 0$.

The matrix H of an Hermitian form $\bar{X}'HX$ is called positive definite or positive semi-definite according as the form is positive definite or positive semi-definite.

V. An Hermitian form is positive definite if and only if there exists a non-singular matrix C such that $H = \bar{C}'C$.

VI. If H is positive definite, every principal minor of H is positive, and conversely.

VII. If H is positive semi-definite, every principal minor of H is non-negative, and conversely.

SOLVED PROBLEM

1. Reduce $\bar{X}'\begin{bmatrix} 1 & 1+2i & 2-3i \\ 1-2i & 5 & -4-2i \\ 2+3i & -4+2i & 13 \end{bmatrix}X$ to canonical form **(18.4)**.

From Problem 7, Chapter 15,

$$\left[\begin{array}{ccc|ccc} 1 & 1+2i & 2-3i & 1 & 0 & 0 \\ 1-2i & 5 & -4-2i & 0 & 1 & 0 \\ 2+3i & -4+2i & 13 & 0 & 0 & 1 \end{array}\right] \overset{HC}{\sim} \left[\begin{array}{ccc|ccc} 1 & 0 & 0 & 1 & 0 & 0 \\ 0 & 1 & 0 & 2/\sqrt{10} & 1/\sqrt{10} & i/\sqrt{10} \\ 0 & 0 & -1 & (-4-4i)/\sqrt{10} & i/\sqrt{10} & 1/\sqrt{10} \end{array}\right]$$

Thus, the non-singular linear transformation

$$X \quad = \quad BY \quad = \quad \begin{bmatrix} 1 & 2/\sqrt{10} & (-4+4i)/\sqrt{10} \\ 0 & 1/\sqrt{10} & -i/\sqrt{10} \\ 0 & -i/\sqrt{10} & 1/\sqrt{10} \end{bmatrix} Y$$

reduces the given Hermitian form to $\bar{y}_1 y_1 + \bar{y}_2 y_2 - \bar{y}_3 y_3$.

SUPPLEMENTARY PROBLEMS

2. Reduce each of the following to canonical form.

(a) $\bar{X}'\begin{bmatrix} 1 & 1+2i \\ 1-2i & 2 \end{bmatrix}X$ (c) $\bar{X}'\begin{bmatrix} 1 & 1-3i & 2-3i \\ 1+3i & 1 & 2+3i \\ 2+3i & 2-3i & 4 \end{bmatrix}X$

(b) $\bar{X}'\begin{bmatrix} 0 & i \\ -i & 0 \end{bmatrix}X$ (d) $\bar{X}'\begin{bmatrix} 1 & 1-i & 3-2i \\ 1+i & 0 & 2-i \\ 3+2i & 2+i & 4 \end{bmatrix}X$

Hint: For (*b*), first multiply the second row of H by i and add to the first row.

Ans. (a) $X = \begin{bmatrix} 1 & (-1-2i)/\sqrt{3} \\ 0 & 1/\sqrt{3} \end{bmatrix}Y$; $\bar{y}_1y_1 - \bar{y}_2y_2$

(b) $X = \dfrac{1}{\sqrt{2}}\begin{bmatrix} 1 & -i \\ -i & 1 \end{bmatrix}Y$; $\bar{y}_1y_1 - \bar{y}_2y_2$

(c) $X = \begin{bmatrix} 1 & (-1+3i)/3 & -1 \\ 0 & 1/3 & -1 \\ 0 & 0 & 1 \end{bmatrix}Y$; $\bar{y}_1y_1 - \bar{y}_2y_2$

(d) $X = \begin{bmatrix} 1 & (-1+i)/\sqrt{2} & (-1+3i)/\sqrt{10} \\ 0 & 1/\sqrt{2} & (-3-2i)/\sqrt{10} \\ 0 & 0 & 2/\sqrt{10} \end{bmatrix}Y$; $\bar{y}_1y_1 - \bar{y}_2y_2 - \bar{y}_3y_3$

3. Obtain the linear transformation $X = BY$ which followed by $Y = IX$ carries (*a*) of Problem 2 into (*b*).

Ans. $X = \dfrac{1}{\sqrt{2}}\begin{bmatrix} 1 & (-1-2i)/\sqrt{3} \\ 0 & 1/\sqrt{3} \end{bmatrix}\begin{bmatrix} 1 & i \\ i & 1 \end{bmatrix}Y$

4. Show that $\bar{X}'\begin{bmatrix} 1 & 1+i & -1 \\ 1-i & 6 & -3+i \\ -1 & -3-i & 11 \end{bmatrix}X$ is positive definite and $\bar{X}'\begin{bmatrix} 1 & 1+i & 1+2i \\ 1-i & 3 & 5 \\ 1-2i & 5 & 10 \end{bmatrix}X$ is positive semi-definite.

5. Prove Theorems **V-VII**.

6. Obtain for Hermitian forms theorems analogous to Theorems **XIX-XXI**, Chapter 17, for quadratic forms.

7. Prove: $\begin{vmatrix} 0 & x_1 & x_2 & \cdots & x_n \\ \bar{x}_1 & h_{11} & h_{12} & \cdots & h_{1n} \\ \bar{x}_2 & h_{21} & h_{22} & \cdots & h_{2n} \\ \cdots & \cdots & \cdots & \cdots & \cdots \\ \bar{x}_n & h_{n1} & h_{n2} & \cdots & h_{nn} \end{vmatrix} = -\sum_{i=1}^{n}\sum_{j=1}^{n} \eta_{ij}\bar{x}_i x_j$ where η_{ij} is the cofactor of h_{ij} in $H = |h_{ij}|$.

Hint: Use (**4.3**).

Chapter 19

The Characteristic Equation of a Matrix

THE PROBLEM. Let $Y = AX$, where $A = [a_{ij}]$, $(i,j = 1,2,\ldots,n)$, be a linear transformation over F. In general, the transformation carries a vector $X = [x_1, x_2, \ldots, x_n]'$ into a vector $Y = [y_1, y_2, \ldots, y_n]'$ whose only connection with X is through the transformation. We shall investigate here the possibility of certain vectors X being carried by the transformation into λX, where λ is either a scalar of F or of some field $\mathfrak{F}$ of which F is a subfield.

Any vector X which by the transformation is carried into λX, that is, any vector X for which

$$(19.1) \qquad AX = \lambda X$$

is called an **invariant vector** under the transformation.

THE CHARACTERISTIC EQUATION. From **(19.1)**, we obtain

$$(19.2) \qquad \lambda X - AX = (\lambda I - A)X = \begin{bmatrix} \lambda - a_{11} & -a_{12} & \cdots & -a_{1n} \\ -a_{21} & \lambda - a_{22} & \cdots & -a_{2n} \\ \cdots & \cdots & \cdots & \cdots \\ -a_{n1} & -a_{n2} & \cdots & \lambda - a_{nn} \end{bmatrix} \begin{bmatrix} x_1 \\ x_2 \\ \vdots \\ x_n \end{bmatrix} = 0$$

The system of homogeneous equations **(19.2)** has non-trivial solutions if and only if

$$(19.3) \qquad |\lambda I - A| = \begin{vmatrix} \lambda - a_{11} & -a_{12} & \cdots & -a_{1n} \\ -a_{21} & \lambda - a_{22} & \cdots & -a_{2n} \\ \cdots & \cdots & \cdots & \cdots \\ -a_{n1} & -a_{n2} & \cdots & \lambda - a_{nn} \end{vmatrix} = 0$$

The expansion of this determinant yields a polynomial $\phi(\lambda)$ of degree n in λ which is known as the **characteristic polynomial** of the transformation or of the matrix A. The equation $\phi(\lambda) = 0$ is called the **characteristic equation** of A and its roots $\lambda_1, \lambda_2, \ldots, \lambda_n$ are called the **characteristic roots** of A. If $\lambda = \lambda_i$ is a characteristic root, then **(19.2)** has non-trivial solutions which are the components of **invariant** or **characteristic vectors** associated with (corresponding to) that root.

Characteristic roots are also known as **latent roots** and **eigenvalues**; characteristic vectors are called **latent vectors** and **eigenvectors**.

Example 1. Determine the characteristic roots and associated invariant vectors, given $A = \begin{bmatrix} 2 & 2 & 1 \\ 1 & 3 & 1 \\ 1 & 2 & 2 \end{bmatrix}$.

The characteristic equation is $\begin{vmatrix} \lambda - 2 & -2 & -1 \\ -1 & \lambda - 3 & -1 \\ -1 & -2 & \lambda - 2 \end{vmatrix} = \lambda^3 - 7\lambda^2 + 11\lambda - 5 = 0$ and the characteristic roots are $\lambda_1 = 5$, $\lambda_2 = 1$, $\lambda_3 = 1$.

When $\lambda = \lambda_1 = 5$, **(19.2)** becomes

$$\begin{bmatrix} 3 & -2 & -1 \\ -1 & 2 & -1 \\ -1 & -2 & 3 \end{bmatrix}\begin{bmatrix} x_1 \\ x_2 \\ x_3 \end{bmatrix} = 0 \qquad \text{or} \qquad \begin{bmatrix} 0 & 1 & -1 \\ 1 & 0 & -1 \\ 0 & 0 & 0 \end{bmatrix}\begin{bmatrix} x_1 \\ x_2 \\ x_3 \end{bmatrix} = 0$$

since $\begin{bmatrix} 3 & -2 & -1 \\ -1 & 2 & -1 \\ -1 & -2 & 3 \end{bmatrix}$ is row equivalent to $\begin{bmatrix} 0 & 1 & -1 \\ 1 & 0 & -1 \\ 0 & 0 & 0 \end{bmatrix}$.

A solution is given by $x_1 = x_2 = x_3 = 1$; hence, associated with the characteristic root $\lambda = 5$ is the one-dimensional vector space spanned by the vector $[1,1,1]'$. Every vector $[k,k,k]'$ of this space is an invariant vector of A.

When $\lambda = \lambda_2 = 1$, **(19.2)** becomes

$$\begin{bmatrix} -1 & -2 & -1 \\ -1 & -2 & -1 \\ -1 & -2 & -1 \end{bmatrix}\begin{bmatrix} x_1 \\ x_2 \\ x_3 \end{bmatrix} = 0 \qquad \text{or} \qquad x_1 + 2x_2 + x_3 = 0$$

Two linearly independent solutions are $(2,-1,0)$ and $(1,0,-1)$. Thus, associated with the characteristic root $\lambda = 1$ is the two-dimensional vector space spanned by $X_1 = [2,-1,0]'$ and $X_2 = [1,0,-1]'$. Every vector $hX_1 + kX_2 = [2h+k, -h, -k]'$ is an invariant vector of A.

See Problems 1-2.

GENERAL THEOREMS. In Problem 3, we prove a special case ($k = 3$) of

I. If $\lambda_1, \lambda_2, \ldots, \lambda_k$ are distinct characteristic roots of a matrix A and if $X_1, X_2, \ldots, X_k$ are non-zero invariant vectors associated respectively with these roots, the X's are linearly independent.

In Problem 4, we prove a special case ($n = 3$) of

II. The kth derivative of $\phi(\lambda) = |\lambda I - A|$, where A is n-square, with respect to λ is $k!$ times the sum of the principal minors of order $n-k$ of the characteristic matrix when $k<n$, is $n!$ when $k=n$, and is 0 when $k>n$.

As a consequence of Theorem **II**, we have

III. If λ_i is an r-fold characteristic root of an n-square matrix A, the rank of $\lambda_i I - A$ is not less than $n-r$ and the dimension of the associated invariant vector space is not greater than r.

See Problem 5.

In particular

III′. If λ_i is a simple characteristic root of an n-square matrix A, the rank of $\lambda_i I - A$ is $n-1$ and the dimension of the associated invariant vector space is 1.

Example 2. For the matrix $A = \begin{bmatrix} 2 & 2 & 1 \\ 1 & 3 & 1 \\ 1 & 2 & 2 \end{bmatrix}$ of Example 1, the characteristic equation is $\phi(\lambda) = (\lambda-5)(\lambda-1)^2 = 0$. The invariant vector $[1,1,1]'$ associated with the characteristic root $\lambda = 5$ and the linearly independent invariant vectors $[2,-1,0]'$ and $[1,0,-1]'$ associated with the multiple root $\lambda = 1$ are a linearly independent set (see Theorem **I**).

The invariant vector space associated with the simple characteristic root $\lambda = 5$ is of dimension 1. The invariant vector space associated with the characteristic root $\lambda = 1$, of multiplicity 2, is of dimension 2 (see Theorems **III** and **III′**).

See also Problem 6.

Since any principal minor of A' is equal to the corresponding principal minor of A, we have by **(19.4)** of Problem 1,

IV. The characteristic roots of A and A' are the same.

Since any principal minor of $\bar{A}$ is the conjugate of the corresponding principal minor of A, we have

V. The characteristic roots of $\bar{A}$ and of $\bar{A}'$ are the conjugates of the characteristic roots of A.

By comparing characteristic equations, we have

VI. If $\lambda_1, \lambda_2, \ldots, \lambda_n$ are the characteristic roots of an n-square matrix A and if k is a scalar, then $k\lambda_1, k\lambda_2, \ldots, k\lambda_n$ are the characteristic roots of kA.

VII. If $\lambda_1, \lambda_2, \ldots, \lambda_n$ are the characteristic roots of an n-square matrix A and if k is a scalar, then $\lambda_1 - k, \lambda_2 - k, \ldots, \lambda_n - k$ are the characteristic roots of $A - kI$.

In Problem 7, we prove

VIII. If α is a characteristic root of a non-singular matrix A, then $|A|/\alpha$ is a characteristic root of adj A.

SOLVED PROBLEMS

1. If A is n-square, show that

$$\textbf{(19.4)} \qquad \phi(\lambda) \;=\; |\lambda I - A| \;=\; \lambda^n + s_1\lambda^{n-1} + s_2\lambda^{n-2} + \ldots + s_{n-1}\lambda + (-1)^n |A|$$

where s_m, $(m = 1, 2, \ldots, n-1)$ is $(-1)^m$ times the sum of all the m-square principal minors of A.

We rewrite $|\lambda I - A|$ in the form

$$\begin{vmatrix} \lambda - a_{11} & 0 - a_{12} & \ldots\ldots & 0 - a_{1n} \\ 0 - a_{21} & \lambda - a_{22} & \ldots\ldots & 0 - a_{2n} \\ \ldots & \ldots & \ldots & \ldots \\ 0 - a_{n1} & 0 - a_{n2} & \ldots\ldots & \lambda - a_{nn} \end{vmatrix}$$

and, each element being a binomial, suppose that the determinant has been expressed as the sum of 2^n determinants in accordance with Theorem **VIII**, Chapter 3. One of these determinants has λ as diagonal elements and zeros elsewhere; its value is λ^n. Another is free of λ; its value is $(-1)^n|A|$. The remaining determinants have m columns, $(m = 1, 2, \ldots, n-1)$, of $-A$ and $n - m$ columns each of which contains just one non-zero element λ.

Consider one of these determinants and suppose that its columns numbered $i_1, i_2, \ldots, i_m$ are columns of $-A$.

After an even number of interchanges (count them) of adjacent rows and of adjacent columns, the determinant becomes

$$(-1)^m \begin{vmatrix} a_{i_1,i_1} & a_{i_1,i_2} & \cdots & a_{i_1,i_m} & \\ a_{i_2,i_1} & a_{i_2,i_2} & \cdots & a_{i_2,i_m} & 0 \\ \cdots & \cdots & \cdots & \cdots & \\ a_{i_m,i_1} & a_{i_m,i_2} & \cdots & a_{i_m,i_m} & \\ \hline & & & & \begin{matrix} \lambda & 0 & \cdots & 0 \end{matrix} \\ \cdots & \cdots & \cdots & \cdots & \begin{matrix} 0 & \lambda & \cdots & 0 \end{matrix} \\ a_{i_n,i_1} & a_{i_n,i_2} & \cdots & a_{i_n,i_m} & \\ & & & & \begin{matrix} 0 & 0 & \cdots & \lambda \end{matrix} \end{vmatrix} = (-1)^m \left| A^{i_1,i_2,\ldots,i_m}_{i_1,i_2,\ldots,i_m} \right| \lambda^{n-m}$$

where $\left| A^{i_1,i_2,\ldots,i_m}_{i_1,i_2,\ldots,i_m} \right|$ is an m-square principal minor of A. Now

$$s_m = (-1)^m \sum_{\rho} \left| A^{i_1,i_2,\ldots,i_m}_{i_1,i_2,\ldots,i_m} \right|$$

as $(i_1, i_2, \ldots, i_m)$ runs through the $\rho = \dfrac{n(n-1)\ldots(n-m+1)}{1\ 2 \ldots m}$ different combinations of $1, 2, \ldots, n$ taken m at a time.

2. Use **(19.4)** of Problem 1 to expand $|\lambda I - A|$, given $A = \begin{bmatrix} 1 & -4 & -1 & 4 \\ 2 & 0 & 5 & -4 \\ -1 & 1 & -2 & 3 \\ -1 & 4 & -1 & 6 \end{bmatrix}$.

We have $s_1 = 1 + 0 - 2 + 6 = 5$

$$s_2 = \begin{vmatrix} 1 & -4 \\ 2 & 0 \end{vmatrix} + \begin{vmatrix} 1 & -1 \\ -1 & -2 \end{vmatrix} + \begin{vmatrix} 1 & -4 \\ -1 & 6 \end{vmatrix} + \begin{vmatrix} 0 & 5 \\ 1 & -2 \end{vmatrix} + \begin{vmatrix} 0 & -4 \\ 4 & 6 \end{vmatrix} + \begin{vmatrix} -2 & 3 \\ -1 & 6 \end{vmatrix}$$

$$= 8 - 3 + 2 - 5 + 16 - 9 = 9$$

$$s_3 = \begin{vmatrix} 1 & -4 & -1 \\ 2 & 0 & 5 \\ -1 & 1 & -2 \end{vmatrix} + \begin{vmatrix} 1 & -4 & -4 \\ 2 & 0 & -4 \\ -1 & 4 & 6 \end{vmatrix} + \begin{vmatrix} 1 & -1 & -4 \\ -1 & -2 & 3 \\ -1 & -1 & 6 \end{vmatrix} + \begin{vmatrix} 0 & 5 & -4 \\ 1 & -2 & 3 \\ 4 & -1 & 6 \end{vmatrix}$$

$$= -3 + 16 - 8 + 2 = 7$$

$$|A| = 2$$

Then $|\lambda I - A| = \lambda^4 - 5\lambda^3 + 9\lambda^2 - 7\lambda + 2.$

3. Let $\lambda_1, X_1;\ \lambda_2, X_2;\ \lambda_3, X_3$ be distinct characteristic roots and associated invariant vectors of A. Show that X_1, X_2, X_3 are linearly independent.

Assume the contrary, that is, assume that there exist scalars a_1, a_2, a_3, not all zero, such that

(i) $$a_1 X_1 + a_2 X_2 + a_3 X_3 = 0$$

Multiply **(i)** by A and recall that $AX_i = \lambda_i X_i$; we have

(ii) $$a_1 AX_1 + a_2 AX_2 + a_3 AX_3 = a_1\lambda_1 X_1 + a_2\lambda_2 X_2 + a_3\lambda_3 X_3 = 0$$

Multiply **(ii)** by A and obtain

(iii) $$a_1\lambda_1^2 X_1 + a_2\lambda_2^2 X_2 + a_3\lambda_3^2 X_3 = 0$$

Now **(i)**, **(ii)**, **(iii)** may be written as

(iv) $$\begin{bmatrix} 1 & 1 & 1 \\ \lambda_1 & \lambda_2 & \lambda_3 \\ \lambda_1^2 & \lambda_2^2 & \lambda_3^2 \end{bmatrix} \begin{bmatrix} a_1 X_1 \\ a_2 X_2 \\ a_3 X_3 \end{bmatrix} = 0$$

By Problem 5, Chapter 3, $B = \begin{vmatrix} 1 & 1 & 1 \\ \lambda_1 & \lambda_2 & \lambda_3 \\ \lambda_1^2 & \lambda_2^2 & \lambda_3^2 \end{vmatrix} \neq 0$; hence, B^{-1} exists. Multiplying **(iv)** by B^{-1}, we have $[a_1X_1, a_2X_2, a_3X_3]' = 0$. But this requires $a_1 = a_2 = a_3 = 0$, contrary to the hypothesis.

Thus, X_1, X_2, X_3 are linearly independent.

4. From $\phi(\lambda) = |\lambda I - A| = \begin{vmatrix} \lambda - a_{11} & -a_{12} & -a_{13} \\ -a_{21} & \lambda - a_{22} & -a_{23} \\ -a_{31} & -a_{32} & \lambda - a_{33} \end{vmatrix}$ we obtain

$$\phi'(\lambda) = \begin{vmatrix} 1 & 0 & 0 \\ -a_{21} & \lambda - a_{22} & -a_{23} \\ -a_{31} & -a_{32} & \lambda - a_{33} \end{vmatrix} + \begin{vmatrix} \lambda - a_{11} & -a_{12} & -a_{13} \\ 0 & 1 & 0 \\ -a_{31} & -a_{32} & \lambda - a_{33} \end{vmatrix} + \begin{vmatrix} \lambda - a_{11} & -a_{12} & -a_{13} \\ -a_{21} & \lambda - a_{22} & -a_{23} \\ 0 & 0 & 1 \end{vmatrix}$$

$$= \begin{vmatrix} \lambda - a_{22} & -a_{23} \\ -a_{32} & \lambda - a_{33} \end{vmatrix} + \begin{vmatrix} \lambda - a_{11} & -a_{13} \\ -a_{31} & \lambda - a_{33} \end{vmatrix} + \begin{vmatrix} \lambda - a_{11} & -a_{12} \\ -a_{21} & \lambda - a_{22} \end{vmatrix}$$

= the sum of the principal minors of $\lambda I - A$ of order two

$$\phi''(\lambda) = \begin{vmatrix} 1 & 0 \\ -a_{32} & \lambda - a_{33} \end{vmatrix} + \begin{vmatrix} \lambda - a_{22} & -a_{23} \\ 0 & 1 \end{vmatrix} + \begin{vmatrix} 1 & 0 \\ -a_{31} & \lambda - a_{33} \end{vmatrix} + \begin{vmatrix} \lambda - a_{11} & -a_{13} \\ 0 & 1 \end{vmatrix}$$
$$+ \begin{vmatrix} 1 & 0 \\ -a_{21} & \lambda - a_{22} \end{vmatrix} + \begin{vmatrix} \lambda - a_{11} & -a_{12} \\ 0 & 1 \end{vmatrix}$$

$$= 2\{(\lambda - a_{11}) + (\lambda - a_{22}) + (\lambda - a_{33})\}$$

= 2! times the sum of the principal minors of $\lambda I - A$ of order one

$$\phi'''(\lambda) = 3!$$

Also $\phi^{(iv)}(\lambda) = \phi^{(v)}(\lambda) = \dots = 0$.

5. Prove: If λ_i is an r-fold characteristic root of an n-square matrix A, the rank of $\lambda_i I - A$ is not less than $n - r$ and the dimension of the associated invariant vector space is not greater than r.

Since λ_i is an r-fold root of $\phi(\lambda) = 0$, $\phi(\lambda_i) = \phi'(\lambda_i) = \phi''(\lambda_i) = \dots = \phi^{(r-1)}(\lambda_i) = 0$ and $\phi^{(r)}(\lambda_i) \neq 0$. Now $\phi^{(r)}(\lambda_i)$ is $r!$ times the sum of the principal minors of order $n - r$ of $\lambda_i I - A$; hence, not every principal minor can vanish and $\lambda_i I - A$ is of rank at least $n - r$. By **(11.2)**, the associated invariant vector space of $\lambda_i I - A$, i.e., its null-space, is of dimension at most r.

6. For the matrix of Problem 2, find the characteristic roots and the associated invariant vector spaces.

The characteristic roots are 1, 1, 1, 2.

For $\lambda = 2$: $\lambda I - A = \begin{bmatrix} 1 & 4 & 1 & 4 \\ -2 & 2 & -5 & 4 \\ 1 & -1 & 4 & -3 \\ 1 & -4 & 1 & -4 \end{bmatrix} \sim \begin{bmatrix} 1 & 0 & 1 & 0 \\ 0 & 0 & 0 & 0 \\ 0 & 0 & 3 & -2 \\ 0 & 1 & 0 & 1 \end{bmatrix}$ is of rank 3; its null-space is of dimension 1. The associated invariant vector space is that spanned by $[2, 3, -2, -3]'$.

For $\lambda = 1$: $\lambda I - A = \begin{bmatrix} 0 & 4 & 1 & 4 \\ -2 & 1 & -5 & 4 \\ 1 & -1 & 3 & -3 \\ 1 & -4 & 1 & -5 \end{bmatrix} \sim \begin{bmatrix} 0 & 0 & 5 & -4 \\ 0 & -1 & 1 & -2 \\ 1 & 0 & 2 & -1 \\ 0 & 0 & 0 & 0 \end{bmatrix}$ is of rank 3; its null-space is of dimension 1. The associated invariant vector space is that spanned by $[3, 6, -4, -5]'$.

7. Prove: If α is a non-zero characteristic root of the non-singular n-square matrix A, then $|A|/\alpha$ is a characteristic root of adj A.

By Problem 1,

$$\text{(i)} \qquad \alpha^n + s_1\alpha^{n-1} + \ldots + s_{n-1}\alpha + (-1)^n|A| \;=\; 0$$

where s_i, $(i = 1, 2, \ldots, n-1)$ is $(-1)^i$ times the sum of all i-square principal minors of A, and

$$|\mu I - \text{adj } A| \;=\; \mu^n + S_1\mu^{n-1} + \ldots + S_{n-1}\mu + (-1)^n|\text{adj } A|$$

where S_j, $(j = 1, 2, \ldots, n-1)$ is $(-1)^j$ times the sum of the j-square principal minors of adj A.

By **(6.4)** and the definitions of s_i and S_j, $S_1 = (-1)^n s_{n-1}$, $S_2 = (-1)^n|A|s_{n-2}$, $\ldots$, $S_{n-1} = (-1)^n|A|^{n-2}s_1$, and $|\text{adj } A| = |A|^{n-1}$; then

$$|\mu I - \text{adj } A| \;=\; (-1)^n\{(-1)^n\mu^n + s_{n-1}\mu^{n-1} + s_{n-2}|A|\mu^{n-2} + \ldots + s_2|A|^{n-3}\mu^2 + s_1|A|^{n-2}\mu + |A|^{n-1}\}$$

and

$$|A|^{1-n}|\mu I - \text{adj } A| \;=\; (-1)^n\{1 + s_1(\frac{\mu}{|A|}) + \ldots + s_{n-1}(\frac{\mu}{|A|})^{n-1} + (-1)^n(\frac{\mu}{|A|})^n|A|\} \;=\; f(\mu)$$

Now

$$f(\frac{|A|}{\alpha}) \;=\; (-1)^n\{1 + s_1(\frac{1}{\alpha}) + \ldots + s_{n-1}(\frac{1}{\alpha})^{n-1} + (-1)^n(\frac{1}{\alpha})^n|A|\}$$

and by **(i)**

$$\alpha^n f(\frac{|A|}{\alpha}) \;=\; (-1)^n\{\alpha^n + s_1\alpha^{n-1} + \ldots + s_{n-1}\alpha + (-1)^n|A|\} \;=\; 0$$

Hence, $|A|/\alpha$ is a characteristic root of adj A.

8. Prove: The characteristic equation of an orthogonal matrix P is a reciprocal equation.

We have

$$\phi(\lambda) \;=\; |\lambda I - P| \;=\; |\lambda PIP' - P| \;=\; |-P\lambda(\frac{1}{\lambda}I - P')| \;=\; \pm\lambda^n|\frac{1}{\lambda}I - P| \;=\; \pm\lambda^n\phi(\frac{1}{\lambda})$$

SUPPLEMENTARY PROBLEMS

9. For the following matrices, determine the characteristic roots and a basis of each of the associated invariant vector spaces.

$(a)\ \begin{bmatrix} 1 & 0 & -1 \\ 1 & 2 & 1 \\ 2 & 2 & 3 \end{bmatrix}$ $\quad (c)\ \begin{bmatrix} 1 & 2 & 2 \\ 0 & 2 & 1 \\ -1 & 2 & 2 \end{bmatrix}$ $\quad (e)\ \begin{bmatrix} -2 & -8 & -12 \\ 1 & 4 & 4 \\ 0 & 0 & 1 \end{bmatrix}$ $\quad (g)\ \begin{bmatrix} 2 & 1 & 1 \\ 1 & 2 & 1 \\ 0 & 0 & 1 \end{bmatrix}$ $\quad (i)\ \begin{bmatrix} 1 & -1 & -1 \\ 1 & -1 & 0 \\ 1 & 0 & -1 \end{bmatrix}$

$(b)\ \begin{bmatrix} 1 & 1 & -2 \\ -1 & 2 & 1 \\ 0 & 1 & -1 \end{bmatrix}$ $\quad (d)\ \begin{bmatrix} 0 & 1 & 0 \\ 0 & 0 & 1 \\ 1 & -3 & 3 \end{bmatrix}$ $\quad (f)\ \begin{bmatrix} -3 & -9 & -12 \\ 1 & 3 & 4 \\ 0 & 0 & 1 \end{bmatrix}$ $\quad (h)\ \begin{bmatrix} 2 & 2 & 0 \\ 2 & 2 & 0 \\ 0 & 0 & 1 \end{bmatrix}$ $\quad (j)\ \begin{bmatrix} 2-i & 0 & i \\ 0 & 1+i & 0 \\ i & 0 & 2-i \end{bmatrix}$

$(k)\ \begin{bmatrix} 3 & 2 & 2 & -4 \\ 2 & 3 & 2 & -1 \\ 1 & 1 & 2 & -1 \\ 2 & 2 & 2 & -1 \end{bmatrix}$ $\quad (l)\ \begin{bmatrix} 5 & 6 & -10 & 7 \\ -5 & -4 & 9 & -6 \\ -3 & -2 & 6 & -4 \\ -3 & -3 & 7 & -5 \end{bmatrix}$ $\quad (m)\ \begin{bmatrix} -1 & -1 & -6 & 3 \\ 1 & -2 & -3 & 0 \\ -1 & 1 & 0 & 1 \\ -1 & -1 & -5 & 3 \end{bmatrix}$

Ans. (a) 1, [1, −1, 0]′; 2, [2, −1, −2]′; 3, [1, −1, −2]′
(b) −1, [1, 0, 1]′; 2, [1, 3, 1]′; 1, [3, 2, 1]′
(c) 1, [1, 1, −1]′; 2, [2, 1, 0]′;
(d) 1, [1, 1, 1]′

(e) 2, $[2,-1,0]'$; 0, $[4,-1,0]'$; 1, $[4,0,-1]'$
(f) 0, $[3,-1,0]'$; 1, $[12,-4,-1]'$
(g) 1, $[1,0,-1]'$, $[0,1,-1]'$; 3, $[1,1,0]'$
(h) 0, $[1,-1,0]'$; 1, $[0,0,1]'$; 4, $[1,1,0]'$
(i) -1, $[0,1,-1]'$; i, $[1+i,1,1]'$; $-i$, $[1-i,1,1]'$
(j) 2, $[1,0,1]'$; $1+i$, $[0,1,0]'$; $2-2i$, $[1,0,-1]'$
(k) 1, $[1,0,-1,0]'$, $[1,-1,0,0]'$; 2, $[-2,4,1,2]'$; 3, $[0,3,1,2]'$
(l) 1, $[1,2,3,2]'$; -1, $[-3,0,1,4]'$
(m) 0, $[2,1,0,1]'$; 1, $[3,0,1,4]'$; -1, $[3,0,1,2]'$

10. Prove: If X is a unit vector and if $AX = \lambda X$ then $X'AX = \lambda$.

11. Prove: The characteristic roots of a diagonal matrix are the elements of its diagonal and the associated invariant vectors are the elementary vectors E_i.

12. Prove Theorems **I** and **VI**.

13. Prove Theorem **VII**.
Hint. If $|\lambda I - A| = (\lambda-\lambda_1)(\lambda-\lambda_2)\dots(\lambda-\lambda_n)$ then $|(\lambda+k)I - A| = (\lambda+k-\lambda_1)(\lambda+k-\lambda_2)\dots(\lambda+k-\lambda_n)$.

14. Prove: The characteristic roots of the direct sum $\text{diag}(A_1, A_2, \dots, A_s)$ are the characteristic roots of $A_1, A_2, \dots, A_s$.

15. Prove: If A and $N = \begin{bmatrix} I_r & 0 \\ 0 & 0 \end{bmatrix}$ are n-square and $r < n$, show that NA and AN have the same characteristic equation.

16. Prove: If the n-square matrix A is of rank r, then at least $n-r$ of its characteristic roots are zero.

17. Prove: If A and B are n-square and A is non-singular, then $A^{-1}B$ and BA^{-1} have the same characteristic roots.

18. For A and B of Problem 17, show that B and $A^{-1}BA$ have the same characteristic roots.

19. Let A be an n-square matrix. Write $|\lambda I - A^{-1}| = |-\lambda A^{-1}(\frac{1}{\lambda}I - A)|$ and conclude that $1/\lambda_1, 1/\lambda_2, \dots, 1/\lambda$ are the characteristic roots of A^{-1}.

20. Prove: The characteristic roots of an orthogonal matrix P are of absolute value 1.
Hint. If λ_i, X_i are a characteristic root and associated invariant vector of P, then $X_i'X_i = (PX_i)'(PX_i) = \bar{\lambda}_i\lambda_i X_i'X_i$.

21. Prove: If $\lambda_i \neq \pm 1$ is a characteristic root and X_i is the associated invariant vector of an orthogonal matrix P, then $X_i'X_i = 0$.

22. Prove: The characteristic roots of a unitary matrix are of absolute value 1.

23. Obtain, using Theorem **II**,

$$\phi(0) = (-1)^n|A|$$
$$\phi'(0) = (-1)^{n-1} \text{ times the sum of the principal minors of order } n-1 \text{ of } A$$
$$\dots\dots\dots\dots\dots\dots\dots\dots$$
$$\phi^{(r)}(0) = (-1)^{n-r}r! \text{ times the sum of the principal minors of order } n-r \text{ of } A$$
$$\dots\dots\dots\dots\dots\dots\dots\dots$$
$$\phi^{(n)}(0) = n!$$

24. Substitute from Problem 23 into

$$\phi(\lambda) = \phi(0) + \phi'(0)\cdot\lambda + \frac{1}{2!}\phi''(0)\cdot\lambda^2 + \dots + \frac{1}{n!}\phi^{(n)}(0)\cdot\lambda^n$$

to obtain **(19.4)**.

Chapter 20

Similarity

TWO n**-SQUARE MATRICES** A and B over F are called **similar** over F if there exists a non-singular matrix R over F such that

(20.1) $$B = R^{-1}AR$$

Example 1. The matrices $A = \begin{bmatrix} 2 & 2 & 1 \\ 1 & 3 & 1 \\ 1 & 2 & 2 \end{bmatrix}$ of Example 1, Chapter 19, and

$$B = R^{-1}AR = \begin{bmatrix} 7 & -3 & -3 \\ -1 & 1 & 0 \\ -1 & 0 & 1 \end{bmatrix} \begin{bmatrix} 2 & 2 & 1 \\ 1 & 3 & 1 \\ 1 & 2 & 2 \end{bmatrix} \begin{bmatrix} 1 & 3 & 3 \\ 1 & 4 & 3 \\ 1 & 3 & 4 \end{bmatrix} = \begin{bmatrix} 5 & 14 & 13 \\ 0 & 1 & 0 \\ 0 & 0 & 1 \end{bmatrix}$$

are similar.

The characteristic equation $(\lambda-5)(\lambda-1)^2 = 0$ of B is also the characteristic equation of A.

An invariant vector of B associated with $\lambda = 5$ is $Y_1 = [1, 0, 0]'$ and it is readily shown that $X_1 = RY_1 = [1, 1, 1]'$ is an invariant vector of A associated with the same characteristic root $\lambda = 5$. The reader will show that $Y_2 = [7, -2, 0]'$ and $Y_3 = [17, -3, -2]'$ are a pair of linearly independent invariant vectors of B associated with $\lambda = 1$ while $X_2 = RY_2$ and $X_3 = RY_3$ are a pair of linearly independent invariant vectors of A associated with the same root $\lambda = 1$.

Example 1 illustrates the following theorems:

I. Two similar matrices have the same characteristic roots.

For a proof, see Problem 1.

II. If Y is an invariant vector of $B = R^{-1}AR$ corresponding to the characteristic root λ_i of B, then $X = RY$ is an invariant vector of A corresponding to the same characteristic root λ_i of A.

For a proof, see Problem 2.

DIAGONAL MATRICES. The characteristic roots of a diagonal matrix $D = \text{diag}(a_1, a_2, \ldots, a_n)$ are simply the diagonal elements.

A diagonal matrix always has n linearly independent invariant vectors. The elementary vectors E_i are such a set since $DE_i = a_i E_i$, $(i = 1, 2, \ldots, n)$.

As a consequence, we have (see Problems 3 and 4 for proofs)

III. Any n-square matrix A, similar to a diagonal matrix, has n linearly independent invariant vectors.

IV. If an n-square matrix A has n linearly independent invariant vectors, it is similar to a diagonal matrix.

See Problem 5.

In Problem 6, we prove

V. Over a field F an n-square matrix A is similar to a diagonal matrix if and only if $\lambda I - A$ factors completely in F and the multiplicity of each λ_i is equal to the dimension of the null-space of $\lambda_i I - A$.

Not every n-square matrix is similar to a diagonal matrix. The matrix of Problem 6, Chapter 19, is an example. There, corresponding to the triple root $\lambda = 1$, the null-space of $\lambda I - A$ is of dimension 1.

We can prove, however,

VI. Every n-square matrix A is similar to a triangular matrix whose diagonal elements are the characteristic roots of A.

See Problems 7-8.

As special cases, we have

VII. If A is any real n-square matrix with real characteristic roots, there exists an orthogonal matrix P such that $P^{-1}AP = P'AP$ is triangular and has as diagonal elements the characteristic roots of A.

See Problems 9-10.

VIII. If A is any n-square matrix with complex elements or a real n-square matrix with complex characteristic roots, there exists a unitary matrix U such that $U^{-1}AU = \bar{U}'AU$ is triangular and has as diagonal elements the characteristic roots of A.

See Problem 11.

The matrices A and $P^{-1}AP$ of Theorem **VII** are called **orthogonally similar**.

The matrices A and $U^{-1}AU$ of Theorem **VIII** are called **unitarily similar**.

DIAGONABLE MATRICES. A matrix A which is similar to a diagonal matrix is called **diagonable**. Theorem **IV** is basic to the study of certain types of diagonable matrices in the next chapter.

SOLVED PROBLEMS

1. Prove: Two similar matrices have the same characteristic roots.

Let A and $B = R^{-1}AR$ be the similar matrices; then

$$(i) \qquad \lambda I - B \;=\; \lambda I - R^{-1}AR \;=\; R^{-1}\lambda IR - R^{-1}AR \;=\; R^{-1}(\lambda I - A)R$$

and

$$|\lambda I - B| \;=\; |R^{-1}|\cdot|\lambda I - A|\cdot|R| \;=\; |\lambda I - A|$$

Thus, A and B have the same characteristic equation and the same characteristic roots.

2. Prove: If Y is an invariant vector of $B = R^{-1}AR$ corresponding to the characteristic root λ_i, then $X = RY$ is an invariant vector of A corresponding to the same characteristic root λ_i of A.

By hypothesis, $BY = \lambda_i Y$ and $RB = AR$; then

$$AX \;=\; ARY \;=\; RBY \;=\; R\lambda_i Y \;=\; \lambda_i RY \;=\; \lambda_i X$$

and X is an invariant vector of A corresponding to the characteristic root λ_i.

3. Prove: Any matrix A which is similar to a diagonal matrix has n linearly independent invariant vectors.

Let $R^{-1}AR = \text{diag}(b_1, b_2, \ldots, b_n) = B$. Now the elementary vectors $E_1, E_2, \ldots, E_n$ are invariant vectors of B. Then, by Theorem **II**, the vectors $X_j = RE_j$ are invariant vectors of A. Since R is non-singular, its column vectors are linearly independent.

4. Prove: If an n-square matrix A has n linearly independent invariant vectors, it is similar to a diagonal matrix.

Let the n linearly independent invariant vectors $X_1, X_2, \ldots, X_n$ be associated with the respective characteristic roots $\lambda_1, \lambda_2, \ldots, \lambda_n$ so that $AX_i = \lambda_i X_i$, $(i = 1, 2, \ldots, n)$. Let $R = [X_1, X_2, \ldots, X_n]$; then

$$AR = [AX_1, AX_2, \ldots, AX_n] = [\lambda_1 X_1, \lambda_2 X_2, \ldots, \lambda_n X_n]$$

$$= [X_1, X_2, \ldots, X_n]\begin{bmatrix} \lambda_1 & 0 & 0 & \ldots & 0 \\ 0 & \lambda_2 & 0 & \ldots & 0 \\ \ldots & \ldots & \ldots & \ldots & \ldots \\ 0 & 0 & 0 & \ldots & \lambda_n \end{bmatrix} = R\,\text{diag}(\lambda_1, \lambda_2, \ldots, \lambda_n)$$

Hence, $R^{-1}AR = \text{diag}(\lambda_1, \lambda_2, \ldots, \lambda_n)$.

5. A set of linearly independent invariant vectors of the matrix A of Example 1, Chapter 19, is

$$X_1 = [1, 1, 1]', \quad X_2 = [2, -1, 0]', \quad X_3 = [1, 0, -1]'$$

Take $R = [X_1, X_2, X_3] = \begin{bmatrix} 1 & 2 & 1 \\ 1 & -1 & 0 \\ 1 & 0 & -1 \end{bmatrix}$; then $R^{-1} = \frac{1}{4}\begin{bmatrix} 1 & 2 & 1 \\ 1 & -2 & 1 \\ 1 & 2 & -3 \end{bmatrix}$ and

$$R^{-1}AR = \frac{1}{4}\begin{bmatrix} 1 & 2 & 1 \\ 1 & -2 & 1 \\ 1 & 2 & -3 \end{bmatrix}\begin{bmatrix} 2 & 2 & 1 \\ 1 & 3 & 1 \\ 1 & 2 & 2 \end{bmatrix}\begin{bmatrix} 1 & 2 & 1 \\ 1 & -1 & 0 \\ 1 & 0 & -1 \end{bmatrix} = \begin{bmatrix} 5 & 0 & 0 \\ 0 & 1 & 0 \\ 0 & 0 & 1 \end{bmatrix}.$$

a diagonal matrix.

6. Prove: Over a field F an n-square matrix A is similar to a diagonal matrix if and only if $\lambda I - A$ factors completely in F and the multiplicity of each λ_i is equal to the dimension of the null-space of $\lambda_i I - A$.

First, suppose that $R^{-1}AR = \text{diag}(\lambda_1, \lambda_2, \ldots, \lambda_n) = B$ and that exactly k of these characteristic roots are equal to λ_i. Then $\lambda_i I - B$ has exactly k zeros in its diagonal and, hence, is of rank $n-k$; its null-space is then of dimension $n-(n-k) = k$. But $\lambda_i I - A = R(\lambda_i I - B)R^{-1}$; thus, $\lambda_i I - A$ has the same rank $n-k$ and nullity k as has $\lambda_i I - B$.

Conversely, let $\lambda_1, \lambda_2, \ldots, \lambda_s$ be the distinct characteristic roots of A with respective multiplicities $r_1, r_2, \ldots, r_s$, where $r_1 + r_2 + \ldots + r_s = n$. Denote by $V_{r_1}, V_{r_2}, \ldots, V_{r_s}$ the associated invariant vector spaces. Take $X_{i1}, X_{i2}, \ldots, X_{ir_i}$ as a basis of the invariant vector space V_{r_i}, $(i = 1, 2, \ldots, s)$. Suppose that there exist scalars a_{ij}, not all zero, such that

(i) $$(a_{11}X_{11} + a_{12}X_{12} + \ldots + a_{1r_1}X_{1r_1}) + (a_{21}X_{21} + a_{22}X_{22} + \ldots + a_{2r_2}X_{2r_2})$$
$$+ \ldots + (a_{s1}X_{s1} + a_{s2}X_{s2} + \ldots + a_{sr_s}X_{sr_s}) = 0$$

Now each vector $Y_i = a_{i1}X_{i1} + a_{i2}X_{i2} + \ldots + a_{ir_i}X_{ir_i}) = 0$, $(i = 1, 2, \ldots, s)$, for otherwise, it is an invariant vector and by Theorem **I** their totality is linearly independent. But this contradicts (**i**); thus, the X's constitute a basis of V_n and A is similar to a diagonal matrix by Theorem **IV**.

7. Prove: Every n-square matrix A is similar to a triangular matrix whose diagonal elements are the characteristic roots of A.

Let the characteristic roots of A be $\lambda_1, \lambda_2, \ldots, \lambda_n$ and let X_1 be an invariant vector of A corresponding to the characteristic root λ_1. Take X_1 as the first column of a non-singular matrix Q_1 whose remaining columns may be any whatever such that $|Q_1| \neq 0$. The first column of AQ_1 is $AX_1 = \lambda_1 X_1$ and the first column of $Q_1^{-1}AQ_1$ is $Q_1^{-1}\lambda_1 X_1$. But this, being the first column of $Q_1^{-1}\lambda_1 Q_1$, is $[\lambda_1, 0, \ldots, 0]'$. Thus,

(i) $$Q_1^{-1}AQ_1 = \begin{bmatrix} \lambda_1 & B_1 \\ 0 & A_1 \end{bmatrix}$$

where A_1 is of order $n-1$.

Since $|\lambda I - Q_1^{-1}AQ_1| = (\lambda-\lambda_1)|\lambda I - A_1|$, and $Q_1^{-1}AQ_1$ and A have the same characteristic roots, it follows that the characteristic roots of A_1 are $\lambda_2, \lambda_3, \ldots, \lambda_n$. If $n = 2$, $A_1 = [\lambda_2]$ and the theorem is proved with $Q = Q_1$.

Otherwise, let X_2 be an invariant vector of A_1 corresponding to the characteristic root λ_2. Take X_2 as the first column of a non-singular matrix Q_2 whose remaining columns may be any whatever such that $|Q_2| \neq 0$. Then

(ii) $$Q_2^{-1}A_1Q_2 = \begin{bmatrix} \lambda_2 & B_2 \\ 0 & A_2 \end{bmatrix}$$

where A_2 is of order $n-2$. If $n = 3$, $A_2 = [\lambda_3]$, and the theorem is proved with $Q = Q_1 \cdot \begin{bmatrix} I_1 & 0 \\ 0 & Q_2 \end{bmatrix}$.

Otherwise, we repeat the procedure and, after $n-1$ steps at most, obtain

(iii) $$Q = Q_1 \cdot \begin{bmatrix} I_1 & 0 \\ 0 & Q_2 \end{bmatrix}\begin{bmatrix} I_2 & 0 \\ 0 & Q_3 \end{bmatrix} \cdots \begin{bmatrix} I_{n-2} & 0 \\ 0 & Q_{n-1} \end{bmatrix}$$

such that $Q^{-1}AQ$ is triangular and has as diagonal elements the characteristic roots of A.

8. Find a non-singular matrix Q such that $Q^{-1}AQ$ is triangular, given

$$A = \begin{bmatrix} 9 & -1 & 8 & -9 \\ 6 & -1 & 5 & -5 \\ -5 & 1 & -4 & 5 \\ 4 & 0 & 5 & -4 \end{bmatrix}$$

Here $|\lambda I - A| = (\lambda^2-1)(\lambda^2-4)$ and the characteristic roots are $1, -1, 2, -2$. Take $[5, 5, -1, 3]'$, an invariant vector corresponding to the characteristic root 1 as the first column of a non-singular matrix Q_1 whose remaining columns are elementary vectors, say

$$Q_1 = \begin{bmatrix} 5 & 0 & 0 & 0 \\ 5 & 1 & 0 & 0 \\ -1 & 0 & 1 & 0 \\ 3 & 0 & 0 & 1 \end{bmatrix}$$

Then

$$Q_1^{-1} = \frac{1}{5}\begin{bmatrix} 1 & 0 & 0 & 0 \\ -5 & 5 & 0 & 0 \\ 1 & 0 & 5 & 0 \\ -3 & 0 & 0 & 5 \end{bmatrix} \quad \text{and} \quad Q_1^{-1}AQ_1 = \frac{1}{5}\begin{bmatrix} 5 & -1 & 8 & -9 \\ 0 & 0 & -15 & 20 \\ 0 & 4 & -12 & 16 \\ 0 & 3 & 1 & 7 \end{bmatrix} = \begin{bmatrix} 1 & B_1 \\ 0 & A_1 \end{bmatrix}$$

A characteristic root of A_1 is -1 and an associated invariant vector is $[4, 0, -1]'$. Take $Q_2 = \begin{bmatrix} 4 & 0 & 0 \\ 0 & 1 & 0 \\ -1 & 0 & 1 \end{bmatrix}$; then

$$Q_2^{-1} = \frac{1}{4}\begin{bmatrix} 1 & 0 & 0 \\ 0 & 4 & 0 \\ 1 & 0 & 4 \end{bmatrix} \quad \text{and} \quad Q_2^{-1}A_1Q_2 = \frac{1}{20}\begin{bmatrix} -20 & -15 & 20 \\ 0 & -48 & 64 \\ 0 & -11 & 48 \end{bmatrix} = \begin{bmatrix} -1 & B_2 \\ 0 & A_2 \end{bmatrix}$$

A characteristic root of A_2 is 2 and an associated invariant vector is $[8, 11]'$. Take $Q_3 = \begin{bmatrix} 8 & 0 \\ 11 & 1 \end{bmatrix}$; then

$$Q_3^{-1} = \frac{1}{8}\begin{bmatrix} 1 & 0 \\ -11 & 8 \end{bmatrix} \quad \text{and} \quad Q_3^{-1}A_2Q_3 = \begin{bmatrix} 2 & 2/5 \\ 0 & -2 \end{bmatrix}$$

Now

$$Q = Q_1 \cdot \begin{bmatrix} I_1 & 0 \\ 0 & Q_2 \end{bmatrix} \cdot \begin{bmatrix} I_2 & 0 \\ 0 & Q_3 \end{bmatrix} = \begin{bmatrix} 5 & 0 & 0 & 0 \\ 5 & 4 & 0 & 0 \\ -1 & 0 & 8 & 0 \\ 3 & -1 & 11 & 1 \end{bmatrix}, \qquad Q^{-1} = \frac{1}{160}\begin{bmatrix} 32 & 0 & 0 & 0 \\ -40 & 40 & 0 & 0 \\ 4 & 0 & 20 & 0 \\ -180 & 40 & -220 & 160 \end{bmatrix}.$$

and

$$Q^{-1}AQ = \begin{bmatrix} 1 & 1 & -7 & -9/5 \\ 0 & -1 & 5 & 1 \\ 0 & 0 & 2 & 2/5 \\ 0 & 0 & 0 & -2 \end{bmatrix}$$

9. If A is any real n-square matrix with real characteristic roots then there exists an orthogonal matrix P such that $P^{-1}AP$ is triangular and has as diagonal elements the characteristic roots of A.

Let $\lambda_1, \lambda_2, \ldots, \lambda_n$ be the characteristic roots of A. Since the roots are real the associated invariant vectors will also be real. As in Problem 7, let Q_1 be formed having an invariant vector corresponding to λ_1 as first column. Using the Gram-Schmidt process, obtain from Q_1 an orthogonal matrix P_1 whose first column is proportional to that of Q_1. Then

$$P_1^{-1}AP_1 = \begin{bmatrix} \lambda_1 & B_1 \\ 0 & A_1 \end{bmatrix}$$

where A_1 is of order $n-1$ and has $\lambda_2, \lambda_3, \ldots, \lambda_n$ as characteristic roots.

Next, form Q_2 having as first column an invariant vector of A_1 corresponding to the root λ_2 and, using the Gram-Schmidt process, obtain an orthogonal matrix P_2. Then

$$P_2^{-1}A_1P_2 = \begin{bmatrix} \lambda_2 & B_2 \\ 0 & A_2 \end{bmatrix}$$

After sufficient repetitions, build the orthogonal matrix

$$P = P_1 \cdot \begin{bmatrix} I_1 & 0 \\ 0 & P_2 \end{bmatrix} \cdots \begin{bmatrix} I_{n-2} & 0 \\ 0 & P_{n-1} \end{bmatrix}$$

for which $P^{-1}AP$ is triangular with the characteristic roots of A as diagonal elements.

10. Find an orthogonal matrix P such that

$$P^{-1}AP = P^{-1}\begin{bmatrix} 2 & 2 & 1 \\ 1 & 3 & 1 \\ 1 & 2 & 2 \end{bmatrix}P$$

is triangular and has the characteristic roots of A as diagonal elements.

From Example 1, Chapter 19, the characteristic roots are 5, 1, 1 and an invariant vector corresponding to $\lambda = 1$ is $[1, 0, -1]'$.

We take $Q_1 = \begin{bmatrix} 1 & 0 & 0 \\ 0 & 1 & 0 \\ -1 & 0 & 1 \end{bmatrix}$ and, using the Gram-Schmidt process, obtain

$$P_1 = \begin{bmatrix} 1/\sqrt{2} & 0 & 1/\sqrt{2} \\ 0 & 1 & 0 \\ -1/\sqrt{2} & 0 & 1/\sqrt{2} \end{bmatrix}$$

an orthogonal matrix whose first column is proportional to $[1, 0, -1]'$.

We find

$$P_1^{-1}AP_1 = \begin{bmatrix} 1/\sqrt{2} & 0 & -1/\sqrt{2} \\ 0 & 1 & 0 \\ 1/\sqrt{2} & 0 & 1/\sqrt{2} \end{bmatrix}\begin{bmatrix} 2 & 2 & 1 \\ 1 & 3 & 1 \\ 1 & 2 & 2 \end{bmatrix}\begin{bmatrix} 1/\sqrt{2} & 0 & 1/\sqrt{2} \\ 0 & 1 & 0 \\ -1/\sqrt{2} & 0 & 1/\sqrt{2} \end{bmatrix} = \begin{bmatrix} 1 & 0 & 0 \\ 0 & 3 & \sqrt{2} \\ 0 & 2\sqrt{2} & 3 \end{bmatrix} = \begin{bmatrix} 1 & B_1 \\ 0 & A_1 \end{bmatrix}$$

Now A_1 has $\lambda = 1$ a characteristic root and $[1, -\sqrt{2}]'$ as associated invariant vector. From $Q_2 = \begin{bmatrix} 1 & 0 \\ -\sqrt{2} & 1 \end{bmatrix}$, we obtain by the Gram-Schmidt process the orthogonal matrix $P_2 = \begin{bmatrix} 1/\sqrt{3} & 2/\sqrt{6} \\ -2/\sqrt{6} & 1/\sqrt{3} \end{bmatrix}$. Then

$$P = P_1 \cdot \begin{bmatrix} I_1 & 0 \\ 0 & P_2 \end{bmatrix} = \begin{bmatrix} 1/\sqrt{2} & -1/\sqrt{3} & 1/\sqrt{6} \\ 0 & 1/\sqrt{3} & 2/\sqrt{6} \\ -1/\sqrt{2} & -1/\sqrt{3} & 1/\sqrt{6} \end{bmatrix}$$

is orthogonal and $P^{-1}AP = \begin{bmatrix} 1 & 0 & 0 \\ 0 & 1 & -\sqrt{2} \\ 0 & 0 & 5 \end{bmatrix}$.

11. Find a unitary matrix U such that $U^{-1}AU$ is triangular and has as diagonal elements the characteristic roots of A, given

$$A = \begin{bmatrix} 5+5i & -1+i & -6-4i \\ -4-6i & 2-2i & 6+4i \\ 2+3i & -1+i & -3-2i \end{bmatrix}$$

The characteristic equation of A is $\lambda(\lambda^2 + (-4-i)\lambda + 5 - i) = 0$ and the characteristic roots are $0, 1-i, 3+2i$. For $\lambda = 0$, take $[1, -1, 1]'$ as associated invariant vector and form $Q_1 = \begin{bmatrix} 1 & 0 & 0 \\ -1 & 1 & 0 \\ 1 & 0 & 1 \end{bmatrix}$.

The Gram-Schmidt process produces the unitary matrix

$$U_1 = \begin{bmatrix} 1/\sqrt{3} & 1/\sqrt{6} & -1/\sqrt{2} \\ -1/\sqrt{3} & 2/\sqrt{6} & 0 \\ 1/\sqrt{3} & 1/\sqrt{6} & 1/\sqrt{2} \end{bmatrix}$$

Now

$$U_1^{-1}AU_1 = \begin{bmatrix} 0 & -2\sqrt{2}(1-i) & -(26+24i)/\sqrt{6} \\ 0 & 1-i & (2+3i)/\sqrt{3} \\ 0 & 0 & 3+2i \end{bmatrix}$$

so that, for this choice of Q_1, the required matrix $U = U_1$.

12. Find an orthogonal matrix P such that $P^{-1}AP$ is triangular and has as diagonal elements the characteristic roots of A, given

$$A = \begin{bmatrix} 3 & -1 & 1 \\ -1 & 5 & -1 \\ 1 & -1 & 3 \end{bmatrix}$$

The characteristic roots are $2, 3, 6$ and the associated invariant vectors may be taken as $[1, 0, -1]'$, $[1, 1, 1]'$, $[1, -2, 1]'$ respectively. Now these three vectors are both linearly independent and mutually orthogonal. Taking

$$P = \begin{bmatrix} 1/\sqrt{2} & 1/\sqrt{3} & 1/\sqrt{6} \\ 0 & 1/\sqrt{3} & -2/\sqrt{6} \\ -1/\sqrt{2} & 1/\sqrt{3} & 1/\sqrt{6} \end{bmatrix}$$

we find $P^{-1}AP = \text{diag}(2, 3, 6)$. This suggests the more thorough study of the real symmetric matrix made in the next chapter.

SUPPLEMENTARY PROBLEMS

13. Find an orthogonal matrix P such that $P^{-1}AP$ is triangular and has as diagonal elements the characteristic roots of A for each of the matrices A of Problem 9(a), (b), (c), (d), Chapter 19.

Ans. (a) $\begin{bmatrix} 1/\sqrt{2} & 1/3\sqrt{2} & 2/3 \\ -1/\sqrt{2} & 1/3\sqrt{2} & 2/3 \\ 0 & -4/3\sqrt{2} & 1/3 \end{bmatrix}$, (c) $\begin{bmatrix} 1/\sqrt{3} & 1/\sqrt{2} & -1/\sqrt{6} \\ 1/\sqrt{3} & 0 & 2/\sqrt{6} \\ -1/\sqrt{3} & 1/\sqrt{2} & 1/\sqrt{6} \end{bmatrix}$

(b) $\begin{bmatrix} 1/\sqrt{2} & 0 & -1/\sqrt{2} \\ 0 & 1 & 0 \\ 1/\sqrt{2} & 0 & 1/\sqrt{2} \end{bmatrix}$, (d) $\begin{bmatrix} 1/\sqrt{3} & -1/\sqrt{2} & -1/\sqrt{6} \\ 1/\sqrt{3} & 0 & 2/\sqrt{6} \\ 1/\sqrt{3} & 1/\sqrt{2} & -1/\sqrt{6} \end{bmatrix}$

14. Explain why the matrices (a) and (b) of Problem 13 are similar to a diagonal matrix while (c) and (d) are not. Examine the matrices (a) – (m) of Problem 9, Chapter 19 and determine those which are similar to a diagonal matrix having the characteristic roots as diagonal elements.

15. For each of the matrices A of Problem 9(i), (j), Chapter 19, find a unitary matrix U such that $U^{-1}AU$ is triangular and has as diagonal elements the characteristic roots of A.

Ans. (i) $\begin{bmatrix} 0 & 1/\sqrt{2} & -(1+i)/2 \\ 1/\sqrt{2} & (1-i)/2\sqrt{2} & \frac{1}{2} \\ -1/\sqrt{2} & (1-i)/2\sqrt{2} & \frac{1}{2} \end{bmatrix}$, (j) $\begin{bmatrix} 1/\sqrt{2} & 0 & -1/\sqrt{2} \\ 0 & 1 & 0 \\ 1/\sqrt{2} & 0 & 1/\sqrt{2} \end{bmatrix}$

16. Prove: If A is real and symmetric and P is orthogonal, then $P^{-1}AP$ is real and symmetric.

17. Make the necessary modification of Problem 9 to prove Theorem **VIII**.

18. Let B_i and C_i be similar matrices for $(i = 1, 2, \ldots, m)$. Show that

$$B = \text{diag}(B_1, B_2, \ldots, B_m) \quad \text{and} \quad C = \text{diag}(C_1, C_2, \ldots, C_m)$$

are similar. Hint. Suppose $C_i = R_i^{-1} B_i R_i$ and form $R = \text{diag}(R_1, R_2, \ldots, R_m)$.

19. Let $B = \text{diag}(B_1, B_2)$ and $C = \text{diag}(B_2, B_1)$. Write $I = \text{diag}(I_1, I_2)$, where the orders of I_1 and I_2 are those of B_1 and B_2 respectively, and define $R = \begin{bmatrix} 0 & I_1 \\ I_2 & 0 \end{bmatrix}$. Show that $R^{-1}BR = C$ to prove B and C are similar.

20. Extend the result of Problem 19 to $B = \text{diag}(B_1, B_2, \ldots, B_m)$ and C any matrix obtained by rearranging the B_i along the diagonal.

21. If A and B are n-square, then AB and BA have the same characteristic roots.
Hint. Let $PAQ = N$; then $PABP^{-1} = NQ^{-1}BP^{-1}$ and $Q^{-1}BAQ = Q^{-1}BP^{-1}N$. See Problem 15, Chapter 19

22. If $A_1, A_2, \ldots, A_s$ are non-singular and of the same order, show that $A_1A_2 \ldots A_s$, $A_2A_3 \ldots A_sA_1$, $A_3 \ldots A_sA_1A_2$, ... have the same characteristic equation.

23. Let $Q^{-1}AQ = B$ where B is triangular and has as diagonal elements the characteristic roots $\lambda_1, \lambda_2, \ldots, \lambda_n$ of A.

(a) Show that $Q^{-1}A^kQ$ is triangular and has as diagonal elements the kth powers of the characteristic roots of A.

(b) Show that $\sum_{i=1}^{n} \lambda_i^k = \text{trace } A^k$.

24. Show that similarity is an equivalence relation.

25. Show that $\begin{bmatrix} 2 & 2 & 1 \\ 1 & 3 & 1 \\ 1 & 2 & 2 \end{bmatrix}$ and $\begin{bmatrix} 2 & 1 & -1 \\ 0 & 2 & -1 \\ -3 & -2 & 3 \end{bmatrix}$ have the same characteristic roots but are not similar.

Chapter 21

Similarity to a Diagonal Matrix

REAL SYMMETRIC MATRICES. The study of real symmetric matrices and Hermitian matrices may be combined but we shall treat them separately here. For real symmetric matrices, we have:

I. The characteristic roots of a real symmetric matrix are all real.

See Problem 1.

II. The invariant vectors associated with distinct characteristic roots of a real symmetric matrix are mutually orthogonal.

See Problem 2.

When A is real and symmetric, each B_i of Problem 9, Chapter 20, is 0; hence,

III. If A is a real n-square symmetric matrix with characteristic roots $\lambda_1, \lambda_2, \ldots, \lambda_n$, then there exists a real orthogonal matrix P such that $P'AP = P^{-1}AP = \text{diag}(\lambda_1, \lambda_2, \ldots, \lambda_n)$.

Theorem **III** implies

IV. If λ_i is a characteristic root of multiplicity r_i of a real symmetric matrix, then there is associated with λ_i an invariant space of dimension r_i.

In terms of a real quadratic form, Theorem **III** becomes

V. Every real quadratic form $q = X'AX$ can be reduced by an orthogonal transformation $X = BY$ to a canonical form

$$\lambda_1 y_1^2 + \lambda_2 y_2^2 + \ldots + \lambda_r y_r^2 \tag{21.1}$$

where r is the rank of A and $\lambda_1, \lambda_2, \ldots, \lambda_r$ are its non-zero characteristic roots.

Thus, the rank of q is the number of non-zero characteristic roots of A while the index is the number of positive characteristic roots or, by Descartes Rule of signs, the number of variations of sign in $|\lambda I - A| = 0$.

VI. A real symmetric matrix is positive definite if and only if all of its characteristic roots are positive.

ORTHOGONAL SIMILARITY. If P is an orthogonal matrix and $B = P^{-1}AP$, then B is said to be **orthogonally similar** to A. Since $P^{-1} = P'$, B is also orthogonally congruent and orthogonally equivalent to A. Theorem **III** may be restated as

VII. Every real symmetric matrix A is orthogonally similar to a diagonal matrix whose diagonal elements are the characteristic roots of A.

See Problem 3.

Let the characteristic roots of the real symmetric matrix A be arranged so that $\lambda_1 \geqq \lambda_2 \geqq \ldots \geqq \lambda_n$. Then $\text{diag}(\lambda_1, \lambda_2, \ldots, \lambda_n)$ is a unique diagonal matrix similar to A. The totality of such diagonal matrices constitutes a canonical set for real symmetric matrices under orthogonal similarity. We have

VIII. Two real symmetric matrices are orthogonally similar if and only if they have the same characteristic roots, that is, if and only if they are similar.

PAIRS OF REAL QUADRATIC FORMS. In Problem 4, we prove

IX. If $X'AX$ and $X'BX$ are real quadratic forms in $(x_1, x_2, \ldots, x_n)$ and if $X'BX$ is positive definite, there exists a real non-singular linear transformation $X = CY$ which carries $X'AX$ into

$$\lambda_1 y_1^2 + \lambda_2 y_2^2 + \ldots + \lambda_n y_n^2$$

and $X'BX$ into

$$y_1^2 + y_2^2 + \ldots + y_n^2$$

where λ_i are the roots of $|\lambda B - A| = 0$.

See also Problems 4-5.

HERMITIAN MATRICES. Paralleling the theorems for real symmetric matrices, we have

X. The characteristic roots of an Hermitian matrix are real.

See Problem 7.

XI. The invariant vectors associated with distinct characteristic roots of an Hermitian matrix are mutually orthogonal.

XII. If H is an n-square Hermitian matrix with characteristic roots $\lambda_1, \lambda_2, \ldots, \lambda_n$, there exists a unitary matrix U such that $\bar{U}'HU = U^{-1}HU = \text{diag}(\lambda_1, \lambda_2, \ldots, \lambda_n)$. The matrix H is called **unitarily similar** to $U^{-1}HU$.

XIII. If λ_i is a characteristic root of multiplicity r_i of the Hermitian matrix H, then there is associated with λ_i an invariant space of dimension r_i.

Let the characteristic roots of the Hermitian matrix H be arranged so that $\lambda_1 \leqq \lambda_2 \leqq \ldots \leqq \lambda_n$. Then $\text{diag}(\lambda_1, \lambda_2, \ldots, \lambda_n)$ is a unique diagonal matrix similar to H. The totality of such diagonal matrices constitutes a canonical set for Hermitian matrices under unitary similarity. There follows

XIV. Two Hermitian matrices are unitarily similar if and only if they have the same characteristic roots, that is, if and only if they are similar.

NORMAL MATRICES. An n-square matrix A is called normal if $A\bar{A}' = \bar{A}'A$. Normal matrices include diagonal, real symmetric, real skew-symmetric, orthogonal, Hermitian, skew-Hermitian, and unitary matrices.

Let A be a normal matrix and U be a unitary matrix, and write $B = \bar{U}'AU$. Then $\bar{B}' = \bar{U}'\bar{A}'U$ and $\bar{B}'B = \bar{U}'\bar{A}'U \cdot \bar{U}'AU = \bar{U}'\bar{A}'AU = \bar{U}'A\bar{A}'U = \bar{U}'AU \cdot \bar{U}'\bar{A}'U = B\bar{B}'$. Thus,

XV. If A is a normal matrix and U is a unitary matrix, then $B = \bar{U}'AU$ is a normal matrix.

In Problem 8, we prove

XVI. If X_i is an invariant vector corresponding to the characteristic root λ_i of a normal matrix A, then X_i is also an invariant vector of $\bar{A}'$ corresponding to the characteristic root $\bar{\lambda}_i$.

In Problem 9, we prove

XVII. A square matrix A is unitarily similar to a diagonal matrix if and only if A is normal.

As a consequence, we have

XVIII. If A is normal, the invariant vectors corresponding to distinct characteristic roots are orthogonal.

See Problem 10.

XIX. If λ_i is a characteristic root of multiplicity r_i of a normal matrix A, the associated invariant vector space has dimension r_i.

XX. Two normal matrices are unitarily similar if and only if they have the same characteristic roots, that is, if and only if they are similar.

SOLVED PROBLEMS

1. Prove: The characteristic roots of an n-square real symmetric matrix A are all real.

Suppose that $h+ik$ is a complex characteristic root of A. Consider

$$B = \{(h+ik)I - A\}\{(h-ik)I - A\} = (hI-A)^2 + k^2 I$$

which is real and singular since $(h+ik)I - A$ is singular. There exists a non-zero real vector X such that $BX = 0$ and, hence,

$$X'BX = X'(hI-A)^2X + k^2X'X = X'(hI-A)'(hI-A)X + k^2X'X = 0$$

The vector $(hI-A)X$ is real; hence, $\{(hI-A)X\}'\{(hI-A)X\} \geqq 0$. Also, $X'X > 0$. Thus, $k=0$ and there are no complex roots.

2. Prove: The invariant vectors associated with distinct characteristic roots of a real symmetric matrix A are mutually orthogonal.

Let X_1 and X_2 be invariant vectors associated respectively with the distinct characteristic roots λ_1 and λ_2 of A. Then

$$AX_1 = \lambda_1 X_1 \quad \text{and} \quad AX_2 = \lambda_2 X_2, \qquad \text{also} \qquad X_2'AX_1 = \lambda_1 X_2'X_1 \quad \text{and} \quad X_1'AX_2 = \lambda_2 X_1'X_2$$

Taking transposes

$$X_1'AX_2 = \lambda_1 X_1'X_2 \quad \text{and} \quad X_2'AX_1 = \lambda_2 X_2'X_1$$

Then $\lambda_1 X_1'X_2 = \lambda_2 X_1'X_2$ and, since $\lambda_1 \neq \lambda_2$, $X_1'X_2 = 0$. Thus, X_1 and X_2 are orthogonal.

3. Find an orthogonal matrix P such that $P^{-1}AP$ is diagonal and has as diagonal elements the characteristic roots of A, given

$$A = \begin{bmatrix} 7 & -2 & 1 \\ -2 & 10 & -2 \\ 1 & -2 & 7 \end{bmatrix}$$

The characteristic equation is

$$\begin{vmatrix} \lambda-7 & 2 & -1 \\ 2 & \lambda-10 & 2 \\ -1 & 2 & \lambda-7 \end{vmatrix} = \lambda^3 - 24\lambda^2 + 180\lambda - 432 = 0$$

and the characteristic roots are 6, 6, 12.

For $\lambda = 6$, we have $\begin{bmatrix} -1 & 2 & -1 \\ 2 & -4 & 2 \\ -1 & 2 & -1 \end{bmatrix}\begin{bmatrix} x_1 \\ x_2 \\ x_3 \end{bmatrix} = 0$ or $x_1 - 2x_2 + x_3 = 0$ and choose as associated invariant vectors the mutually orthogonal pair $X_1 = [1, 0, -1]'$ and $X_2 = [1, 1, 1]'$. When $\lambda = 12$, we take $X_3 = [1, -2, 1]'$ as associated invariant vector.

Using the normalized form of these vectors as columns of P, we have

$$P = \begin{bmatrix} 1/\sqrt{2} & 1/\sqrt{3} & 1/\sqrt{6} \\ 0 & 1/\sqrt{3} & -2/\sqrt{6} \\ -1/\sqrt{2} & 1/\sqrt{3} & 1/\sqrt{6} \end{bmatrix}$$

It is left as an exercise to show that $P^{-1}AP = \text{diag}(6, 6, 12)$.

4. Prove: If $X'AX$ and $X'BX$ are real quadratic forms in $(x_1, x_2, \ldots, x_n)$ and if $X'BX$ is positive definite, there exists a real non-singular linear transformation $X = CY$ which carries $X'AX$ into $\lambda_1 y_1^2 + \lambda_2 y_2^2 + \ldots + \lambda_n y_n^2$ and $X'BX$ into $y_1^2 + y_2^2 + \ldots + y_n^2$, where $\lambda_1, \lambda_2, \ldots, \lambda_n$ are the roots of $|\lambda B - A| = 0$.

By Theorem **VII** there exists an orthogonal transformation $X = GV$ which carries $X'BX$ into

$$\text{(i)} \qquad V'(G'BG)V \;=\; \mu_1 v_1^2 + \mu_2 v_2^2 + \ldots + \mu_n v_n^2$$

where $\mu_1, \mu_2, \ldots, \mu_n$ are the characteristic roots (all positive) of B.

Let $H = \text{diag}(1/\sqrt{\mu_1}, 1/\sqrt{\mu_2}, \ldots, 1/\sqrt{\mu_n})$. Then $V = HW$ carries (i) into

$$\text{(ii)} \qquad W'(H'G'BGH)W \;=\; w_1^2 + w_2^2 + \ldots + w_n^2$$

Now for the real quadratic form $W'(H'G'AGH)W$ there exists an orthogonal transformation $W = KY$ which carries it into

$$Y'(K'H'G'AGHK)Y \;=\; \lambda_1 y_1^2 + \lambda_2 y_2^2 + \ldots + \lambda_n y_n^2$$

where $\lambda_1, \lambda_2, \ldots, \lambda_n$ are the characteristic roots of $H'G'AGH$. Thus, there exists a real non-singular transformation $X = CY = GHKY$ which carries $X'AX$ into $\lambda_1 y_1^2 + \lambda_2 y_2^2 + \ldots + \lambda_n y_n^2$ and $X'BX$ into

$$Y'(K'H'G'BGHK)Y \;=\; Y'(K^{-1}IK)Y \;=\; y_1^2 + y_2^2 + \ldots + y_n^2$$

Since for all values of λ,

$$\begin{aligned} K'H'G'(\lambda B - A)GHK \;&=\; \lambda K'H'G'BGHK - K'H'G'AGHK \;=\; \text{diag}(\lambda, \lambda, \ldots, \lambda) - \text{diag}(\lambda_1, \lambda_2, \ldots, \lambda_n) \\ &=\; \text{diag}(\lambda - \lambda_1, \lambda - \lambda_2, \ldots, \lambda - \lambda_n) \end{aligned}$$

it follows that $\lambda_1, \lambda_2, \ldots, \lambda_n$ are the roots of $|\lambda B - A| = 0$

5. From Problem 3, the linear transformation

$$X \;=\; (GH)W \;=\; \begin{bmatrix} 1/\sqrt{2} & 1/\sqrt{3} & 1/\sqrt{6} \\ 0 & 1/\sqrt{3} & -2/\sqrt{6} \\ -1/\sqrt{2} & 1/\sqrt{3} & 1/\sqrt{6} \end{bmatrix} \begin{bmatrix} 1/\sqrt{6} & 0 & 0 \\ 0 & 1/\sqrt{6} & 0 \\ 0 & 0 & 1/2\sqrt{3} \end{bmatrix} W$$

$$= \begin{bmatrix} 1/2\sqrt{3} & 1/3\sqrt{2} & 1/6\sqrt{2} \\ 0 & 1/3\sqrt{2} & -1/3\sqrt{2} \\ -1/2\sqrt{3} & 1/3\sqrt{2} & 1/6\sqrt{2} \end{bmatrix} W$$

carries $q = X'BX = X' \begin{bmatrix} 7 & -2 & 1 \\ -2 & 10 & -2 \\ 1 & -2 & 7 \end{bmatrix} X$ into $W'IW$.

The same transformation carries

$$X'AX \;=\; X' \begin{bmatrix} 3 & -1 & 1 \\ -1 & 5 & -1 \\ 1 & -1 & 3 \end{bmatrix} X \quad \text{into} \quad W' \begin{bmatrix} 1/3 & 0 & 0 \\ 0 & \frac{1}{2} & 0 \\ 0 & 0 & \frac{1}{2} \end{bmatrix} W$$

Since this is a diagonal matrix, the transformation $W = KY$ of Problem 4 is the identity transformation $W = IY$.

Thus, the real linear transformation $X = CY = (GH)Y$ carries the positive definite quadratic form $X'BX$ into $y_1^2 + y_2^2 + y_3^2$ and the quadratic form $X'AX$ into $\frac{1}{3}y_1^2 + \frac{1}{2}y_2^2 + \frac{1}{2}y_3^2$. It is left as an exercise to show that $|\lambda B - A| = 36(3\lambda - 1)(2\lambda - 1)^2$.

6. Prove: Every non-singular real matrix A can be written as $A = CP$ where C is positive definite symmetric and P is orthogonal.

Since A is non-singular, AA' is positive definite symmetric by Theorem **X**, Chapter 17. Then there exists an orthogonal matrix Q such that $Q^{-1}AA'Q = \text{diag}(k_1, k_2, \ldots, k_n) = B$ with each $k_i > 0$. Define $B_1 = \text{diag}(\sqrt{k_1}, \sqrt{k_2}, \ldots, \sqrt{k_n})$ and $C = QB_1Q^{-1}$. Now C is positive definite symmetric and

$$C^2 = QB_1Q^{-1}QB_1Q^{-1} = QB_1^2Q^{-1} = QBQ^{-1} = AA'$$

Define $P = C^{-1}A$. Then $PP' = C^{-1}AA'C^{-1} = C^{-1}C^2C^{-1} = I$ and P is orthogonal. Thus $A = CP$ with C positive definite symmetric and P is orthogonal as required.

7. Prove: The characteristic roots of an Hermitian matrix are real.

Let λ_i be a characteristic root of the Hermitian matrix H. Then there exists a non-zero vector X_i such that $HX_i = \lambda_i X_i$. Now $\bar{X}_i'HX_i = \lambda_i\bar{X}_i'X_i$ is real and different from zero and so also is the conjugate transpose $\bar{X}_i'HX_i = \bar{\lambda}_i\bar{X}_i'X_i$. Thus, $\bar{\lambda}_i = \lambda_i$ and λ_i is real.

8. Prove: If X_i is an invariant vector corresponding to a characteristic root λ_i of a normal matrix A, then X_i is an invariant vector of $\bar{A}'$ corresponding to the characteristic root $\bar{\lambda}_i$.

Since A is normal,

$$\begin{aligned}(\lambda I - A)(\overline{\lambda I - A})' &= (\lambda I - A)(\bar{\lambda}I - \bar{A}') = \lambda\bar{\lambda}I - \lambda\bar{A}' - \bar{\lambda}A + A\bar{A}' \\ &= \bar{\lambda}\lambda I - \lambda\bar{A}' - \bar{\lambda}A + \bar{A}'A = (\overline{\lambda I - A})'(\lambda I - A)\end{aligned}$$

so that $\lambda I - A$ is normal. By hypothesis, $BX_i = (\lambda_i I - A)X_i = 0$; then

$$(\overline{BX_i})'(BX_i) = \bar{X}_i'\bar{B}'\cdot BX_i = \bar{X}_i'B\cdot\bar{B}'X_i = (\bar{B}'X_i)'(\bar{B}'X_i) = 0 \quad \text{and} \quad \bar{B}'X_i = (\bar{\lambda}_i I - \bar{A}')X_i = 0$$

Thus, X_i is an invariant vector of $\bar{A}'$ corresponding to the characteristic root $\bar{\lambda}_i$.

9. Prove: An n-square matrix A is unitarily similar to a diagonal matrix if and only if A is normal.

Suppose A is normal. By Theorem **VIII**, Chapter 20, there exists a unitary matrix U such that

$$\bar{U}'AU = \begin{bmatrix} \lambda_1 & b_{12} & b_{13} & \cdots & b_{1,n-1} & b_{1n} \\ 0 & \lambda_2 & b_{23} & \cdots & b_{2,n-1} & b_{2n} \\ \cdots & \cdots & \cdots & \cdots & \cdots & \cdots \\ 0 & 0 & 0 & \cdots & \lambda_{n-1} & b_{n-1,n} \\ 0 & 0 & 0 & \cdots & 0 & \lambda_n \end{bmatrix} = B$$

By Theorem **XV**, B is normal so that $\bar{B}'B = B\bar{B}'$. Now the element in the first row and first column of $\bar{B}'B$ is $\bar{\lambda}_1\lambda_1$ while the corresponding element of $B\bar{B}'$ is

$$\lambda_1\bar{\lambda}_1 + b_{12}\bar{b}_{12} + b_{13}\bar{b}_{13} + \ldots + b_{1n}\bar{b}_{1n}$$

Since these elements are equal and since each $b_{1j}\bar{b}_{1j} \geqq 0$, we conclude that each $b_{1j} = 0$. Continuing with the corresponding elements in the second row and second column, ..., we conclude that every b_{ij} of B is zero. Thus, $B = \text{diag}(\lambda_1, \lambda_2, \ldots, \lambda_n)$. Conversely, let A be diagonal; then A is normal.

10. Prove: If A is normal, the invariant vectors corresponding to distinct characteristic roots are orthogonal.

Let λ_1, X_1 and λ_2, X_2 be distinct characteristic roots and associated invariant vectors of A. Then $AX_1 = \lambda_1 X_1$, $AX_2 = \lambda_2 X_2$ and, by Problem 8, $\bar{A}'X_1 = \bar{\lambda}_1 X_1$, $\bar{A}'X_2 = \bar{\lambda}_2 X_2$. Now $\bar{X}_2' A X_1 = \lambda_1 \bar{X}_2' X_1$ and, taking the conjugate transpose, $\bar{X}_1' \bar{A}' X_2 = \bar{\lambda}_1 \bar{X}_1' X_2$. But $\bar{X}_1' \bar{A}' X_2 = \bar{\lambda}_2 \bar{X}_1' X_2$. Thus, $\bar{\lambda}_1 \bar{X}_1' X_2 = \bar{\lambda}_2 \bar{X}_1' X_2$ and, since $\bar{\lambda}_1 \neq \bar{\lambda}_2$, $\bar{X}_1' X_2 = 0$ as required.

11. Consider the conic $x_1^2 - 12x_1x_2 - 4x_2^2 = 40$ or

$$\text{(i)} \qquad X'AX \quad = \quad X'\begin{bmatrix} 1 & -6 \\ -6 & -4 \end{bmatrix} X \quad = \quad 40$$

referred to rectangular coordinate axes OX_1 and OX_2.

The characteristic equation of A is

$$|\lambda I - A| \quad = \quad \begin{vmatrix} \lambda - 1 & 6 \\ 6 & \lambda + 4 \end{vmatrix} \quad = \quad (\lambda - 5)(\lambda + 8) \quad = \quad 0$$

For the characteristic roots $\lambda_1 = 5$ and $\lambda_2 = -8$, take $[3, -2]'$ and $[2, 3]'$ respectively as associated invariant vectors. Now form the orthogonal matrix $P = \begin{bmatrix} 3/\sqrt{13} & 2/\sqrt{13} \\ -2/\sqrt{13} & 3/\sqrt{13} \end{bmatrix}$ whose columns are the two vectors after normalization. The transformation $X = PY$ reduces (i) to

$$Y'\begin{bmatrix} 3/\sqrt{13} & -2/\sqrt{13} \\ 2/\sqrt{13} & 3/\sqrt{13} \end{bmatrix}\begin{bmatrix} 1 & -6 \\ -6 & -4 \end{bmatrix}\begin{bmatrix} 3/\sqrt{13} & 2/\sqrt{13} \\ -2/\sqrt{13} & 3/\sqrt{13} \end{bmatrix} Y \quad = \quad Y'\begin{bmatrix} 5 & 0 \\ 0 & -8 \end{bmatrix} Y \quad = \quad 5y_1^2 - 8y_2^2 \quad = \quad 40$$

The conic is an hyperbola.

Aside from the procedure, this is the familiar rotation of axes in plane analytic geometry to effect the elimination of the cross-product term in the equation of a conic. Note that by Theorem **VII** the result is known as soon as the characteristic roots are found.

12. One problem of solid analytic geometry is that of reducing, by translation and rotation of axes, the equation of a quadric surface to simplest form. The main tasks are to locate the centre and to determine the principal directions, i.e., the directions of the axes after rotation. Without attempting to justify the steps, we show here the role of two matrices in such a reduction of the equation of a central quadric.

Consider the surface $3x^2 + 2xy + 2xz + 4yz - 2x - 14y + 2z - 9 = 0$ and the symmetric matrices

$$A \quad = \quad \begin{bmatrix} 3 & 1 & 1 \\ 1 & 0 & 2 \\ 1 & 2 & 0 \end{bmatrix} \qquad \text{and} \quad B \quad = \quad \begin{bmatrix} 3 & 1 & 1 & -1 \\ 1 & 0 & 2 & -7 \\ 1 & 2 & 0 & 1 \\ -1 & -7 & 1 & -9 \end{bmatrix}$$

formed respectively from the terms of degree two and from all the terms.

The characteristic equation of A is

$$|\lambda I - A| \quad = \quad \begin{vmatrix} \lambda - 3 & -1 & -1 \\ -1 & \lambda & -2 \\ -1 & -2 & \lambda \end{vmatrix} \quad = \quad 0$$

The characteristic roots and associated unit invariant vectors are :

$$\lambda_1 = 1,\ v_1 = \left[\frac{1}{\sqrt{3}}, \frac{-1}{\sqrt{3}}, \frac{-1}{\sqrt{3}}\right]'; \qquad \lambda_2 = 4,\ v_2 = \left[\frac{2}{\sqrt{6}}, \frac{1}{\sqrt{6}}, \frac{1}{\sqrt{6}}\right]'; \qquad \lambda_3 = -2,\ v_3 = \left[0, \frac{-1}{\sqrt{2}}, \frac{1}{\sqrt{2}}\right]'$$

Using only the elementary row transformations $H_j(k)$ and $H_{ij}(k)$, where $j \neq 4$,

$$B = \begin{bmatrix} B_1 \\ B_2 \end{bmatrix} = \begin{bmatrix} 3 & 1 & 1 & -1 \\ 1 & 0 & 2 & -7 \\ 1 & 2 & 0 & 1 \\ \hline -1 & -7 & 1 & -9 \end{bmatrix} \sim \begin{bmatrix} 0 & 0 & 1 & -4 \\ 0 & 1 & 0 & 0 \\ 1 & 0 & 0 & 1 \\ \hline 0 & 0 & 0 & -4 \end{bmatrix} = \begin{bmatrix} D_1 \\ D_2 \end{bmatrix}$$

Considering B_1 as the augmented matrix of the system of equations $\begin{cases} 3x + y + z - 1 = 0 \\ x + 2z - 7 = 0 \\ x + 2y + 1 = 0 \end{cases}$ we find from D_1 the solution $x = -1,\ y = 0,\ z = 4$ or $C(-1,0,4)$. From D_2, we have $d = -4$.

The rank of A is 3 and the rank of B is 4; the quadric has as centre $C(-1,0,4)$. The required reduced equation is

$$\lambda_1 X^2 + \lambda_2 Y^2 + \lambda_3 Z^2 + d = X^2 + 4Y^2 - 2Z^2 - 4 = 0$$

The equations of translation are $x = x' - 1,\ y = y',\ z = z' + 4$.

The principal directions are v_1, v_2, v_3. Denote by E the inverse of $[v_1, v_2, v_3]$. The equations of the rotation of axes to the principal directions are

$$\begin{bmatrix} x' \\ y' \\ z' \end{bmatrix} = [X\ Y\ Z] \cdot E = [X\ Y\ Z] \begin{bmatrix} 1/\sqrt{3} & -1/\sqrt{3} & -1/\sqrt{3} \\ 2/\sqrt{6} & 1/\sqrt{6} & 1/\sqrt{6} \\ 0 & -1/\sqrt{2} & 1/\sqrt{2} \end{bmatrix}$$

SUPPLEMENTARY PROBLEMS

13. For each of the following real symmetric matrices A, find an orthogonal matrix P such that $P^{-1}AP$ is diagonal and has as diagonal elements the characteristic roots of A.

(a) $\begin{bmatrix} 2 & 0 & -1 \\ 0 & 2 & 0 \\ -1 & 0 & 2 \end{bmatrix}$, (b) $\begin{bmatrix} 2 & 0 & 1 \\ 0 & 3 & 0 \\ 1 & 0 & 2 \end{bmatrix}$, (c) $\begin{bmatrix} 2 & -4 & 2 \\ -4 & 2 & -2 \\ 2 & -2 & -1 \end{bmatrix}$, (d) $\begin{bmatrix} 3 & 2 & 2 \\ 2 & 2 & 0 \\ 2 & 0 & 4 \end{bmatrix}$, (e) $\begin{bmatrix} 4 & -1 & 1 \\ -1 & 4 & -1 \\ 1 & -1 & 4 \end{bmatrix}$

Ans. (a) $\begin{bmatrix} 1/\sqrt{2} & 0 & 1/\sqrt{2} \\ 0 & 1 & 0 \\ 1/\sqrt{2} & 0 & -1/\sqrt{2} \end{bmatrix}$, (b) $\begin{bmatrix} 1/\sqrt{2} & 1/\sqrt{6} & 1/\sqrt{3} \\ 0 & -2/\sqrt{6} & 1/\sqrt{3} \\ -1/\sqrt{2} & 1/\sqrt{6} & 1/\sqrt{3} \end{bmatrix}$, (c) $\begin{bmatrix} 2/3 & 1/3 & 2/3 \\ -2/3 & 2/3 & 1/3 \\ 1/3 & 2/3 & -2/3 \end{bmatrix}$

(d) $\begin{bmatrix} 2/3 & 1/3 & 2/3 \\ -2/3 & 2/3 & 1/3 \\ -1/3 & -2/3 & 2/3 \end{bmatrix}$, (e) $\begin{bmatrix} 1/\sqrt{6} & 1/\sqrt{2} & 1/\sqrt{3} \\ 2/\sqrt{6} & 0 & -1/\sqrt{3} \\ 1/\sqrt{6} & -1/\sqrt{2} & 1/\sqrt{3} \end{bmatrix}$

14. Find a linear transformation which reduces $X'BX$ to $y_1^2 + y_2^2 + y_3^2$ and $X'AX$ to $\lambda_1 y_1^2 + \lambda_2 y_2^2 + \lambda_3 y_3^2$, where λ_i are the roots of $|\lambda B - A| = 0$, given

(a) $A = \begin{bmatrix} 7 & -2 & 1 \\ -2 & 10 & -2 \\ 1 & -2 & 7 \end{bmatrix}$, $B = \begin{bmatrix} 2 & 0 & 1 \\ 0 & 3 & 0 \\ 1 & 0 & 2 \end{bmatrix}$ (b) $A = \begin{bmatrix} 7 & -4 & -4 \\ -4 & 1 & -8 \\ -4 & -8 & 1 \end{bmatrix}$, $B = \begin{bmatrix} 2 & 2 & 2 \\ 2 & 5 & 4 \\ 2 & 4 & 5 \end{bmatrix}$

Ans. (a) $\begin{bmatrix} 1/\sqrt{2} & 1/3\sqrt{2} & 1/3 \\ 0 & -2/3\sqrt{2} & 1/3 \\ -1/\sqrt{2} & 1/3\sqrt{2} & 1/3 \end{bmatrix}$, (b) $\begin{bmatrix} 2/3 & 2/3 & 1/3\sqrt{10} \\ -2/3 & 1/3 & 2/3\sqrt{10} \\ 1/3 & -2/3 & 2/3\sqrt{10} \end{bmatrix}$

15. Prove Theorem **IV**.
Hint. If $P^{-1}AP = \text{diag}(\lambda_1, \lambda_1, \ldots, \lambda_1, \lambda_{r+1}, \lambda_{r+2}, \ldots, \lambda_n)$, then $P^{-1}(\lambda_1 I - A)P = \text{diag}(0, 0, \ldots, 0, \lambda_1 - \lambda_{r+1}, \lambda_1 - \lambda_{r+2}, \ldots, \lambda_1 - \lambda_n)$ is of rank $n-r$.

16. Modify the proof in Problem 2 to prove Theorem **XI**.

17. Prove Theorems **XII**, **XIII**, and **XIX**.

18. Identify each locus:

(a) $20x_1^2 - 24x_1x_2 + 27x_2^2 = 369$, (c) $108x_1^2 - 312x_1x_2 + 17x_2^2 = 900$,

(b) $3x_1^2 + 2x_1x_2 + 3x_2^2 = 4$, (d) $x_1^2 + 2x_1x_2 + x_2^2 = 8$

19. Let A be real and skew-symmetric. Prove:

(a) Each characteristic root of A is either zero or a pure imaginary.
(b) $I+A$ is non-singular, $I-A$ is non-singular.
(c) $B = (I+A)^{-1}(I-A)$ is orthogonal. (See Problem 35, Chapter 13.)

20. Prove: If A is normal and non-singular so also is A^{-1}.

21. Prove: If A is normal, then A is similar to A'.

22. Prove: A square matrix A is normal if and only if it can be expressed as $H+iK$, where H and K are commutative Hermitian matrices.

23. If A is n-square with characteristic roots $\lambda_1, \lambda_2, \ldots, \lambda_n$, then A is normal if and only if the characteristic roots of $A\bar{A}'$ are $\lambda_1\bar{\lambda}_1, \lambda_2\bar{\lambda}_2, \ldots, \lambda_n\bar{\lambda}_n$.
Hint. Write $U^{-1}AU = T = [t_{ij}]$, where U is unitary and T is triangular. Now $\text{tr}(T\bar{T}') = \text{tr}(A\bar{A}')$ requires $t_{ij} = 0$ for $i \neq j$.

24. Prove: If A is non-singular, then $A\bar{A}'$ is positive definite Hermitian. Restate the theorem when A is real and non-singular.

25. Prove: If A and B are n-square and normal and if A and $\bar{B}'$ commute, then AB and BA are normal.

26. Let the characteristic function of the n-square matrix A be

$$\phi(\lambda) = (\lambda-\lambda_1)^{r_1}(\lambda-\lambda_2)^{r_2}\cdots(\lambda-\lambda_s)^{r_s}$$

and suppose there exists a non-singular matrix P such that

$$(1) \qquad P^{-1}AP = \text{diag}(\lambda_1 I_{r_1}, \lambda_2 I_{r_2}, \ldots, \lambda_s I_{r_s})$$

Define by B_i, $(i = 1, 2, \ldots, s)$ the n-square matrix $\text{diag}(0, 0, \ldots, 0, I_{r_i}, 0, \ldots, 0)$ obtained by replacing λ_i by 1 and λ_j, $(j \neq i)$, by 0 in the right member of (1) and define

$$E_i = PB_iP^{-1}, \quad (i = 1, 2, \ldots, s)$$

Show that

(a) $P^{-1}AP = \lambda_1 B_1 + \lambda_2 B_2 + \ldots + \lambda_s B_s$
(b) $A = \lambda_1 E_1 + \lambda_2 E_2 + \ldots + \lambda_s E_s$
(c) Every E_i is idempotent.
(d) $E_i E_j = 0$ for $i \neq j$.
(e) $E_1 + E_2 + \ldots + E_s = I$
(f) The rank of E_i is the multiplicity of the characteristic root λ_i.
(g) $(\lambda_i I - A)E_i = 0$, $(i = 1, 2, \ldots, s)$
(h) If $p(x)$ is any polynomial in x, then $p(A) = p(\lambda_1)E_1 + p(\lambda_2)E_2 + \ldots + p(\lambda_s)E_s$.
Hint. Establish $A^2 = \lambda_1^2 E_1 + \lambda_2^2 E_2 + \ldots + \lambda_s^2 E_s$, $A^3 = \lambda_1^3 E_1 + \lambda_2^3 E_2 + \ldots + \lambda_s^3 E_s$, ...

(*i*) Each E_i is a polynomial in A.
Hint. Define $f(\lambda) = (\lambda-\lambda_1)(\lambda-\lambda_2)\dots(\lambda-\lambda_s)$ and $f_i(\lambda) = f(\lambda)/(\lambda-\lambda_i)$, $(i = 1, 2, \dots, s)$. Then $f_i(A) = f_i(\lambda_i)E_i$.

(*j*) A matrix B commutes with A if and only if it commutes with every E_i.
Hint. If B commutes with A it commutes with every polynomial in A.

(*k*) If A is normal, then each E_i is Hermitian.

(*l*) If A is non-singular, then

$$A^{-1} = \lambda_1^{-1}E_1 + \lambda_2^{-1}E_2 + \dots + \lambda_s^{-1}E_s$$

(*m*) If A is positive definite Hermitian, then

$$H = A^{1/2} = \sqrt{\lambda_1}\,E_1 + \sqrt{\lambda_2}\,E_2 + \dots + \sqrt{\lambda_s}\,E_s$$

is positive definite Hermitian.

(*n*) Equation (*b*) is called the **spectral decomposition** of A. Show that it is unique.

27. (*a*) Obtain the spectral decomposition

$$A = \begin{bmatrix} 24 & -20 & 10 \\ -20 & 24 & -10 \\ 10 & -10 & 9 \end{bmatrix} = 49\begin{bmatrix} 4/9 & -4/9 & 2/9 \\ -4/9 & 4/9 & -2/9 \\ 2/9 & -2/9 & 1/9 \end{bmatrix} + 4\begin{bmatrix} 5/9 & 4/9 & -2/9 \\ 4/9 & 5/9 & 2/9 \\ -2/9 & 2/9 & 8/9 \end{bmatrix}$$

(*b*) Obtain $A^{-1} = \dfrac{1}{196}\begin{bmatrix} 29 & 20 & -10 \\ 20 & 29 & 10 \\ -10 & 10 & 44 \end{bmatrix}$.

(*c*) Obtain $A^{1/2} = \begin{bmatrix} 38/9 & -20/9 & 10/9 \\ -20/9 & 38/9 & -10/9 \\ 10/9 & -10/9 & 23/9 \end{bmatrix}$

28. Prove: If A is normal and commutes with B, then $\bar{A}'$ and B commute.
Hint. Use Problem 26(*j*).

29. Prove: If A is non-singular then there exists a unitary matrix U and a positive definite Hermitian matrix H such that $A = HU$.
Hint. Define H by $H^2 = A\bar{A}'$ and $U = H^{-1}A$.

30. Prove: If A is non-singular, then A is normal if and only if H and U of Problem 29 commute.

31. Prove: The square matrix A is similar to a diagonal matrix if and only if there exists a positive definite Hermitian matrix H such that $H^{-1}AH$ is normal.

32. Prove: A real symmetric (Hermitian) matrix is idempotent if and only if its characteristic roots are 0's and 1's.

33. Prove: If A is real symmetric (Hermitian) and idempotent, then $r_A = \operatorname{tr} A$.

34. Let A be normal, $B = I + A$ be non-singular, and $C = B^{-1}\bar{B}'$.
Prove: (*a*) A and $(\bar{B}')^{-1}$ commute, (*b*) C is unitary.

35. Prove: If H is Hermitian, then $(I+iH)^{-1}(I-iH)$ is unitary.

36. If A is n-square, the set of numbers $\bar{X}'AX$ where X is a unit vector is called the **field of values** of A. Prove:

(*a*) The characteristic roots of A are in its field of values.

(*b*) Every diagonal element of A and every diagonal element of $U^{-1}AU$, where U is unitary, is in the field of values of A.

(*c*) If A is real symmetric (Hermitian), every element in its field of values is real.

(*d*) If A is real symmetric (Hermitian), its field of values is the set of reals $\lambda_1 \leqq \lambda \leqq \lambda_n$, where λ_1 is the least and λ_n is the greatest characteristic root of A.

Chapter 22

Polynomials Over a Field

POLYNOMIAL DOMAIN OVER F. Let λ denote an abstract symbol (indeterminate) which is assumed to be commutative with itself and with the elements of a field F. The expression

$$(22.1) \qquad f(\lambda) = a_n\lambda^n + a_{n-1}\lambda^{n-1} + \cdots + a_1\lambda + a_0\lambda^0$$

where the a_i are in F is called a **polynomial in** λ **over** F.

If every $a_i = 0$, **(22.1)** is called the **zero polynomial** and we write $f(\lambda) = 0$. If $a_n \neq 0$, **(22.1)** is said to be of degree n and a_n is called its **leading coefficient**. The polynomial $f(\lambda) = a_0\lambda^0 = a_0 \neq 0$ is said to be of degree zero; the degree of the zero polynomial is not defined.

If $a_n = 1$ in **(22.1)**, the polynomial is called **monic**.

Two polynomials in λ which contain, apart from terms with zero coefficients, the same terms are said to be **equal**.

The totality of polynomials **(22.1)** is called a **polynomial domain** $F[\lambda]$ over F.

SUM AND PRODUCT. Regarding the individual polynomials of $F[\lambda]$ as elements of a number system, the polynomial domain has most but not all of the properties of a field. For example

$$f(\lambda) + g(\lambda) = g(\lambda) + f(\lambda) \qquad \text{and} \qquad f(\lambda)\cdot g(\lambda) = g(\lambda)\cdot f(\lambda)$$

If $f(\lambda)$ is of degree m and $g(\lambda)$ is of degree n,

(i) $f(\lambda) + g(\lambda)$ is of degree m when $m>n$, of degree at most m when $m=n$, and of degree n when $m<n$.

(ii) $f(\lambda)\cdot g(\lambda)$ is of degree $m+n$.

If $f(\lambda) \neq 0$ while $f(\lambda)\cdot g(\lambda) = 0$, then $g(\lambda) = 0$.

If $g(\lambda) \neq 0$ and $h(\lambda)\cdot g(\lambda) = k(\lambda)\cdot g(\lambda)$, then $h(\lambda) = k(\lambda)$.

QUOTIENTS. In Problem 1, we prove

I. If $f(\lambda)$ and $g(\lambda) \neq 0$ are polynomials in $F[\lambda]$, then there exist unique polynomials $h(\lambda)$ and $r(\lambda)$ in $F[\lambda]$, where $r(\lambda)$ is either the zero polynomial or is of degree less than that of $g(\lambda)$, such that

$$(22.2) \qquad f(\lambda) = h(\lambda)\cdot g(\lambda) + r(\lambda)$$

Here, $r(\lambda)$ is called the **remainder** in the division of $f(\lambda)$ by $g(\lambda)$. If $r(\lambda) = 0$, $g(\lambda)$ is said to divide $f(\lambda)$ while $g(\lambda)$ and $h(\lambda)$ are called **factors** of $f(\lambda)$.

Let $f(\lambda) = h(\lambda)\cdot g(\lambda)$. When $g(\lambda)$ is of degree zero, that is, when $g(\lambda) = c$, a constant, the factorization is called **trivial**. A non-constant polynomial over F is called **irreducible** over F if its only factorization is trivial.

Example 1. Over the rational field λ^2-3 is irreducible; over the real field it is factorable as $(\lambda+\sqrt{3})(\lambda-\sqrt{3})$. Over the real field (and hence over the rational field) λ^2+4 is irreducible; over the complex field it is factorable as $(\lambda+2i)(\lambda-2i)$.

THE REMAINDER THEOREM. Let $f(\lambda)$ be any polynomial and $g(\lambda) = \lambda - a$. Then **(22.2)** becomes

$$f(\lambda) = h(\lambda)\cdot(\lambda-a) + r \tag{22.3}$$

where r is free of λ. By **(22.3)**, $f(a) = r$, and we have

II. When $f(\lambda)$ is divided by $\lambda-a$ until a remainder free of λ is obtained, that remainder is $f(a)$.

III. A polynomial $f(\lambda)$ has $\lambda-a$ as a factor if and only if $f(a) = 0$.

GREATEST COMMON DIVISOR. If $h(\lambda)$ divides both $f(\lambda)$ and $g(\lambda)$, it is called a **common divisor** of $f(\lambda)$ and $g(\lambda)$.

A polynomial $d(\lambda)$ is called the **greatest common divisor** of $f(\lambda)$ and $g(\lambda)$ if

(i) $d(\lambda)$ is monic,

(ii) $d(\lambda)$ is a common divisor of $f(\lambda)$ and $g(\lambda)$,

(iii) every common divisor of $f(\lambda)$ and $g(\lambda)$ is a divisor of $d(\lambda)$.

In Problem 2, we prove

IV. If $f(\lambda)$ and $g(\lambda)$ are polynomials in $F[\lambda]$, not both the zero polynomial, they have a unique greatest common divisor $d(\lambda)$ and there exist polynomials $h(\lambda)$ and $k(\lambda)$ in $F[\lambda]$ such that

$$d(\lambda) = h(\lambda)\cdot f(\lambda) + k(\lambda)\cdot g(\lambda) \tag{22.4}$$

See also Problem 3.

When the only common divisors of $f(\lambda)$ and $g(\lambda)$ are constants, their greatest common divisor is $d(\lambda) = 1$.

Example 2. The greatest common divisor of $f(\lambda) = (\lambda^2+4)(\lambda^2+3\lambda+5)$ and $g(\lambda) = (\lambda^2-1)(\lambda^2+3\lambda+5)$ is $\lambda^2+3\lambda+5$, and **(22.4)** is

$$\lambda^2 + 3\lambda + 5 = \tfrac{1}{5}f(\lambda) - \tfrac{1}{5}g(\lambda)$$

We have also $(1-\lambda^2)\cdot f(\lambda) + (\lambda^2+4)\cdot g(\lambda) = 0$. This illustrates

V. If the greatest common divisor of $f(\lambda)$ of degree $n>0$ and $g(\lambda)$ of degree $m>0$ is not 1, there exist non-zero polynomials $a(\lambda)$ of degree $<m$ and $b(\lambda)$ of degree $<n$ such that

$$a(\lambda)\cdot f(\lambda) + b(\lambda)\cdot g(\lambda) = 0$$

and conversely. See Problem 4.

RELATIVELY PRIME POLYNOMIALS. Two polynomials are called **relatively prime** if their greatest common divisor is 1.

VI. If $g(\lambda)$ is irreducible in $F[\lambda]$ and $f(\lambda)$ is any polynomial of $F[\lambda]$, then either $g(\lambda)$ divides $f(\lambda)$ or $g(\lambda)$ is relatively prime to $f(\lambda)$.

VII. If $g(\lambda)$ is irreducible but divides $f(\lambda)\cdot h(\lambda)$, it divides at least one of $f(\lambda)$ and $h(\lambda)$.

VIII. If $f(\lambda)$ and $g(\lambda)$ are relatively prime and if each divides $h(\lambda)$, so also does $f(\lambda)\cdot g(\lambda)$.

UNIQUE FACTORIZATION. In Problem 5, we prove

IX. Every non-zero polynomial $f(\lambda)$ of $F[\lambda]$ can be written as

$$f(\lambda) \quad = \quad c\cdot q_1(\lambda)\cdot q_2(\lambda)\ \ldots\ q_r(\lambda) \tag{22.5}$$

where $c \neq 0$ is a constant and the $q_i(\lambda)$ are monic irreducible polynomials of $F[\lambda]$.

SOLVED PROBLEMS

1. Prove: If $f(\lambda)$ and $g(\lambda)\neq 0$ are polynomials in $F[\lambda]$, there exist unique polynomials $h(\lambda)$ and $r(\lambda)$ in $F[\lambda]$, where $r(\lambda)$ is either the zero polynomial or is of degree less than that of $g(\lambda)$, such that

$$f(\lambda) \quad = \quad h(\lambda)\cdot g(\lambda) + r(\lambda) \tag{i}$$

Let

$$f(\lambda) \quad = \quad a_n\lambda^n + a_{n-1}\lambda^{n-1} + \cdots + a_1\lambda + a_0$$

and

$$g(\lambda) \quad = \quad b_m\lambda^m + b_{m-1}\lambda^{m-1} + \cdots + b_1\lambda + b_0, \qquad b_m \neq 0$$

Clearly, the theorem is true if $f(\lambda) = 0$ or if $n < m$. Suppose that $n \geq m$; then

$$f(\lambda) - \frac{a_n}{b_m}\lambda^{n-m}g(\lambda) \quad = \quad f_1(\lambda) \quad = \quad c_p\lambda^p + c_{p-1}\lambda^{p-1} + \cdots + c_0$$

is either the zero polynomial or is of degree less than that of $f(\lambda)$.

If $f_1(\lambda) = 0$ or is of degree less than that of $g(\lambda)$, we have proved the theorem with $h(\lambda) = \frac{a_n}{b_m}\lambda^{n-m}$ and $r(\lambda) = f_1(\lambda)$. Otherwise, we form

$$f(\lambda) - \frac{a_n}{b_m}\lambda^{n-m}g(\lambda) - \frac{c_p}{b_m}\lambda^{p-m}g(\lambda) \quad = \quad f_2(\lambda)$$

Again, if $f_2(\lambda) = 0$ or is of degree less than that of $g(\lambda)$, we have proved the theorem. Otherwise, we repeat the process. Since in each step, the degree of the remainder (assumed $\neq 0$) is reduced, we eventually reach a remainder $r(\lambda) = f_s(\lambda)$ which is either the zero polynomial or is of degree less than that of $g(\lambda)$.

To prove uniqueness, suppose

$$f(\lambda) \;=\; h(\lambda)\cdot g(\lambda) + r(\lambda) \qquad \text{and} \qquad f(\lambda) \;=\; k(\lambda)\cdot g(\lambda) + s(\lambda)$$

where the degrees of $r(\lambda)$ and $s(\lambda)$ are less than that of $g(\lambda)$. Then

$$h(\lambda)\cdot g(\lambda) + r(\lambda) \quad = \quad k(\lambda)\cdot g(\lambda) + s(\lambda)$$

and

$$[k(\lambda) - h(\lambda)]g(\lambda) \quad = \quad r(\lambda) - s(\lambda)$$

Now $r(\lambda) - s(\lambda)$ is of degree less than m while, unless $k(\lambda) - h(\lambda) = 0$, $[k(\lambda) - h(\lambda)]g(\lambda)$ is of degree equal to or greater than m. Thus, $k(\lambda) - h(\lambda) = 0$, $r(\lambda) - s(\lambda) = 0$ so that $k(\lambda) = h(\lambda)$ and $r(\lambda) = s(\lambda)$. Then both $h(\lambda)$ and $r(\lambda)$ are unique.

2. Prove: If $f(\lambda)$ and $g(\lambda)$ are polynomials in $F[\lambda]$, not both zero, they have a unique greatest common divisor $d(\lambda)$ and there exist polynomials $h(\lambda)$ and $k(\lambda)$ in F such that

(a) $$d(\lambda) = h(\lambda)\cdot f(\lambda) + k(\lambda)\cdot g(\lambda)$$

If, say, $f(\lambda) = 0$, then $d(\lambda) = b_m^{-1} g(\lambda)$ where b_m is the leading coefficient of $g(\lambda)$ and we have (a) with $h(\lambda) = 1$ and $k(\lambda) = b_m^{-1}$.

Suppose next that the degree of $g(\lambda)$ is not greater than that of $f(\lambda)$. By Theorem **I**, we have

(**i**) $$f(\lambda) = q_1(\lambda)\cdot g(\lambda) + r_1(\lambda)$$

where $r_1(\lambda) = 0$ or is of degree less than that of $g(\lambda)$. If $r_1(\lambda) = 0$, then $d(\lambda) = b_m^{-1} g(\lambda)$ and we have (a) with $h(\lambda) = 0$ and $k(\lambda) = b_m^{-1}$.

If $r_1(\lambda) \neq 0$, we have

(**ii**) $$g(\lambda) = q_2(\lambda)\cdot r_1(\lambda) + r_2(\lambda)$$

where $r_2(\lambda) = 0$ or is of degree less than that of $r_1(\lambda)$. If $r_2(\lambda) = 0$, we have from (**i**)

$$r_1(\lambda) = f(\lambda) - q_1(\lambda)\cdot g(\lambda)$$

and from it obtain (a) by dividing by the leading coefficient of $r_1(\lambda)$.

If $r_2(\lambda) \neq 0$, we have

(**iii**) $$r_1(\lambda) = q_3(\lambda)\cdot r_2(\lambda) + r_3(\lambda)$$

where $r_3(\lambda) = 0$ or is of degree less than that of $r_2(\lambda)$. If $r_3(\lambda) = 0$, we have from (**i**) and (**ii**)

$$\begin{aligned} r_2(\lambda) &= g(\lambda) - q_2(\lambda)\cdot r_1(\lambda) = g(\lambda) - q_2(\lambda)[f(\lambda) - q_1(\lambda)\cdot g(\lambda)] \\ &= -q_2(\lambda)\cdot f(\lambda) + [1 + q_1(\lambda)\cdot q_2(\lambda)]g(\lambda) \end{aligned}$$

and from it obtain (a) by dividing by the leading coefficient of $r_2(\lambda)$.

Continuing the process under the assumption that each new remainder is different from 0, we have, in general,

(**iv**) $$r_i(\lambda) = q_{i+2}(\lambda)\cdot r_{i+1}(\lambda) + r_{i+2}(\lambda)$$

moreover, the process must conclude with

(**v**) $$r_{s-2}(\lambda) = q_s(\lambda)\cdot r_{s-1}(\lambda) + r_s(\lambda), \qquad r_s(\lambda) \neq 0$$

and

(**vi**) $$r_{s-1}(\lambda) = q_{s+1}(\lambda)\cdot r_s(\lambda)$$

By (**vi**), $r_s(\lambda)$ divides $r_{s-1}(\lambda)$, and by (**v**), also divides $r_{s-2}(\lambda)$. From (**iv**), we have

$$r_{s-3}(\lambda) = q_{s-1}(\lambda)\cdot r_{s-2}(\lambda) + r_{s-1}(\lambda)$$

so that $r_s(\lambda)$ divides $r_{s-3}(\lambda)$. Thus, by retracing the steps leading to (**vi**), we conclude that $r_s(\lambda)$ divides both $f(\lambda)$ and $g(\lambda)$. If the leading coefficient of $r_s(\lambda)$ is c, then $d(\lambda) = c^{-1} r_s(\lambda)$.

From (**i**) $r_1(\lambda) = f(\lambda) - q_1(\lambda)\cdot g(\lambda) = h_1(\lambda)\cdot f(\lambda) + k_1(\lambda)\cdot g(\lambda)$ and substituting in (**ii**)

$$r_2(\lambda) = -q_2(\lambda)\cdot f(\lambda) + [1 + q_1(\lambda)\cdot q_2(\lambda)]g(\lambda) = h_2(\lambda)\cdot f(\lambda) + k_2(\lambda)\cdot g(\lambda)$$

From (**iii**), $r_3(\lambda) = r_1(\lambda) - q_3(\lambda)\cdot r_2(\lambda)$. Substituting for $r_1(\lambda)$ and $r_2(\lambda)$, we have

$$\begin{aligned} r_3(\lambda) &= [1 + q_2(\lambda)\cdot q_3(\lambda)]f(\lambda) + [-q_1(\lambda) - q_3(\lambda) - q_1(\lambda)\cdot q_2(\lambda)\cdot q_3(\lambda)]g(\lambda) \\ &= h_3(\lambda)\cdot f(\lambda) + k_3(\lambda)\cdot g(\lambda) \end{aligned}$$

Continuing, we obtain finally,

$$r_s(\lambda) = h_s(\lambda)\cdot f(\lambda) + k_s(\lambda)\cdot g(\lambda)$$

Then $d(\lambda) = c^{-1} r_s(\lambda) = c^{-1} h_s(\lambda)\cdot f(\lambda) + c^{-1} k_s(\lambda)\cdot g(\lambda) = h(\lambda)\cdot f(\lambda) + k(\lambda)\cdot g(\lambda)$ as required.

The proof that $d(\lambda)$ is unique is left as an exercise.

3. Find the greatest common divisor $d(\lambda)$ of

$$f(\lambda) = 3\lambda^5 + 7\lambda^4 + 11\lambda + 6 \quad \text{and} \quad g(\lambda) = \lambda^4 + 2\lambda^3 - \lambda^2 - \lambda + 2$$

and express $d(\lambda)$ in the form of Theorem **III**.

We find

(i) $f(\lambda) = (3\lambda+1)g(\lambda) + (\lambda^3+4\lambda^2+6\lambda+4)$

(ii) $g(\lambda) = (\lambda-2)(\lambda^3+4\lambda^2+6\lambda+4) + (\lambda^2+7\lambda+10)$

(iii) $\lambda^3+4\lambda^2+6\lambda+4 = (\lambda-3)(\lambda^2+7\lambda+10) + (17\lambda+34)$

and

(iv) $\lambda^2 + 7\lambda + 10 = (\frac{1}{17}\lambda+\frac{5}{17})(17\lambda+34)$

The greatest common divisor is $\frac{1}{17}(17\lambda+34) = \lambda + 2$.

From **(iii)**,

$$17\lambda + 34 = (\lambda^3 + 4\lambda^2 + 6\lambda + 4) - (\lambda - 3)(\lambda^2 + 7\lambda + 10)$$

Substituting for $\lambda^2 + 7\lambda + 10$ from **(ii)**

$$\begin{aligned} 17\lambda + 34 &= (\lambda^3 + 4\lambda^2 + 6\lambda + 4) - (\lambda - 3)[g(\lambda) - (\lambda-2)(\lambda^3+4\lambda^2+6\lambda+4)] \\ &= (\lambda^2 - 5\lambda + 7)(\lambda^3 + 4\lambda^2 + 6\lambda + 4) - (\lambda - 3)g(\lambda) \end{aligned}$$

and for $\lambda^3 + 4\lambda^2 + 6\lambda + 4$ from **(i)**

$$17\lambda + 34 = (\lambda^2 - 5\lambda + 7)f(\lambda) + (-3\lambda^3 + 14\lambda^2 - 17\lambda - 4)g(\lambda)$$

Then

$$\lambda + 2 = \frac{1}{17}(\lambda^2 - 5\lambda + 7)\cdot f(\lambda) + \frac{1}{17}(-3\lambda^3 + 14\lambda^2 - 17\lambda - 4)\cdot g(\lambda)$$

4. Prove: If the greatest common divisor of $f(\lambda)$ of degree $n>0$ and $g(\lambda)$ of degree $m>0$ is not 1, there exist non-zero polynomials $a(\lambda)$ of degree $<m$ and $b(\lambda)$ of degree $<n$ such that

$$(a) \qquad a(\lambda)\cdot f(\lambda) + b(\lambda)\cdot g(\lambda) = 0$$

and conversely.

Let the greatest common divisor of $f(\lambda)$ and $g(\lambda)$ be $d(\lambda) \neq 1$; then

$$f(\lambda) = d(\lambda)\cdot f_1(\lambda) \quad \text{and} \quad g(\lambda) = d(\lambda)\cdot g_1(\lambda)$$

where $f_1(\lambda)$ is of degree $<n$ and $g_1(\lambda)$ is of degree $<m$. Now

$$g_1(\lambda)\cdot f(\lambda) = g_1(\lambda)\cdot d(\lambda)\cdot f_1(\lambda) = g(\lambda)\cdot f_1(\lambda)$$

and

$$g_1(\lambda)\cdot f(\lambda) + [-f_1(\lambda)\cdot g(\lambda)] = 0$$

Thus, taking $a(\lambda) = g_1(\lambda)$ and $b(\lambda) = -f_1(\lambda)$, we have (a).

Conversely, suppose $f(\lambda)$ and $g(\lambda)$ are relatively prime and (a) holds. Then by Theorem **IV** there exist polynomials $h(\lambda)$ and $k(\lambda)$ such that

$$h(\lambda)\cdot f(\lambda) + k(\lambda)\cdot g(\lambda) = 1$$

Then, using (a),

$$\begin{aligned} a(\lambda) &= a(\lambda)\cdot h(\lambda)\cdot f(\lambda) + a(\lambda)\cdot k(\lambda)\cdot g(\lambda) \\ &= -b(\lambda)\cdot h(\lambda)\cdot g(\lambda) + a(\lambda)\cdot k(\lambda)\cdot g(\lambda) \end{aligned}$$

and $g(\lambda)$ divides $a(\lambda)$. But this is impossible; hence, if (a) holds, $f(\lambda)$ and $g(\lambda)$ cannot be relatively prime.

5. Prove: Every non-zero polynomial $f(\lambda)$ in $F[\lambda]$ can be written as

$$f(\lambda) = c \cdot q_1(\lambda) \cdot q_2(\lambda) \ \ldots \ q_r(\lambda)$$

where $c \neq 0$ is a constant and the $q_i(\lambda)$ are monic irreducible polynomials in $F[\lambda]$.

Write

(i) $$f(\lambda) = a_n \cdot f_1(\lambda)$$

where a_n is the leading coefficient of $f(\lambda)$. If $f_1(\lambda)$ is irreducible, then (i) satisfies the conditions of the theorem. Otherwise, there is a factorization

(ii) $$f(\lambda) = a_n \cdot g(\lambda) \cdot h(\lambda)$$

If $g(\lambda)$ and $h(\lambda)$ are irreducible, then **(ii)** satisfies the conditions of the theorem. Otherwise, further factorization leads to a set of monic irreducible factors.

To prove uniqueness, suppose that

$$a_n \cdot q_1(\lambda) \cdot q_2(\lambda) \ \ldots \ q_r(\lambda) \qquad \text{and} \qquad a_n \cdot p_1(\lambda) \cdot p_2(\lambda) \ \ldots \ p_s(\lambda)$$

are two factorizations with $r < s$. Since $q_1(\lambda)$ divides $p_1(\lambda) \cdot p_2(\lambda) \ \ldots \ p_s(\lambda)$, it must divide some one of the $p_i(\lambda)$ which, by a change in numbering, may be taken as $p_1(\lambda)$. Since $p_1(\lambda)$ is monic and irreducible, $q_1(\lambda) = p_1(\lambda)$. Then $q_2(\lambda)$ divides $p_2(\lambda) \cdot p_3(\lambda) \ \ldots \ p_s(\lambda)$ and, after a repetition of the argument above, $q_2(\lambda) = p_2(\lambda)$. Eventually, we have $q_i(\lambda) = p_i(\lambda)$ for $i = 1, 2, \ldots, r$ and $p_{r+1}(\lambda) \cdot p_{r+2}(\lambda) \ldots p_s(\lambda) = 1$. Since the latter equality is impossible, $r = s$ and uniqueness is established.

SUPPLEMENTARY PROBLEMS

6. Give an example in which the degree of $f(\lambda) + g(\lambda)$ is less than the degree of either $f(\lambda)$ or $g(\lambda)$.

7. Prove Theorem **III**.

8. Prove: If $f(\lambda)$ divides $g(\lambda)$ and $h(\lambda)$, it divides $g(\lambda) \pm h(\lambda)$.

9. Find a necessary and sufficient condition that the two non-zero polynomials $f(\lambda)$ and $g(\lambda)$ in $F[\lambda]$ divide each other.

10. For each of the following, express the greatest common divisor in the form of Theorem **IV**.

(a) $f(\lambda) = 2\lambda^5 - \lambda^3 + 2\lambda^2 - 6\lambda - 4$, $\quad g(\lambda) = \lambda^4 - \lambda^3 - \lambda^2 + 2\lambda - 2$

(b) $f(\lambda) = \lambda^4 - \lambda^3 - 3\lambda^2 - 11\lambda + 6$, $\quad g(\lambda) = \lambda^3 - 2\lambda^2 - 2\lambda - 3$

(c) $f(\lambda) = 2\lambda^5 + 5\lambda^4 + 4\lambda^3 - \lambda^2 - \lambda + 1$, $\quad g(\lambda) = \lambda^3 + 2\lambda^2 + 2\lambda + 1$

(d) $f(\lambda) = 3\lambda^4 - 4\lambda^3 + \lambda^2 - 5\lambda + 6$, $\quad g(\lambda) = \lambda^2 + 2\lambda + 2$

Ans. (a) $\lambda^2 - 2 = -\frac{1}{3}(\lambda - 1) f(\lambda) + \frac{1}{3}(2\lambda^2 + 1) g(\lambda)$

(b) $\lambda - 3 = -\frac{1}{13}(\lambda + 4) f(\lambda) + \frac{1}{13}(\lambda^2 + 5\lambda + 5) g(\lambda)$

(c) $\lambda + 1 = \frac{1}{13}(\lambda + 4) f(\lambda) + \frac{1}{13}(-2\lambda^3 - 9\lambda^2 - 2\lambda + 9) g(\lambda)$

(d) $1 = \frac{1}{102}(5\lambda + 2) f(\lambda) + \frac{1}{102}(-15\lambda^3 + 44\lambda^2 - 55\lambda + 45) g(\lambda)$

11. Prove Theorem **VI**.

Hint. Let $d(\lambda)$ be the greatest common divisor of $f(\lambda)$ and $g(\lambda)$; then $g(\lambda) = d(\lambda) \cdot h(\lambda)$ and either $d(\lambda)$ or $h(\lambda)$ is a constant.

12. Prove Theorems **VII** and **VIII**.

13. Prove: If $f(\lambda)$ is relatively prime to $g(\lambda)$ and divides $g(\lambda)\cdot a(\lambda)$, it divides $a(\lambda)$.

14. The *least common multiple* of $f(\lambda)$ and $g(\lambda)$ is a monic polynomial which is a multiple of both $f(\lambda)$ and $g(\lambda)$, and is of minimum degree. Find the greatest common divisor and the least common multiple of

(*a*) $f(\lambda) = \lambda^3 - 1, \quad g(\lambda) = \lambda^2 - 1$
(*b*) $f(\lambda) = (\lambda-1)(\lambda+1)^2(\lambda+2), \quad g(\lambda) = (\lambda+1)(\lambda+2)^3(\lambda-3)$

Ans. (*a*) g.c.d. $= \lambda - 1$; l.c.m. $= (\lambda^2-1)(\lambda^2+\lambda+1)$
(*b*) g.c.d. $= (\lambda+1)(\lambda+2)$; l.c.m. $= (\lambda-1)(\lambda+1)^2(\lambda+2)^3(\lambda-3)$

15. Given $A = \begin{bmatrix} 1 & 2 & 2 \\ 2 & 1 & 2 \\ 2 & 2 & 1 \end{bmatrix}$, show

(*a*) $\phi(\lambda) = \lambda^3 - 3\lambda^2 - 9\lambda - 5$ and $\phi(A) = A^3 - 3A^2 - 9A - 5I = 0$
(*b*) $m(A) = 0$, when $m(\lambda) = \lambda^2 - 4\lambda - 5$.

16. What property of a field is not satisfied by a polynomial domain?

17. The scalar c is called a **root** of the polynomial $f(\lambda)$ if $f(c) = 0$. Prove: The scalar c is a root of $f(\lambda)$ if and only if $\lambda - c$ is a factor of $f(\lambda)$.

18. Suppose $f(\lambda) = (\lambda - c)^k g(\lambda)$. (*a*) Show that c is a root of multiplicity $k-1$ of $f'(\lambda)$. (*b*) Show that c is a root of multiplicity $k > 1$ of $f(\lambda)$ if and only if c is a root of both $f(\lambda)$ and $f'(\lambda)$.

19. Take $f(\lambda)$ and $g(\lambda)$, not both 0, in $F[\lambda]$ with greatest common divisor $d(\lambda)$. Let K be any field containing F. Show that if $D(\lambda)$ is the greatest common divisor of $f(\lambda)$ and $g(\lambda)$ considered in $K[\lambda]$, then $D(\lambda) = d(\lambda)$.
Hint: Let $d(\lambda) = h(\lambda)\cdot f(\lambda) + k(\lambda)\cdot g(\lambda)$, $f(\lambda) = s(\lambda)\cdot D(\lambda)$, $g(\lambda) = t(\lambda)\cdot D(\lambda)$, and $D(\lambda) = c(\lambda)\cdot d(\lambda)$.

20. Prove: An n-square matrix A is normal if $\bar{A}'$ can be expressed as a polynomial

$$a_s A^s + a_{s-1} A^{s-1} + \dots + a_1 A + a_0 I$$

in A.

Chapter 23

Lambda Matrices

DEFINITIONS. Let $F[\lambda]$ be a polynomial domain consisting of all polynomials in λ with coefficients in F. A non-zero $m \times n$ matrix over $F[\lambda]$

$$(23.1) \qquad A(\lambda) = [a_{ij}(\lambda)] = \begin{bmatrix} a_{11}(\lambda) & a_{12}(\lambda) & \ldots & a_{1n}(\lambda) \\ a_{21}(\lambda) & a_{22}(\lambda) & \ldots & a_{2n}(\lambda) \\ \ldots & \ldots & \ldots & \ldots \\ a_{m1}(\lambda) & a_{m2}(\lambda) & \ldots & a_{mn}(\lambda) \end{bmatrix}$$

is called a λ-matrix.

Let p be the maximum degree in λ of the polynomials $a_{ij}(\lambda)$ of **(23.1)**. Then $A(\lambda)$ can be written as a **matrix polynomial** of degree p in λ,

$$(23.2) \qquad A(\lambda) = A_p\lambda^p + A_{p-1}\lambda^{p-1} + \ldots + A_1\lambda + A_0$$

where the A_i are $m \times n$ matrices over F.

Example 1.

$$A(\lambda) = \begin{bmatrix} \lambda^2+\lambda+1 & \lambda^4+2\lambda^3+3\lambda^2+5 \\ \lambda^3-4 & \lambda^3-3\lambda^2 \end{bmatrix}$$

$$= \begin{bmatrix} 0 & 1 \\ 0 & 0 \end{bmatrix}\lambda^4 + \begin{bmatrix} 0 & 2 \\ 1 & 1 \end{bmatrix}\lambda^3 + \begin{bmatrix} 1 & 3 \\ 0 & -3 \end{bmatrix}\lambda^2 + \begin{bmatrix} 1 & 0 \\ 0 & 0 \end{bmatrix}\lambda + \begin{bmatrix} 1 & 5 \\ -4 & 0 \end{bmatrix}$$

is a λ-matrix or matrix polynomial of degree four.

If $A(\lambda)$ is n-square, it is called singular or non-singular according as $|A(\lambda)|$ is or is not zero. Further, $A(\lambda)$ is called **proper** or **improper** according as A_p is non-singular or singular. The matrix polynomial of Example 1 is non-singular and improper.

OPERATIONS WITH λ-MATRICES. Consider the two n-square λ-matrices or matrix polynomials over $F(\lambda)$

$$(23.3) \qquad A(\lambda) = A_p\lambda^p + A_{p-1}\lambda^{p-1} + \ldots + A_1\lambda + A_0$$

and

$$(23.4) \qquad B(\lambda) = B_q\lambda^q + B_{q-1}\lambda^{q-1} + \ldots + B_1\lambda + B_0$$

The matrices **(23.3)** and **(23.4)** are said to be equal, $A(\lambda) = B(\lambda)$, provided $p = q$ and $A_i = B_i$, $(i = 0, 1, 2, \ldots, p)$.

The sum $A(\lambda) + B(\lambda)$ is a λ-matrix $C(\lambda)$ obtained by adding corresponding elements of the two λ-matrices.

The product $A(\lambda) \cdot B(\lambda)$ is a λ-matrix or matrix polynomial of degree at most $p+q$. If either $A(\lambda)$ or $B(\lambda)$ is non-singular, the degree of $A(\lambda) \cdot B(\lambda)$ and also of $B(\lambda) \cdot A(\lambda)$ is exactly $p+q$.

The equality **(23.3)** is not disturbed when λ is replaced throughout by any scalar k of F. For example, putting $\lambda = k$ in **(23.3)** yields

$$A(k) = A_p k^p + A_{p-1}k^{p-1} + \ldots + A_1 k + A_0$$

However, when λ is replaced by an n-square matrix C, two results can be obtained due to the fact that, in general, two n-square matrices do not commute. We define

(23.5) $$A_R(C) = A_p C^p + A_{p-1}C^{p-1} + \ldots + A_1 C + A_0$$

and

(23.6) $$A_L(C) = C^p A_p + C^{p-1}A_{p-1} + \ldots + CA_1 + A_0$$

called respectively the **right and left functional values** of $A(\lambda)$.

Example 2. Let $A(\lambda) = \begin{bmatrix} \lambda^2 & \lambda+1 \\ \lambda-2 & \lambda^2+2 \end{bmatrix} = \begin{bmatrix} 1 & 0 \\ 0 & 1 \end{bmatrix}\lambda^2 + \begin{bmatrix} 0 & 1 \\ 1 & 0 \end{bmatrix}\lambda + \begin{bmatrix} 0 & 1 \\ -2 & 2 \end{bmatrix}$ and $C = \begin{bmatrix} 1 & 2 \\ 3 & 4 \end{bmatrix}$

Then $$A_R(C) = \begin{bmatrix} 1 & 0 \\ 0 & 1 \end{bmatrix}\begin{bmatrix} 1 & 2 \\ 3 & 4 \end{bmatrix}^2 + \begin{bmatrix} 0 & 1 \\ 1 & 0 \end{bmatrix}\begin{bmatrix} 1 & 2 \\ 3 & 4 \end{bmatrix} + \begin{bmatrix} 0 & 1 \\ -2 & 2 \end{bmatrix} = \begin{bmatrix} 10 & 15 \\ 14 & 26 \end{bmatrix}$$

and

$$A_L(C) = \begin{bmatrix} 1 & 2 \\ 3 & 4 \end{bmatrix}^2\begin{bmatrix} 1 & 0 \\ 0 & 1 \end{bmatrix} + \begin{bmatrix} 1 & 2 \\ 3 & 4 \end{bmatrix}\begin{bmatrix} 0 & 1 \\ 1 & 0 \end{bmatrix} + \begin{bmatrix} 0 & 1 \\ -2 & 2 \end{bmatrix} = \begin{bmatrix} 9 & 12 \\ 17 & 27 \end{bmatrix}$$

See Problem 1.

DIVISION. In Problem 2, we prove

I. If $A(\lambda)$ and $B(\lambda)$ are matrix polynomials **(23.3)** and **(23.4)** and if B_q is non-singular, then there exist unique matrix polynomials $Q_1(\lambda)$, $R_1(\lambda)$; $Q_2(\lambda)$, $R_2(\lambda)$, where $R_1(\lambda)$ and $R_2(\lambda)$ are either zero or of degree less than that of $B(\lambda)$, such that

(23.7) $$A(\lambda) = Q_1(\lambda)\cdot B(\lambda) + R_1(\lambda)$$

and

(23.8) $$A(\lambda) = B(\lambda)\cdot Q_2(\lambda) + R_2(\lambda)$$

If $R_1(\lambda) = 0$, $B(\lambda)$ is called a **right divisor** of $A(\lambda)$; if $R_2(\lambda) = 0$, $B(\lambda)$ is called a **left divisor** of $A(\lambda)$.

Example 3. If $A(\lambda) = \begin{bmatrix} \lambda^4+\lambda^2+\lambda-1 & \lambda^3+\lambda^2+\lambda+2 \\ 2\lambda^3-\lambda & 2\lambda^2+2\lambda \end{bmatrix}$ and $B(\lambda) = \begin{bmatrix} \lambda^2+1 & 1 \\ \lambda & \lambda^2+\lambda \end{bmatrix}$, then

$$A(\lambda) = \begin{bmatrix} \lambda^2-1 & \lambda-1 \\ 2\lambda & 2 \end{bmatrix}\begin{bmatrix} \lambda^2+1 & 1 \\ \lambda & \lambda^2+\lambda \end{bmatrix} + \begin{bmatrix} 2\lambda & 2\lambda+3 \\ -5\lambda & -2\lambda \end{bmatrix} = Q_1(\lambda)\cdot B(\lambda) + R_1(\lambda)$$

and

$$A(\lambda) = \begin{bmatrix} \lambda^2+1 & 1 \\ \lambda & \lambda^2+\lambda \end{bmatrix}\begin{bmatrix} \lambda^2 & \lambda+1 \\ \lambda-1 & 1 \end{bmatrix} = B(\lambda)\cdot Q_2(\lambda)$$

Here, $B(\lambda)$ is a left divisor of $A(\lambda)$.

See Problem 3.

A matrix polynomial of the form

(23.9) $$B(\lambda) = b_q\lambda^q\cdot I_n + b_{q-1}\lambda^{q-1}\cdot I_n + \ldots + b_1\lambda\cdot I_n + b_0 I_n = b(\lambda)\cdot I_n$$

is called **scalar**. A scalar matrix polynomial $B(\lambda) = b(\lambda)\cdot I_n$ commutes with every n-square matrix polynomial.

If in **(23.7)** and **(23.8)**, $B(\lambda) = b(\lambda)\cdot I$, then

(23.10) $$A(\lambda) = Q_1(\lambda)\cdot B(\lambda) + R_1(\lambda) = B(\lambda)\cdot Q_1(\lambda) + R_1(\lambda)$$

Example 4. Let $A(\lambda) = \begin{bmatrix} \lambda^2+2\lambda & \lambda+1 \\ \lambda^2-1 & 2\lambda+1 \end{bmatrix}$ and $B(\lambda) = (\lambda+2)I_2$. Then

$$A(\lambda) = \begin{bmatrix} \lambda & 1 \\ \lambda-2 & 2 \end{bmatrix}\begin{bmatrix} \lambda+2 & 0 \\ 0 & \lambda+2 \end{bmatrix} + \begin{bmatrix} 0 & -1 \\ 3 & -3 \end{bmatrix} = Q_1(\lambda)\cdot B(\lambda) + R_1(\lambda)$$

and

$$A(\lambda) = \begin{bmatrix} \lambda+2 & 0 \\ 0 & \lambda+2 \end{bmatrix}\begin{bmatrix} \lambda & 1 \\ \lambda-2 & 2 \end{bmatrix} + \begin{bmatrix} 0 & -1 \\ 3 & -3 \end{bmatrix} = B(\lambda)\cdot Q_1(\lambda) + R_1(\lambda)$$

If $R_1(\lambda) = 0$ in **(23.10)**, then $A(\lambda) = b(\lambda)\cdot I\cdot Q_1(\lambda)$ and we have

II. A matrix polynomial $A(\lambda) = [a_{ij}(\lambda)]$ of degree n is divisible by a scalar matrix polynomial $B(\lambda) = b(\lambda)\cdot I_n$ if and only if every $a_{ij}(\lambda)$ is divisible by $b(\lambda)$.

THE REMAINDER THEOREM. Let $A(\lambda)$ be the λ-matrix of **(23.3)** and let $B = [b_{ij}]$ be an n-square matrix over F. Since $\lambda I - B$ is non-singular, we may write

(23.11) $$A(\lambda) = Q_1(\lambda)\cdot(\lambda I - B) + R_1$$

and

(23.12) $$A(\lambda) = (\lambda I - B)\cdot Q_2(\lambda) + R_2$$

where R_1 and R_2 are free of λ. It can be shown

III. If the matrix polynomial $A(\lambda)$ of **(23.3)** is divided by $\lambda I - B$, where $B = [b_{ij}]$ is n-square, until remainders R_1 and R_2, free of λ, are obtained, then

$$R_1 = A_R(B) = A_p B^p + A_{p-1}B^{p-1} + \ldots + A_1 B + A_0$$

and

$$R_2 = A_L(B) = B^p A_p + B^{p-1}A_{p-1} + \ldots + BA_1 + A_0$$

Example 5. Let $A(\lambda) = \begin{bmatrix} \lambda^2 & \lambda+1 \\ \lambda-2 & \lambda^2+2 \end{bmatrix}$ and $\lambda I - B = \begin{bmatrix} \lambda-1 & -2 \\ -3 & \lambda-4 \end{bmatrix}$. Then

$$A(\lambda) = \begin{bmatrix} \lambda+1 & 3 \\ 4 & \lambda+4 \end{bmatrix}\begin{bmatrix} \lambda-1 & -2 \\ -3 & \lambda-4 \end{bmatrix} + \begin{bmatrix} 10 & 15 \\ 14 & 26 \end{bmatrix} = Q_1(\lambda)\cdot(\lambda I - B) + R_1$$

and

$$A(\lambda) = \begin{bmatrix} \lambda-1 & -2 \\ -3 & \lambda-4 \end{bmatrix}\begin{bmatrix} \lambda+1 & 3 \\ 4 & \lambda+4 \end{bmatrix} + \begin{bmatrix} 9 & 12 \\ 17 & 27 \end{bmatrix} = (\lambda I - B)Q_2(\lambda) + R_2$$

From Example 2, $R_1 = A_R(B)$ and $R_2 = A_L(B)$ in accordance with Theorem **III**.

When $A(\lambda)$ is a scalar matrix polynomial

$$A(\lambda) = f(\lambda)\cdot I = a_p I\lambda^p + a_{p-1}I\lambda^{p-1} + \ldots + a_1 I\lambda + a_0 I$$

the remainders in **(23.11)** and **(23.12)** are identical so that

$$R_1 = R_2 = a_p B^p + a_{p-1}B^{p-1} + \ldots + a_1 B + a_0 I$$

and we have

IV. If a scalar matrix polynomial $f(\lambda)\cdot I_n$ is divided by $\lambda I_n - B$ until a remainder R, free of λ, is obtained, then $R = f(B)$.

As a consequence, we have

V. A scalar matrix polynomial $f(\lambda)\cdot I_n$ is divisible by $\lambda I_n - B$ if and only if $f(B) = 0$.

CAYLEY-HAMILTON THEOREM. Consider the n-square matrix $A = [a_{ij}]$ having characteristic matrix $\lambda I - A$ and characteristic equation $\phi(\lambda) = |\lambda I - A| = 0$. By **(6.2)**,

$$(\lambda I - A)\cdot \text{adj}(\lambda I - A) = \phi(\lambda)\cdot I$$

Then $\phi(\lambda)\cdot I$ is divisible by $\lambda I - A$ and, by Theorem **V**, $\phi(A) = 0$. Thus,

VI. Every square matrix $A = [a_{ij}]$ satisfies its characteristic equation $\phi(\lambda) = 0$.

Example 6. The characteristic equation of $A = \begin{bmatrix} 2 & 2 & 1 \\ 1 & 3 & 1 \\ 1 & 2 & 2 \end{bmatrix}$ is $\lambda^3 - 7\lambda^2 + 11\lambda - 5 = 0$. Now

$$A^2 = \begin{bmatrix} 7 & 12 & 6 \\ 6 & 13 & 6 \\ 6 & 12 & 7 \end{bmatrix}, \qquad A^3 = \begin{bmatrix} 32 & 62 & 31 \\ 31 & 63 & 31 \\ 31 & 62 & 32 \end{bmatrix},$$

and

$$\begin{bmatrix} 32 & 62 & 31 \\ 31 & 63 & 31 \\ 31 & 62 & 32 \end{bmatrix} - 7\begin{bmatrix} 7 & 12 & 6 \\ 6 & 13 & 6 \\ 6 & 12 & 7 \end{bmatrix} + 11\begin{bmatrix} 2 & 2 & 1 \\ 1 & 3 & 1 \\ 1 & 2 & 2 \end{bmatrix} - 5\begin{bmatrix} 1 & 0 & 0 \\ 0 & 1 & 0 \\ 0 & 0 & 1 \end{bmatrix} = 0$$

See Problem 4.

SOLVED PROBLEMS

1. For the matrix $A(\lambda) = \begin{bmatrix} \lambda^2+\lambda & \lambda^2 \\ \lambda+1 & 1 \end{bmatrix}$, compute $A_R(C)$ and $A_L(C)$ when $C = \begin{bmatrix} 1 & -1 \\ 0 & 2 \end{bmatrix}$.

$$A(\lambda) = \begin{bmatrix} 1 & 1 \\ 0 & 0 \end{bmatrix}\lambda^2 + \begin{bmatrix} 1 & 0 \\ 1 & 0 \end{bmatrix}\lambda + \begin{bmatrix} 0 & 0 \\ 1 & 1 \end{bmatrix} \quad \text{and} \quad C^2 = \begin{bmatrix} 1 & -3 \\ 0 & 4 \end{bmatrix}; \text{ then}$$

$$A_R(C) = \begin{bmatrix} 1 & 1 \\ 0 & 0 \end{bmatrix}\begin{bmatrix} 1 & -3 \\ 0 & 4 \end{bmatrix} + \begin{bmatrix} 1 & 0 \\ 1 & 0 \end{bmatrix}\begin{bmatrix} 1 & -1 \\ 0 & 2 \end{bmatrix} + \begin{bmatrix} 0 & 0 \\ 1 & 1 \end{bmatrix}$$

$$= \begin{bmatrix} 1 & 1 \\ 0 & 0 \end{bmatrix} + \begin{bmatrix} 1 & -1 \\ 1 & -1 \end{bmatrix} + \begin{bmatrix} 0 & 0 \\ 1 & 1 \end{bmatrix} = \begin{bmatrix} 2 & 0 \\ 2 & 0 \end{bmatrix}$$

and

$$A_L(C) = \begin{bmatrix} 1 & -3 \\ 0 & 4 \end{bmatrix}\begin{bmatrix} 1 & 1 \\ 0 & 0 \end{bmatrix} + \begin{bmatrix} 1 & -1 \\ 0 & 2 \end{bmatrix}\begin{bmatrix} 1 & 0 \\ 1 & 0 \end{bmatrix} + \begin{bmatrix} 0 & 0 \\ 1 & 1 \end{bmatrix} = \begin{bmatrix} 1 & 1 \\ 3 & 1 \end{bmatrix}$$

2. Prove: If $A(\lambda)$ and $B(\lambda)$ are the λ-matrices **(23.3)** and **(23.4)** and if B_q is non-singular, then there exist unique polynomial matrices $Q_1(\lambda)$, $R_1(\lambda)$; $Q_2(\lambda)$, $R_2(\lambda)$, where $R_1(\lambda)$ and $R_2(\lambda)$ are either zero or of degree less than that of $B(\lambda)$, such that

(i) $$A(\lambda) = Q_1(\lambda)\cdot B(\lambda) + R_1(\lambda)$$

and

(ii) $$A(\lambda) = B(\lambda)\cdot Q_2(\lambda) + R_2(\lambda)$$

If $p < q$, then **(i)** holds with $Q_1(\lambda) = 0$ and $R_1(\lambda) = A(\lambda)$. Suppose that $p \geqq q$; then

$$A(\lambda) - A_p B_q^{-1} B(\lambda)\lambda^{p-q} = C(\lambda)$$

where $C(\lambda)$ is either zero or of degree at most $p-1$.

If $C(\lambda)$ is zero or of degree less than q, we have **(i)** with

$$Q_1(\lambda) = A_p B_q^{-1}\lambda^{p-q} \qquad \text{and} \qquad R_1(\lambda) = C(\lambda)$$

If $C(\lambda) = C_s\lambda^s + \dots$ where $s > q$, form

$$A(\lambda) - A_p B_q^{-1} B(\lambda)\lambda^{p-q} - C_s B_q^{-1} B(\lambda)\lambda^{s-q} = D(\lambda)$$

If $D(\lambda)$ is either zero or of degree less than q, we have **(i)** with

$$Q_1(\lambda) = A_p B_q^{-1}\lambda^{p-q} + C_s B_q^{-1}\lambda^{s-q} \qquad \text{and} \qquad R_1(\lambda) = D(\lambda)$$

otherwise, we continue the process. Since this results in a sequence of matrix polynomials $C(\lambda)$, $D(\lambda)$, ... of decreasing degrees, we ultimately reach a matrix polynomial which is either zero or of degree less than q and we have **(i)**.

To obtain **(ii)**, begin with

$$A(\lambda) - B(\lambda) B_q^{-1} A_p \lambda^{p-q}$$

This derivation together with the proof of uniqueness will be left as an exercise. See Problem 1, Chapter 22.

3. Given $A(\lambda) = \begin{bmatrix} \lambda^4 + 2\lambda^3 - 1 & \lambda^3 - \lambda - 1 \\ \lambda^3 + \lambda^2 + 1 & \lambda^3 + 1 \end{bmatrix}$ and $B(\lambda) = \begin{bmatrix} 2\lambda^2 - \lambda & -\lambda^2 + \lambda - 1 \\ -\lambda^2 + 2 & \lambda^2 - \lambda \end{bmatrix}$

find matrices $Q_1(\lambda)$, $R_1(\lambda)$; $Q_2(\lambda)$, $R_2(\lambda)$ such that

(*a*) $A(\lambda) = Q_1(\lambda)\cdot B(\lambda) + R_1(\lambda)$, (*b*) $A(\lambda) = B(\lambda)\cdot Q_2(\lambda) + R_2(\lambda)$ as in Problem 2.

We have

$$A(\lambda) = \begin{bmatrix} 1 & 0 \\ 0 & 0 \end{bmatrix}\lambda^4 + \begin{bmatrix} 2 & 1 \\ 1 & 1 \end{bmatrix}\lambda^3 + \begin{bmatrix} 0 & 0 \\ 1 & 0 \end{bmatrix}\lambda^2 + \begin{bmatrix} 0 & -1 \\ 0 & 0 \end{bmatrix}\lambda + \begin{bmatrix} -1 & -1 \\ 1 & 1 \end{bmatrix}$$

and

$$B(\lambda) = \begin{bmatrix} 2 & -1 \\ -1 & 1 \end{bmatrix}\lambda^2 + \begin{bmatrix} -1 & 1 \\ 0 & -1 \end{bmatrix}\lambda + \begin{bmatrix} 0 & -1 \\ 2 & 0 \end{bmatrix}$$

Here, $B_2 = \begin{bmatrix} 2 & -1 \\ -1 & 1 \end{bmatrix}$ and $B_2^{-1} = \begin{bmatrix} 1 & 1 \\ 1 & 2 \end{bmatrix}$.

(*a*) We compute

$$A(\lambda) - A_4 B_2^{-1} B(\lambda)\lambda^2 = \begin{bmatrix} 3 & 1 \\ 1 & 1 \end{bmatrix}\lambda^3 + \begin{bmatrix} -2 & 1 \\ 1 & 0 \end{bmatrix}\lambda^2 + \begin{bmatrix} 0 & -1 \\ 0 & 0 \end{bmatrix}\lambda + \begin{bmatrix} -1 & -1 \\ 1 & 1 \end{bmatrix} = C(\lambda)$$

$$C(\lambda) - C_3 B_2^{-1} B(\lambda)\lambda = \begin{bmatrix} 2 & 2 \\ 3 & 1 \end{bmatrix}\lambda^2 + \begin{bmatrix} -10 & 3 \\ -6 & 2 \end{bmatrix}\lambda + \begin{bmatrix} -1 & -1 \\ 1 & 1 \end{bmatrix} = D(\lambda)$$

and

$$D(\lambda) - D_2 B_2^{-1} B(\lambda) = \begin{bmatrix} -6 & 5 \\ -2 & 3 \end{bmatrix}\lambda + \begin{bmatrix} -13 & 3 \\ -9 & 5 \end{bmatrix} = \begin{bmatrix} -6\lambda - 13 & 5\lambda + 3 \\ -2\lambda - 9 & 3\lambda + 5 \end{bmatrix} = R_1(\lambda)$$

Then $Q_1(\lambda) = (A_4\lambda^2 + C_3\lambda + D_2) B_2^{-1} = \begin{bmatrix} 1 & 1 \\ 0 & 0 \end{bmatrix}\lambda^2 + \begin{bmatrix} 4 & 5 \\ 2 & 3 \end{bmatrix}\lambda + \begin{bmatrix} 4 & 6 \\ 4 & 5 \end{bmatrix}$

$$= \begin{bmatrix} \lambda^2 + 4\lambda + 4 & \lambda^2 + 5\lambda + 6 \\ 2\lambda + 4 & 3\lambda + 5 \end{bmatrix}$$

(*b*) We compute

$$A(\lambda) - B(\lambda) B_2^{-1} A_4\lambda^2 = \begin{bmatrix} 2 & 1 \\ 2 & 1 \end{bmatrix}\lambda^3 + \begin{bmatrix} 1 & 0 \\ -1 & 0 \end{bmatrix}\lambda^2 + \begin{bmatrix} 0 & -1 \\ 0 & 0 \end{bmatrix}\lambda + \begin{bmatrix} -1 & -1 \\ 1 & 1 \end{bmatrix} = E(\lambda)$$

$$E(\lambda) - B(\lambda) B_2^{-1} E_3\lambda = \begin{bmatrix} -1 & -1 \\ 5 & 3 \end{bmatrix}\lambda^2 + \begin{bmatrix} 6 & 2 \\ -8 & -4 \end{bmatrix}\lambda + \begin{bmatrix} -1 & -1 \\ 1 & 1 \end{bmatrix} = F(\lambda)$$

and

$$F(\lambda) - B(\lambda) B_2^{-1} F_2 = \begin{bmatrix} 1 & -1 \\ 1 & 1 \end{bmatrix}\lambda + \begin{bmatrix} 8 & 4 \\ -7 & -3 \end{bmatrix} = \begin{bmatrix} \lambda + 8 & -\lambda + 4 \\ \lambda - 7 & \lambda - 3 \end{bmatrix} = R_2(\lambda)$$

Then $$Q_2(\lambda) = B_2^{-1}(A_4\lambda^2 + E_3\lambda + F_2) = \begin{bmatrix}1 & 0\\1 & 0\end{bmatrix}\lambda^2 + \begin{bmatrix}4 & 2\\6 & 3\end{bmatrix}\lambda + \begin{bmatrix}4 & 2\\9 & 5\end{bmatrix} = \begin{bmatrix}\lambda^2+4\lambda+4 & 2\lambda+2\\ \lambda^2+6\lambda+9 & 3\lambda+5\end{bmatrix}$$

4. Given $A = \begin{bmatrix}1 & 1 & 2\\3 & 1 & 1\\2 & 3 & 1\end{bmatrix}$, use the fact that A satisfies its characteristic equation to compute A^3 and A^4; also, since A is non-singular, to compute A^{-1} and A^{-2}.

$$|\lambda I - A| = \begin{vmatrix}\lambda-1 & -1 & -2\\ -3 & \lambda-1 & -1\\ -2 & -3 & \lambda-1\end{vmatrix} = \lambda^3 - 3\lambda^2 - 7\lambda - 11 = 0$$

Then

$$A^3 = 3A^2 + 7A + 11I = 3\begin{bmatrix}8 & 8 & 5\\8 & 7 & 8\\13 & 8 & 8\end{bmatrix} + 7\begin{bmatrix}1 & 1 & 2\\3 & 1 & 1\\2 & 3 & 1\end{bmatrix} + 11\begin{bmatrix}1 & 0 & 0\\0 & 1 & 0\\0 & 0 & 1\end{bmatrix} = \begin{bmatrix}42 & 31 & 29\\45 & 39 & 31\\53 & 45 & 42\end{bmatrix}$$

$$A^4 = 3A^3 + 7A^2 + 11A = 3\begin{bmatrix}42 & 31 & 29\\45 & 39 & 31\\53 & 45 & 42\end{bmatrix} + 7\begin{bmatrix}8 & 8 & 5\\8 & 7 & 8\\13 & 8 & 8\end{bmatrix} + 11\begin{bmatrix}1 & 1 & 2\\3 & 1 & 1\\2 & 3 & 1\end{bmatrix} = \begin{bmatrix}193 & 160 & 144\\224 & 177 & 160\\272 & 224 & 193\end{bmatrix}$$

From $11I = -7A - 3A^2 + A^3$, we have

$$A^{-1} = \frac{1}{11}\{-7I - 3A + A^2\} = \frac{1}{11}\left\{-7\begin{bmatrix}1 & 0 & 0\\0 & 1 & 0\\0 & 0 & 1\end{bmatrix} - 3\begin{bmatrix}1 & 1 & 2\\3 & 1 & 1\\2 & 3 & 1\end{bmatrix} + \begin{bmatrix}8 & 8 & 5\\8 & 7 & 8\\13 & 8 & 8\end{bmatrix}\right\} = \frac{1}{11}\begin{bmatrix}-2 & 5 & -1\\-1 & -3 & 5\\7 & -1 & -2\end{bmatrix}$$

$$A^{-2} = \frac{1}{11}\{-7A^{-1} - 3I + A\} = \frac{1}{121}\left\{-7\begin{bmatrix}-2 & 5 & -1\\-1 & -3 & 5\\7 & -1 & -2\end{bmatrix} - 33\begin{bmatrix}1 & 0 & 0\\0 & 1 & 0\\0 & 0 & 1\end{bmatrix} + 11\begin{bmatrix}1 & 1 & 2\\3 & 1 & 1\\2 & 3 & 1\end{bmatrix}\right\} = \frac{1}{121}\begin{bmatrix}-8 & -24 & 29\\40 & -1 & -24\\-27 & 40 & -8\end{bmatrix}$$

5. Let $\lambda_1, \lambda_2, \ldots, \lambda_n$ be the characteristic roots of an n-square matrix A and let $h(x)$ be a polynomial of degree p in x. Prove that $|h(A)| = h(\lambda_1)\cdot h(\lambda_2)\ldots h(\lambda_n)$.

We have

(i) $$|\lambda I - A| = (\lambda-\lambda_1)(\lambda-\lambda_2)\ldots(\lambda-\lambda_n)$$

Let

(ii) $$h(x) = c(s_1 - x)(s_2 - x)\ldots(s_p - x)$$

Then

$$h(A) = c(s_1I - A)(s_2I - A)\ldots(s_pI - A)$$

and

$$\begin{aligned}
|h(A)| &= c^{p}|s_1 I - A|\cdot|s_2 I - A|\ldots|s_p I - A| \\
&= \{c(s_1-\lambda_1)(s_1-\lambda_2)\ldots(s_1-\lambda_n)\} \\
&\qquad \cdot\{c(s_2-\lambda_1)(s_2-\lambda_2)\ldots(s_2-\lambda_n)\}\ldots\{c(s_p-\lambda_1)(s_p-\lambda_2)\ldots(s_p-\lambda_n)\} \\
&= \{c(s_1-\lambda_1)(s_2-\lambda_1)\ldots(s_p-\lambda_1)\} \\
&\qquad \cdot\{c(s_1-\lambda_2)(s_2-\lambda_2)\ldots(s_p-\lambda_2)\}\ldots\{c(s_1-\lambda_n)(s_2-\lambda_n)\ldots(s_p-\lambda_n)\} \\
&= h(\lambda_1)\,h(\lambda_2)\ldots h(\lambda_n)
\end{aligned}$$

using **(ii)**.

SUPPLEMENTARY PROBLEMS

6. Given $A(\lambda) = \begin{bmatrix} \lambda^2+2\lambda & \lambda \\ \lambda^2+1 & \lambda-1 \end{bmatrix}$ and $B(\lambda) = \begin{bmatrix} \lambda^2 & \lambda^2+\lambda \\ \lambda+1 & \lambda \end{bmatrix}$, compute:

(*a*) $A(\lambda) + B(\lambda) = \begin{bmatrix} 2\lambda^2+2\lambda & \lambda^2+2\lambda \\ \lambda^2+\lambda+2 & 2\lambda-1 \end{bmatrix}$

(*b*) $A(\lambda) - B(\lambda) = \begin{bmatrix} 2\lambda & -\lambda^2 \\ \lambda^2-\lambda & -1 \end{bmatrix}$

(*c*) $A(\lambda)\cdot B(\lambda) = \begin{bmatrix} \lambda^4+2\lambda^3+\lambda^2+\lambda & \lambda^4+3\lambda^3+3\lambda^2 \\ \lambda^4+2\lambda^2-1 & \lambda^4+\lambda^3+2\lambda^2 \end{bmatrix}$

(*d*) $B(\lambda)\cdot A(\lambda) = \begin{bmatrix} 2\lambda^4+3\lambda^3+\lambda^2+\lambda & 2\lambda^3-\lambda \\ 2\lambda^3+3\lambda^2+3\lambda & 2\lambda^2 \end{bmatrix}$

7. Given $A(\lambda) = \begin{bmatrix} \lambda^2+1 & \lambda^2 \\ \lambda+2 & \lambda-1 \end{bmatrix}$, $B(\lambda) = \begin{bmatrix} \lambda^2 & \lambda \\ -\lambda^2 & 1 \end{bmatrix}$, and $C = \begin{bmatrix} 1 & 0 \\ 2 & -1 \end{bmatrix}$, compute:

$A_R(C) = \begin{bmatrix} 2 & 1 \\ 5 & -2 \end{bmatrix}$, $B_R(C) = \begin{bmatrix} 3 & -1 \\ -1 & 1 \end{bmatrix}$, $A_R(C)\cdot B_R(C) = \begin{bmatrix} 5 & -1 \\ 17 & -7 \end{bmatrix}$, $B_R(C)\cdot A_R(C) = \begin{bmatrix} 1 & 5 \\ 3 & -3 \end{bmatrix}$,

$P_R(C) = \begin{bmatrix} 5 & -1 \\ 9 & -3 \end{bmatrix}$, $Q_R(C) = \begin{bmatrix} 3 & 3 \\ 3 & -3 \end{bmatrix}$;

$A_L(C) = \begin{bmatrix} 2 & 1 \\ 1 & -2 \end{bmatrix}$, $B_L(C) = \begin{bmatrix} 1 & 1 \\ -1 & 3 \end{bmatrix}$, $A_L(C)\cdot B_L(C) = \begin{bmatrix} 1 & 5 \\ 3 & -5 \end{bmatrix}$, $B_L(C)\cdot A_L(C) = \begin{bmatrix} 3 & -1 \\ 1 & -7 \end{bmatrix}$,

$P_L(C) = \begin{bmatrix} 1 & 3 \\ 3 & 1 \end{bmatrix}$, $Q_L(C) = \begin{bmatrix} 5 & 1 \\ 3 & -5 \end{bmatrix}$.

where $P(\lambda) = A(\lambda)\cdot B(\lambda)$ and $Q(\lambda) = B(\lambda)\cdot A(\lambda)$.

8. If $A(\lambda)$ and $B(\lambda)$ are proper n-square λ-matrices of respective degrees p and q, and if $C(\lambda)$ is any non-zero λ-matrix, show that the degree of the triple product in any order is at least $p+q$.

9. For each pair of matrices $A(\lambda)$ and $B(\lambda)$, find matrices $Q_1(\lambda)$, $R_1(\lambda)$; $Q_2(\lambda)$, $R_2(\lambda)$ satisfying **(23.7)** and **(23.8)**.

(a) $A(\lambda) = \begin{bmatrix} \lambda^2+2\lambda & \lambda \\ \lambda^2+1 & \lambda^2-\lambda \end{bmatrix}$, $B(\lambda) = \begin{bmatrix} \lambda & \lambda \\ 1 & -\lambda \end{bmatrix}$

(b) $A(\lambda) = \begin{bmatrix} -\lambda^2+\lambda & -\lambda^2+2\lambda+2 \\ -\lambda^2+2\lambda-1 & 1 \end{bmatrix}$, $B(\lambda) = \begin{bmatrix} \lambda-1 & -1 \\ 0 & \lambda-2 \end{bmatrix}$

(c) $A(\lambda) = \begin{bmatrix} \lambda^3-2\lambda+1 & \lambda^4+\lambda^2+7\lambda-2 & 5\lambda^2+2\lambda+4 \\ \lambda^4+3\lambda^2-3\lambda-1 & 3\lambda^3+2\lambda+2 & 4\lambda^2+6\lambda+1 \\ 2\lambda^3-\lambda+2 & \lambda^3+2\lambda^2 & \lambda^3+\lambda^2+8\lambda-4 \end{bmatrix}$,

$B(\lambda) = \begin{bmatrix} \lambda^2+1 & 1 & 3\lambda-1 \\ 2\lambda & \lambda^2 & \lambda+1 \\ \lambda-2 & 2\lambda & \lambda^2 \end{bmatrix}$

(d) $A(\lambda) = \begin{bmatrix} 3\lambda^4+\lambda^2-1 & \lambda^3-1 & \lambda^2-\lambda \\ \lambda^3-\lambda^2+1 & \lambda^4+\lambda^2+2 & \lambda-1 \\ \lambda^2+\lambda & \lambda+1 & 2\lambda^4+\lambda-2 \end{bmatrix}$, $B(\lambda) = \begin{bmatrix} \lambda^2-2\lambda-1 & \lambda^2+1 & \lambda+1 \\ \lambda+1 & \lambda^2+\lambda & \lambda^2-2\lambda \\ \lambda & \lambda-2 & \lambda^2+\lambda-1 \end{bmatrix}$

Ans. (a) $Q_1(\lambda) = \begin{bmatrix} \lambda+1 & \lambda \\ \lambda & 1 \end{bmatrix}$, $R_1(\lambda) = 0$; $Q_2(\lambda) = \begin{bmatrix} 2\lambda & \lambda-1 \\ -\lambda+2 & -\lambda+2 \end{bmatrix}$, $R_2(\lambda) = \begin{bmatrix} 0 & 0 \\ 1 & 1 \end{bmatrix}$

(b) $Q_1(\lambda) = \begin{bmatrix} -\lambda & -\lambda-1 \\ -\lambda+1 & -1 \end{bmatrix}$, $R_1(\lambda) = 0$; $Q_2(\lambda) = \begin{bmatrix} -\lambda-1 & -\lambda+1 \\ -\lambda & 0 \end{bmatrix}$, $R_2(\lambda) = \begin{bmatrix} -1 & 3 \\ -1 & 1 \end{bmatrix}$

(c) $Q_1(\lambda) = \begin{bmatrix} -\lambda+1 & \lambda^2+3 & -\lambda+7 \\ \lambda^2-1 & 3\lambda+5 & -3\lambda+2 \\ 2\lambda-3 & \lambda & \lambda-6 \end{bmatrix}$, $R_1(\lambda) = \begin{bmatrix} -16\lambda+14 & -6\lambda-3 & -5\lambda+2 \\ -21\lambda+4 & -2\lambda+3 & \lambda-5 \\ 5\lambda-7 & 10\lambda+3 & 18\lambda-7 \end{bmatrix}$

$Q_2(\lambda) = \begin{bmatrix} \lambda-1 & \lambda^2 & 2 \\ \lambda^2+1 & \lambda & 3 \\ -1 & 2 & \lambda+1 \end{bmatrix}$, $R_2(\lambda) = 0$

(d) $Q_1(\lambda) = \begin{bmatrix} 3\lambda^2+6\lambda+31 & -3\lambda^2-5\lambda-16 & 3\lambda^2-7\lambda+8 \\ \lambda-3 & \lambda^2-\lambda-1 & -\lambda^2+4\lambda-7 \\ -2\lambda-1 & 7 & 2\lambda^2-2\lambda-1 \end{bmatrix}$

$R_1(\lambda) = \begin{bmatrix} 81\lambda+46 & -12\lambda-16 & -85\lambda-23 \\ 4\lambda-1 & 15\lambda-9 & 12\lambda-5 \\ -9\lambda-8 & -7\lambda & 17\lambda-2 \end{bmatrix}$

$Q_2(\lambda) = \begin{bmatrix} 3\lambda^2+5\lambda+31 & -\lambda^2-\lambda-4 & 2\lambda^2-4\lambda+3 \\ \lambda-14 & \lambda^2 & -2\lambda^2+6\lambda-6 \\ -3\lambda-2 & 3 & 2\lambda^2-2\lambda-2 \end{bmatrix}$

$R_2(\lambda) = \begin{bmatrix} 71\lambda+46 & -12\lambda-8 & -\lambda+11 \\ -26\lambda-30 & 11\lambda+6 & 4\lambda-4 \\ -15\lambda-30 & 2\lambda+4 & 16\lambda-16 \end{bmatrix}$

10. Verify in Problem 9(*b*) that $R_1(\lambda) = A_R(C)$ and $R_2(\lambda) = A_L(C)$ where $B(\lambda) = \lambda I - C$.

11. Given $B(\lambda) = \begin{bmatrix} \lambda^2+1 & 3\lambda+1 \\ \lambda-2 & \lambda^2-3\lambda+2 \end{bmatrix}$ and $C(\lambda) = \begin{bmatrix} \lambda+2 & \lambda \\ \lambda-3 & \lambda+1 \end{bmatrix}$

(a) compute $A(\lambda) = B(\lambda)\cdot C(\lambda)$

(b) find $Q(\lambda)$ and $R(\lambda)$ of degree at most one such that $A(\lambda) = Q(\lambda)\cdot B(\lambda) + R(\lambda)$.

Ans. $\begin{bmatrix} \lambda^3+5\lambda^2-7\lambda-1 & \lambda^3+3\lambda^2+5\lambda+1 \\ \lambda^3-5\lambda^2+11\lambda-10 & \lambda^3-\lambda^2-3\lambda+2 \end{bmatrix} = \begin{bmatrix} \lambda+4 & \lambda+3 \\ \lambda-6 & \lambda-1 \end{bmatrix} B(\lambda) + \begin{bmatrix} -9\lambda+1 & -\lambda-9 \\ 13\lambda-6 & 9\lambda+10 \end{bmatrix}$

12. Given $A = \begin{bmatrix} 1 & 2 \\ 1 & 1 \end{bmatrix}$, compute as in Problem 4

$$A^2 = \begin{bmatrix} 3 & 4 \\ 2 & 3 \end{bmatrix}, \quad A^3 = \begin{bmatrix} 7 & 10 \\ 5 & 7 \end{bmatrix}, \quad A^4 = \begin{bmatrix} 17 & 24 \\ 12 & 17 \end{bmatrix}$$

$$A^{-1} = \begin{bmatrix} -1 & 2 \\ 1 & -1 \end{bmatrix}, \quad A^{-2} = \begin{bmatrix} 3 & -4 \\ -2 & 3 \end{bmatrix}, \quad A^{-3} = \begin{bmatrix} -7 & 10 \\ 5 & -7 \end{bmatrix}$$

13. Prove: If A and B are similar matrices and $g(\lambda)$ is any scalar polynomial, then $g(A)$ and $g(B)$ are similar.
Hint. Show first that A^k and B^k are similar for any positive integer k.

14. Prove: If $B = \text{diag}(B_1, B_2, \ldots, B_m)$ and $g(\lambda)$ is any scalar polynomial, then

$$g(B) = \text{diag}\left(g(B_1), g(B_2), \ldots, g(B_m)\right)$$

15. Prove Theorem **III**.
Hint. Verify: $\lambda I - B$ divides $A(\lambda) - A_R(B)$.

16. The matrix C is called a **root** of the scalar matrix polynomial $B(\lambda)$ of (**23.9**) if $B(C) = 0$. Prove: The matrix C is a root of $B(\lambda)$ if and only if the characteristic matrix of C divides $B(\lambda)$.

17. Prove: If $\lambda_1, \lambda_2, \ldots, \lambda_n$ are the characteristic roots of A and if $f(A)$ is any scalar polynomial in A, then the characteristic roots of $f(A)$ are $f(\lambda_1), f(\lambda_2), \ldots, f(\lambda_n)$.
Hint. Write $\lambda - f(x) = c(x_1-x)(x_2-x)\ldots(x_s-x)$ so that $|\lambda I - f(A)| = c^n|x_1I-A|\cdot|x_2I-A|\ldots|x_sI-A|$.
Now use $|x_iI-A| = (x_i-\lambda_1)(x_i-\lambda_2)\ldots(x_i-\lambda_n)$ and $c(x_1-\lambda_j)(x_2-\lambda_j)\ldots(x_s-\lambda_j) = \lambda - f(\lambda_j)$.

18. Find the characteristic roots of $f(A) = A^2 - 2A + 3$, given $A = \begin{bmatrix} 1 & -1 & 0 \\ 2 & 3 & 2 \\ 1 & 1 & 2 \end{bmatrix}$

19. Obtain the theorem of Problem 5 as a corollary of Problem 17.

20. Prove: If X is an invariant vector of A of Problem 17, then X is an invariant vector of $f(A)$.

21. Let $A(t) = [a_{ij}(t)]$ where the $a_{ij}(t)$ are real polynomials in the real variable t. Take

$$A(t) = \begin{bmatrix} t^2+t+1 & t^4+2t^3+3t^2+5 \\ t^3-4 & t^3-3t^2 \end{bmatrix} = \begin{bmatrix} 0 & 1 \\ 0 & 0 \end{bmatrix} t^4 + \begin{bmatrix} 0 & 2 \\ 1 & 1 \end{bmatrix} t^3 + \begin{bmatrix} 1 & 3 \\ 0 & -3 \end{bmatrix} t^2 + \begin{bmatrix} 1 & 0 \\ 0 & 0 \end{bmatrix} t + \begin{bmatrix} 1 & 5 \\ -4 & 0 \end{bmatrix}$$

and differentiate the last member as if it were a polynomial with constant coefficients to suggest the definition

$$\frac{d}{dt}A(t) = \left[\frac{d}{dt}a_{ij}(t)\right]$$

22. Derive formulas for:

(a) $\frac{d}{dt}\{A(t) + B(t)\}$; (b) $\frac{d}{dt}\{cA(t)\}$, where c is a constant or $c = [c_{ij}]$; (c) $\frac{d}{dt}\{A(t)\cdot B(t)\}$; (d) $\frac{d}{dt}A^{-1}(t)$.

Hint. For (c), write $A(t)\cdot B(t) = C(t) = [c_{ij}(t)]$ and differentiate $c_{ij}(t) = \sum_{k=1}^{n} a_{ik}(t)\, b_{kj}(t)$. For (d), use $A(t)\cdot A^{-1}(t) = I$.

Chapter 24

Smith Normal Form

BY AN ELEMENTARY TRANSFORMATION on a λ-matrix $A(\lambda)$ over $F[\lambda]$ is meant:

(1) The interchange of the ith and jth row, denoted by H_{ij}; the interchange of the ith and jth column, denoted by K_{ij}.

(2) The multiplication of the ith row by a non-zero constant k, denoted by $H_i(k)$; the multiplication of the ith column by a non-zero constant k, denoted by $K_i(k)$.

(3) The addition to the ith row of the product of $f(\lambda)$, any polynomial of $F[\lambda]$, and the jth row, denoted by $H_{ij}(f(\lambda))$; the addition to the ith column of the product of $f(\lambda)$ and the jth column, denoted by $K_{ij}(f(\lambda))$.

These are the elementary transformations of Chapter 5 except that in (3) the word scalar has been replaced by polynomial. An elementary transformation and the elementary matrix obtained by performing the elementary transformation on I will again be denoted by the same symbol. Also, a row transformation on $A(\lambda)$ is effected by multiplying it on the left by the appropriate H and a column transformation is effected by multiplying $A(\lambda)$ on the right by the appropriate K.

Paralleling Chapter 5, we have

I. Every elementary matrix in $F[\lambda]$ has an inverse which in turn is an elementary matrix in $F[\lambda]$.

II. If $|A(\lambda)| = k \neq 0$, with k in F, $A(\lambda)$ is a product of elementary matrices.

III. The rank of a λ-matrix is invariant under elementary transformations.

Two n-square λ-matrices $A(\lambda)$ and $B(\lambda)$ with elements in $F[\lambda]$ are called **equivalent** provided there exist $P(\lambda) = H_s \ldots H_2 \cdot H_1$ and $Q(\lambda) = K_1 \cdot K_2 \ldots K_t$ such that

(24.1) $$B(\lambda) = P(\lambda) \cdot A(\lambda) \cdot Q(\lambda)$$

Thus,

IV. Equivalent $m \times n$ λ-matrices have the same rank.

THE CANONICAL SET. In Problems 1 and 2, we prove

V. Let $A(\lambda)$ and $B(\lambda)$ be equivalent matrices of rank r; then the greatest common divisor of all s-square minors of $A(\lambda)$, $s \leqq r$, is also the greatest common divisor of all s-square minors of $B(\lambda)$.

In Problem 3, we prove

VI. Every λ-matrix $A(\lambda)$ of rank r can be reduced by elementary transformations to the **Smith normal form**

$$\textbf{(24.2)} \qquad N(\lambda) \;=\; \begin{bmatrix} f_1(\lambda) & 0 & \cdots & 0 & \cdots & 0 \\ 0 & f_2(\lambda) & \cdots & 0 & \cdots & 0 \\ \cdots & \cdots & \cdots & \cdots & \cdots & \cdots \\ 0 & 0 & \cdots & f_r(\lambda) & \cdots & 0 \\ 0 & 0 & \cdots & 0 & \cdots & 0 \\ \cdots & \cdots & \cdots & \cdots & \cdots & \cdots \\ 0 & 0 & \cdots & 0 & \cdots & 0 \end{bmatrix}$$

where each $f_i(\lambda)$ is monic and $f_i(\lambda)$ divides $f_{i+1}(\lambda)$, $(i = 1, 2, \ldots, r-1)$.

When a λ-matrix $A(\lambda)$ of rank r has been reduced to **(24.2)**, the greatest common divisor of all s-square minors of $A(\lambda)$, $s \leqq r$, is the greatest common divisor of all s-square minors of $N(\lambda)$ by Theorem **V**. Since in $N(\lambda)$ each $f_i(\lambda)$ divides $f_{i+1}(\lambda)$, the greatest common divisor of all s-square minors of $N(\lambda)$ and thus of $A(\lambda)$ is

$$\textbf{(24.3)} \qquad g_s(\lambda) \;=\; f_1(\lambda)\cdot f_2(\lambda)\cdot \ldots \cdot f_s(\lambda), \qquad (s = 1, 2, \ldots, r)$$

Suppose $A(\lambda)$ has been reduced to

$$N(\lambda) \;=\; \text{diag}\,(f_1(\lambda),\, f_2(\lambda),\, \ldots,\, f_r(\lambda),\, 0,\, \ldots,\, 0)$$

and to

$$N_1(\lambda) \;=\; \text{diag}\,(h_1(\lambda),\, h_2(\lambda),\, \ldots,\, h_r(\lambda),\, 0,\, \ldots,\, 0)$$

By **(24.3)**,

$$g_s(\lambda) \;=\; f_1(\lambda)\cdot f_2(\lambda)\cdot \ldots \cdot f_s(\lambda) \;=\; h_1(\lambda)\cdot h_2(\lambda)\cdot \ldots \cdot h_s(\lambda)$$

Now $g_1(\lambda) = f_1(\lambda) = h_1(\lambda)$, $g_2(\lambda) = f_1(\lambda)\cdot f_2(\lambda) = h_1(\lambda)\cdot h_2(\lambda)$ so that $f_2(\lambda) = h_2(\lambda)$, ...; in general, if we define $g_0(\lambda) = 1$, then

$$\textbf{(24.4)} \qquad g_s(\lambda)/g_{s-1}(\lambda) \;=\; f_s(\lambda) \;=\; h_s(\lambda), \qquad (s = 1, 2, \ldots, r)$$

and we have

VII. The matrix $N(\lambda)$ of **(24.2)** is uniquely determined by the given matrix $A(\lambda)$.

Thus, the Smith normal matrices are a canonical set for equivalence over $F[\lambda]$.

Example 1. Consider $\quad A(\lambda) \;=\; \begin{bmatrix} \lambda+2 & \lambda+1 & \lambda+3 \\ \lambda^3+2\lambda^2+\lambda & \lambda^3+\lambda^2+\lambda & 2\lambda^3+3\lambda^2+\lambda \\ \lambda^2+3\lambda+2 & \lambda^2+2\lambda+1 & 3\lambda^2+6\lambda+3 \end{bmatrix}$

It is readily found that the greatest common divisor of the one-row minors (elements) of $A(\lambda)$ is $g_1(\lambda) = 1$, the greatest common divisor of the two-row minors of $A(\lambda)$ is $g_2(\lambda) = \lambda$, and $g_3(\lambda) = \frac{1}{2}|A(\lambda)| = \lambda^3 + \lambda^2$. Then, by **(24.4)**,

$$f_1(\lambda) \;=\; g_1(\lambda) \;=\; 1, \qquad f_2(\lambda) \;=\; g_2(\lambda)/g_1(\lambda) \;=\; \lambda, \qquad f_3(\lambda) \;=\; g_3(\lambda)/g_2(\lambda) \;=\; \lambda^2+\lambda$$

and the Smith normal form of $A(\lambda)$ is

$$N(\lambda) \;=\; \begin{bmatrix} 1 & 0 & 0 \\ 0 & \lambda & 0 \\ 0 & 0 & \lambda^2+\lambda \end{bmatrix}$$

For another reduction, see Problem 4.

INVARIANT FACTORS. The polynomials $f_1(\lambda), f_2(\lambda), \ldots, f_r(\lambda)$ in the diagonal of the Smith normal form of $A(\lambda)$ are called **invariant factors** of $A(\lambda)$. If $f_k(\lambda) = 1$, $k \leqq r$, then $f_1(\lambda) = f_2(\lambda) = \ldots = f_k(\lambda) = 1$ and each is called a **trivial** invariant factor.

As a consequence of Theorem **VII**, we have

VIII. Two n-square λ-matrices over $F[\lambda]$ are equivalent over $F[\lambda]$ if and only if they have the same invariant factors.

ELEMENTARY DIVISORS. Let $A(\lambda)$ be an n-square λ-matrix over $F[\lambda]$ and let its invariant factors be expressed as

$$f_i(\lambda) = \{p_1(\lambda)\}^{q_{i1}}\{p_2(\lambda)\}^{q_{i2}}\ldots\{p_s(\lambda)\}^{q_{is}}, \; (i = 1, 2, \ldots, r) \tag{24.5}$$

where $p_1(\lambda), p_2(\lambda), \ldots, p_s(\lambda)$ are distinct monic, irreducible polynomials of $F[\lambda]$. Some of the q_{ij} may be zero and the corresponding factor may be suppressed; however, since $f_i(\lambda)$ divides $f_{i+1}(\lambda)$, $q_{i+1,j} \geq q_{ij}$, $(i = 1, 2, \ldots, r-1;\; j = 1, 2, \ldots, s)$.

The factors $\{p_j(\lambda)\}^{q_{ij}} \neq 1$ which appear in **(24.5)** are called **elementary divisors** over $F[\lambda]$ of $A(\lambda)$.

Example 2. Suppose a 10-square λ-matrix $A(\lambda)$ over the rational field has the Smith normal form

$$\left[\begin{array}{ccccc|c} 1 & 0 & 0 & 0 & 0 & \\ 0 & 1 & 0 & 0 & 0 & \\ 0 & 0 & (\lambda-1)(\lambda^2+1) & 0 & 0 & 0 \\ 0 & 0 & 0 & (\lambda-1)(\lambda^2+1)^2\lambda & 0 & \\ 0 & 0 & 0 & 0 & (\lambda-1)^2(\lambda^2+1)^2\lambda^2(\lambda^2-3) & \\ \hline & & 0 & & & 0 \end{array}\right]$$

The rank is 5. The invariant factors are

$$f_1(\lambda) = 1, \quad f_2(\lambda) = 1, \quad f_3(\lambda) = (\lambda-1)(\lambda^2+1),$$
$$f_4(\lambda) = (\lambda-1)(\lambda^2+1)^2\lambda, \quad f_5(\lambda) = (\lambda-1)^2(\lambda^2+1)^2\lambda^2(\lambda^2-3)$$

The elementary divisors are

$$(\lambda-1)^2, \; \lambda-1, \; \lambda-1, \; (\lambda^2+1)^2, \; (\lambda^2+1)^2, \; (\lambda^2+1), \; \lambda^2, \; \lambda, \; \lambda^2-3$$

Note that the elementary divisors are not necessarily distinct; in the listing each elementary divisor appears as often as it appears in the invariant factors.

Example 3. (*a*) Over the real field the invariant factors of $A(\lambda)$ of Example 2 are unchanged but the elementary divisors are

$$(\lambda-1)^2, \; \lambda-1, \; \lambda-1, \; (\lambda^2+1)^2, \; (\lambda^2+1)^2, \; (\lambda^2+1), \; \lambda^2, \; \lambda, \; \lambda-\sqrt{3}, \; \lambda+\sqrt{3}$$

since λ^2-3 can be factored.

(*b*) Over the complex field the invariant factors remain unchanged but the elementary divisors are

$$(\lambda-1)^2, \; \lambda-1, \; \lambda-1, \; (\lambda+i)^2, \; (\lambda+i)^2, \; \lambda+i, \; (\lambda-i)^2,$$
$$(\lambda-i)^2, \; \lambda-i, \; \lambda^2, \; \lambda, \; \lambda-\sqrt{3}, \; \lambda+\sqrt{3}$$

The invariant factors of a λ-matrix determine its rank and its elementary divisors; conversely, the rank and elementary divisors determine the invariant factors.

Example 4. Let the elementary divisors of the 6-square λ-matrix $A(\lambda)$ of rank 5 be

$$\lambda^3, \; \lambda^2, \; \lambda, \; (\lambda-1)^2, \; (\lambda-1)^2, \; \lambda-1, \; (\lambda+1)^2, \; \lambda+1$$

Find the invariant factors and write the Smith canonical form.

To form $f_5(\lambda)$, form the lowest common multiple of the elementary divisors, i.e.,

$$f_5(\lambda) = \lambda^3(\lambda-1)^2(\lambda+1)^2$$

To form $f_4(\lambda)$, remove the elementary divisors used in $f_5(\lambda)$ from the original list and form the lowest common multiple of those remaining, i.e.,

$$f_4(\lambda) = \lambda^2(\lambda-1)^2(\lambda+1)$$

Repeating, $f_3(\lambda) = \lambda(\lambda-1)$. Now the elementary divisors are exhausted; then $f_2(\lambda) = f_1(\lambda) = 1$. The Smith canonical form is

$$N(\lambda) = \begin{bmatrix} 1 & 0 & 0 & 0 & 0 & 0 \\ 0 & 1 & 0 & 0 & 0 & 0 \\ 0 & 0 & \lambda(\lambda-1) & 0 & 0 & 0 \\ 0 & 0 & 0 & \lambda^2(\lambda-1)^2(\lambda+1) & 0 & 0 \\ 0 & 0 & 0 & 0 & \lambda^3(\lambda-1)^2(\lambda+1)^2 & 0 \\ 0 & 0 & 0 & 0 & 0 & 0 \end{bmatrix}$$

Since the invariant factors of a λ-matrix are invariant under elementary transformations, so also are the elementary divisors. Thus,

IX. Two n-square λ-matrices over $F[\lambda]$ are equivalent over $F[\lambda]$ if and only if they have the same rank and the same elementary divisors.

SOLVED PROBLEMS

1. Prove: If $P(\lambda)$ is a product of elementary matrices, then the greatest common divisor of all s-square minors of $P(\lambda)\cdot A(\lambda)$ is also the greatest common divisor of all s-square minors of $A(\lambda)$.

It is necessary only to consider $P(\lambda)\cdot A(\lambda)$ where $P(\lambda)$ is each of the three types of elementary matrices H.

Let $R(\lambda)$ be an s-square minor of $A(\lambda)$ and let $S(\lambda)$ be the s-square minor of $P(\lambda)\cdot A(\lambda)$ having the same position as $R(\lambda)$. Consider $P(\lambda) = H_{ij}$; its effect on $A(\lambda)$ is either **(i)** to leave $R(\lambda)$ unchanged, **(ii)** to interchange two rows of $R(\lambda)$, or **(iii)** to interchange a row of $R(\lambda)$ with a row not in $R(\lambda)$. In the case of **(i)**, $S(\lambda) = R(\lambda)$; in the case of **(ii)**, $S(\lambda) = -R(\lambda)$; in the case of **(iii)**, $S(\lambda)$ is except possibly for sign another s-square minor of $A(\lambda)$.

Consider $P(\lambda) = H_i(k)$; then either $S(\lambda) = R(\lambda)$ or $S(\lambda) = kR(\lambda)$.

Finally, consider $P(\lambda) = H_{ij}(f(\lambda))$. Its effect on $A(\lambda)$ is either **(i)** to leave $R(\lambda)$ unchanged, **(ii)** to increase one of the rows of $R(\lambda)$ by $f(\lambda)$ times another row of $R(\lambda)$, or **(iii)** to increase one of the rows of $R(\lambda)$ by $f(\lambda)$ times a row not of $R(\lambda)$. In the case of **(i)** and **(ii)**, $S(\lambda) = R(\lambda)$; in the case of **(iii)**,

$$S(\lambda) = R(\lambda) \pm f(\lambda)\cdot T(\lambda)$$

where $T(\lambda)$ is an s-square minor of $A(\lambda)$.

Thus, any s-square minor of $P(\lambda)\cdot A(\lambda)$ is a linear combination of s-square minors of $A(\lambda)$. If $g(\lambda)$ is the greatest common divisor of all s-square minors of $A(\lambda)$ and $g_1(\lambda)$ is the greatest common divisor of all s-square minors of $P(\lambda)\cdot A(\lambda)$, then $g(\lambda)$ divides $g_1(\lambda)$. Let $B(\lambda) = P(\lambda)\cdot A(\lambda)$.

Now $A(\lambda) = P^{-1}(\lambda)\cdot B(\lambda)$ and $P^{-1}(\lambda)$ is a product of elementary matrices. Thus, $g_1(\lambda)$ divides $g(\lambda)$ and $g_1(\lambda) = g(\lambda)$.

2. Prove: If $P(\lambda)$ and $Q(\lambda)$ are products of elementary matrices, then the greatest common divisor of all s-square minors of $P(\lambda)\cdot A(\lambda)\cdot Q(\lambda)$ is also the greatest common divisor of all s-square minors of $A(\lambda)$.

Let $B(\lambda) = P(\lambda)\cdot A(\lambda)$ and $C(\lambda) = B(\lambda)\cdot Q(\lambda)$. Since $C'(\lambda) = Q'(\lambda)\cdot B'(\lambda)$ and $Q'(\lambda)$ is a product of elementary matrices, the greatest common divisor of all s-square minors of $C'(\lambda)$ is the greatest common divisor of all s-square minors of $B'(\lambda)$. But the greatest common divisor of all s-square minors of $C'(\lambda)$ is the greatest common divisor of all s-square minors of $C(\lambda)$ and the same is true for $B'(\lambda)$ and $B(\lambda)$. Thus, the greatest common divisor of all s-square minors of $C(\lambda) = P(\lambda)\cdot A(\lambda)\cdot Q(\lambda)$ is the greatest common divisor of all s-square minors of $A(\lambda)$.

3. Prove: Every λ-matrix $A(\lambda) = [a_{ij}(\lambda)]$ of rank r can be reduced by elementary transformations to the Smith normal form

$$N(\lambda) = \begin{bmatrix} f_1(\lambda) & 0 & \ldots & 0 & \ldots & 0 \\ 0 & f_2(\lambda) & \ldots & 0 & \ldots & 0 \\ \cdots & \cdots & \cdots & \cdots & \cdots & \cdots \\ 0 & 0 & \ldots & f_r(\lambda) & \ldots & 0 \\ 0 & 0 & \ldots & 0 & \ldots & 0 \\ \cdots & \cdots & \cdots & \cdots & \cdots & \cdots \\ 0 & 0 & \ldots & 0 & \ldots & 0 \end{bmatrix}$$

where each $f_i(\lambda)$ is monic and $f_i(\lambda)$ divides $f_{i+1}(\lambda)$, $(i = 1, 2, \ldots, r-1)$.

The theorem is true for $A(\lambda) = 0$. Suppose $A(\lambda) \neq 0$; then there is an element $a_{ij}(\lambda) \neq 0$ of minimum degree. By means of a transformation of type 2, this element may be made monic and, by the proper interchanges of rows and of columns, can be brought into the (1,1)-position in the matrix to become the new $a_{11}(\lambda)$.

(*a*) Suppose $a_{11}(\lambda)$ divides every other element of $A(\lambda)$. Then by transformations of type 3, $A(\lambda)$ can be reduced to

$$\textbf{(i)} \qquad \begin{bmatrix} f_1(\lambda) & 0 \\ 0 & B(\lambda) \end{bmatrix}$$

where $f_1(\lambda) = a_{11}(\lambda)$.

(*b*) Suppose that $a_{11}(\lambda)$ does not divide every element of $A(\lambda)$. Let $a_{1j}(\lambda)$ be an element in the first row which is not divisible by $a_{11}(\lambda)$. By Theorem **I**, Chapter 23, we can write

$$a_{1j}(\lambda) = q(\lambda)\,a_{11}(\lambda) + r_{1j}(\lambda)$$

where $r_{1j}(\lambda)$ is of degree less than that of $a_{11}(\lambda)$. From the jth column subtract the product of $q(\lambda)$ and the first column so that the element in the first row and jth column is now $r_{1j}(\lambda)$. By a transformation of type 2, replace this element by one which is monic and, by an interchange of columns bring it into the (1,1)-position as the new $a_{11}(\lambda)$. If now $a_{11}(\lambda)$ divides every element of $A(\lambda)$, we proceed to obtain **(i)**. Otherwise, after a finite number of repetitions of the above procedure, we obtain a matrix in which every element in the first row and the first column is divisible by the element occupying the (1,1)-position.

If this element divides every element of $A(\lambda)$, we proceed to obtain **(i)**. Otherwise, suppose $a_{ij}(\lambda)$ is not divisible by $a_{11}(\lambda)$. Let $a_{i1}(\lambda) = q_{i1}(\lambda)\cdot a_{11}(\lambda)$ and $a_{1j}(\lambda) = q_{1j}(\lambda)\cdot a_{11}(\lambda)$. From the ith row subtract the product of $q_{i1}(\lambda)$ and the first row. This replaces $a_{i1}(\lambda)$ by 0 and $a_{ij}(\lambda)$ by $a_{ij}(\lambda) - q_{i1}(\lambda)\cdot a_{1j}(\lambda)$. Now add the ith row to the first. This leaves $a_{11}(\lambda)$ unchanged but replaces $a_{1j}(\lambda)$ by

$$a_{ij}(\lambda) - q_{i1}(\lambda)\cdot a_{1j}(\lambda) + a_{1j}(\lambda) = a_{ij}(\lambda) + q_{1j}(\lambda)\{1 - q_{i1}(\lambda)\}a_{11}(\lambda)$$

Since this is not divisible by $a_{11}(\lambda)$, we divide it by $a_{11}(\lambda)$ and as before obtain a new replacement (the remainder) for $a_{11}(\lambda)$. This procedure is continued so long as the monic polynomial last selected as $a_{11}(\lambda)$ does not divide every element of the matrix. After a finite number of steps we must obtain an $a_{11}(\lambda)$ which does divide every element and then reach **(i)**.

Next, we treat $B(\lambda)$ in the same manner and reach

$$\begin{bmatrix} f_1(\lambda) & 0 & 0 \\ 0 & f_2(\lambda) & 0 \\ 0 & 0 & C(\lambda) \end{bmatrix}$$

Ultimately, we have the Smith normal form.

Since $f_1(\lambda)$ is a divisor of every element of $B(\lambda)$ and $f_2(\lambda)$ is the greatest common divisor of the elements of $B(\lambda)$, $f_1(\lambda)$ divides $f_2(\lambda)$. Similarly, it is found that each $f_i(\lambda)$ divides $f_{i+1}(\lambda)$.

4. Reduce

$$A(\lambda) = \begin{bmatrix} \lambda+2 & \lambda+1 & \lambda+3 \\ \lambda^3+2\lambda^2+\lambda & \lambda^3+\lambda^2+\lambda & 2\lambda^3+3\lambda^2+\lambda \\ \lambda^2+3\lambda+2 & \lambda^2+2\lambda+1 & 3\lambda^2+6\lambda+3 \end{bmatrix}$$

to its Smith normal form.

It is not necessary to follow the procedure of Problem 3 here. The element $f_1(\lambda)$ of the Smith normal form is the greatest common divisor of the elements of $A(\lambda)$; clearly this is 1. We proceed at once to obtain such an element in the (1,1)-position and then obtain (i) of Problem 3. After subtracting the second column from the first, we obtain

$$A(\lambda) \sim \begin{bmatrix} 1 & \lambda+1 & \lambda+3 \\ \lambda^2 & \lambda^3+\lambda^2+\lambda & 2\lambda^3+3\lambda^2+\lambda \\ \lambda+1 & \lambda^2+2\lambda+1 & 3\lambda^2+6\lambda+3 \end{bmatrix} \sim \begin{bmatrix} 1 & \lambda+1 & \lambda+3 \\ 0 & \lambda & \lambda^3+\lambda \\ 0 & 0 & 2\lambda^2+2\lambda \end{bmatrix}$$

$$\sim \begin{bmatrix} 1 & 0 & 0 \\ 0 & \lambda & \lambda^3+\lambda \\ 0 & 0 & 2\lambda^2+2\lambda \end{bmatrix} = \begin{bmatrix} 1 & 0 \\ 0 & B(\lambda) \end{bmatrix}$$

Now the greatest common divisor of the elements of $B(\lambda)$ is λ. Then

$$\begin{bmatrix} 1 & 0 & 0 \\ 0 & \lambda & \lambda^3+\lambda \\ 0 & 0 & 2\lambda^2+2\lambda \end{bmatrix} \sim \begin{bmatrix} 1 & 0 & 0 \\ 0 & \lambda & 0 \\ 0 & 0 & 2\lambda^2+2\lambda \end{bmatrix} \sim \begin{bmatrix} 1 & 0 & 0 \\ 0 & \lambda & 0 \\ 0 & 0 & \lambda^2+\lambda \end{bmatrix}$$

and this is the required form.

5. Reduce

$$A(\lambda) = \begin{bmatrix} \lambda & \lambda-1 & \lambda+2 \\ \lambda^2+\lambda & \lambda^2 & \lambda^2+2\lambda \\ \lambda^2-2\lambda & \lambda^2-3\lambda+2 & \lambda^2+\lambda-3 \end{bmatrix}$$

to its Smith normal form.

We find

$$A(\lambda) \sim \begin{bmatrix} 1 & \lambda-1 & \lambda+2 \\ \lambda & \lambda^2 & \lambda^2+2\lambda \\ \lambda-2 & \lambda^2-3\lambda+2 & \lambda^2+\lambda-3 \end{bmatrix} \sim \begin{bmatrix} 1 & \lambda-1 & \lambda+2 \\ 0 & \lambda & 0 \\ 0 & 0 & \lambda+1 \end{bmatrix} \sim \begin{bmatrix} 1 & 0 & 0 \\ 0 & \lambda & 0 \\ 0 & 0 & \lambda+1 \end{bmatrix}$$

$$\sim \begin{bmatrix} 1 & 0 & 0 \\ 0 & \lambda & -\lambda-1 \\ 0 & 0 & \lambda+1 \end{bmatrix} \sim \begin{bmatrix} 1 & 0 & 0 \\ 0 & -1 & -\lambda-1 \\ 0 & \lambda+1 & \lambda+1 \end{bmatrix} \sim \begin{bmatrix} 1 & 0 & 0 \\ 0 & 1 & \lambda+1 \\ 0 & 0 & -\lambda^2-\lambda \end{bmatrix} \sim \begin{bmatrix} 1 & 0 & 0 \\ 0 & 1 & 0 \\ 0 & 0 & \lambda(\lambda+1) \end{bmatrix}$$

using the elementary transformations $K_{12}(-1)$; $H_{21}(-\lambda)$, $H_{31}(-\lambda+2)$; $K_{21}(-\lambda+1)$, $K_{31}(-\lambda-2)$; $H_{23}(-1)$; $K_{23}(1)$; $H_{32}(\lambda+1)$, $H_2(-1)$; $K_{32}(-\lambda-1)$, $K_3(-1)$.

SUPPLEMENTARY PROBLEMS

6. Show that $H_{ij}K_{ij} = H_i(k)K_i(1/k) = H_{ij}(f(\lambda)) \cdot K_{ji}(-f(\lambda)) = I$.

7. Prove: An n-square λ-matrix $A(\lambda)$ is a product of elementary matrices if and only if $|A(\lambda)|$ is a non-zero constant.

8. Prove: An n-square λ-matrix $A(\lambda)$ may be reduced to I by elementary transformations if and only if $|A(\lambda)|$ is a non-zero constant.

9. Prove: A λ-matrix $A(\lambda)$ over $F[\lambda]$ has an inverse with elements in $F[\lambda]$ if and only if $A(\lambda)$ is a product of elementary matrices.

10. Obtain matrices $P(\lambda)$ and $Q(\lambda)$ such that $P(\lambda)\cdot A(\lambda)\cdot Q(\lambda) = I$ and then obtain

$$A(\lambda)^{-1} = Q(\lambda)\cdot P(\lambda)$$

given

$$A(\lambda) = \begin{bmatrix} \lambda+1 & 0 & 1 \\ 1 & \lambda+1 & \lambda \\ 2 & \lambda+2 & \lambda+1 \end{bmatrix}$$

Hint. See Problem 6, Chapter 5. *Ans.* $\begin{bmatrix} 1 & \lambda+2 & -\lambda-1 \\ \lambda-1 & \lambda^2+2\lambda-1 & -\lambda^2-\lambda+1 \\ -\lambda & -\lambda^2-3\lambda-2 & \lambda^2+2\lambda+1 \end{bmatrix}$

11. Reduce each of the following to its Smith normal form:

(*a*) $\begin{bmatrix} \lambda & \lambda & \lambda-1 \\ \lambda^2+\lambda & \lambda^2+2\lambda & \lambda^2-1 \\ 2\lambda^2-2\lambda & \lambda^2-2\lambda & 2\lambda^2-3\lambda+2 \end{bmatrix} \sim \begin{bmatrix} 1 & 0 & 0 \\ 0 & \lambda & 0 \\ 0 & 0 & \lambda^2 \end{bmatrix}$

(*b*) $\begin{bmatrix} \lambda^2+1 & \lambda^3+\lambda & 2\lambda^3-\lambda^2+\lambda \\ \lambda-1 & \lambda^2+1 & \lambda^2-2\lambda+1 \\ \lambda^2 & \lambda^3 & 2\lambda^3-\lambda^2+1 \end{bmatrix} \sim \begin{bmatrix} 1 & 0 & 0 \\ 0 & \lambda+1 & 0 \\ 0 & 0 & \lambda^3+1 \end{bmatrix}$

(*c*) $\begin{bmatrix} \lambda+1 & 2\lambda-2 & \lambda-2 & \lambda^2 \\ \lambda^2+\lambda+1 & 2\lambda^2-2\lambda+1 & \lambda^2-2\lambda & \lambda^3 \\ \lambda^2-\lambda-2 & 3\lambda^2-7\lambda+4 & 2\lambda^2-5\lambda+4 & \lambda^3-2\lambda^2 \\ \lambda^3+\lambda^2 & 2\lambda^3-2\lambda^2 & \lambda^3-2\lambda^2 & \lambda^3 \end{bmatrix} \sim \begin{bmatrix} 1 & 0 & 0 & 0 \\ 0 & 1 & 0 & 0 \\ 0 & 0 & \lambda^2-\lambda & 0 \\ 0 & 0 & 0 & \lambda^4-\lambda^3 \end{bmatrix}$

(*d*) $\begin{bmatrix} \lambda^2+2\lambda+1 & \lambda^2+\lambda & \lambda^3+\lambda^2+\lambda-1 & \lambda^2+\lambda \\ \lambda^2+\lambda+1 & \lambda^2+1 & \lambda^3 & \lambda^2-1 \\ \lambda^2+\lambda & \lambda^2 & \lambda^3+\lambda-1 & \lambda^2 \\ \lambda^3+\lambda^2 & \lambda^3 & \lambda^4 & \lambda^3+\lambda^2-1 \end{bmatrix} \sim \begin{bmatrix} 1 & 0 & 0 & 0 \\ 0 & 1 & 0 & 0 \\ 0 & 0 & \lambda-1 & 0 \\ 0 & 0 & 0 & \lambda^2-1 \end{bmatrix}$

(*e*) $\begin{bmatrix} \lambda^2+1 & \lambda^2+3\lambda+3 & \lambda^2+4\lambda-2 & \lambda^2+3 \\ \lambda-2 & \lambda-1 & \lambda+2 & \lambda-2 \\ 3\lambda+1 & 4\lambda+3 & 2\lambda+2 & 3\lambda+2 \\ \lambda^2+2\lambda & \lambda^2+6\lambda+4 & \lambda^2+6\lambda-1 & \lambda^2+2\lambda+3 \end{bmatrix} \sim I_4$

(*f*) $\begin{bmatrix} \lambda^2 & 0 & 0 \\ 0 & \lambda^2-2\lambda+1 & 0 \\ 0 & 0 & \lambda+1 \end{bmatrix} \sim \begin{bmatrix} 1 & 0 & 0 \\ 0 & 1 & 0 \\ 0 & 0 & \lambda^2(\lambda-1)^2(\lambda+1) \end{bmatrix}$

12. Obtain the elementary divisors over the rational field, the real field, and the complex field for each of the matrices of Problem 11.

13. The following polynomials are non-trivial invariant factors of a matrix. Find its elementary divisors in the real field.

(*a*) $\lambda^2-\lambda,\ \lambda^3-\lambda^2,\ \lambda^6-2\lambda^5+\lambda^4$

(*b*) $\lambda+1,\ \lambda^2-1,\ (\lambda^2-1)^2,\ (\lambda^2-1)^3$

(*c*) $\lambda,\ \lambda^3+\lambda,\ \lambda^7-\lambda^6+2\lambda^5-2\lambda^4+\lambda^3-\lambda^2$

(*d*) $\lambda,\ \lambda^3+\lambda,\ \lambda^5+2\lambda^3+\lambda,\ \lambda^6+\lambda^5+2\lambda^4+2\lambda^3+\lambda^2+\lambda$

Ans. (*a*) $\lambda^4,\ \lambda^2,\ \lambda,\ (\lambda-1)^2,\ \lambda-1,\ \lambda-1$

(*b*) $\lambda+1,\ \lambda+1,\ (\lambda+1)^2,\ (\lambda+1)^3,\ \lambda-1,\ (\lambda-1)^2,\ (\lambda-1)^3$

(*c*) $\lambda,\ \lambda,\ \lambda^2,\ \lambda^2+1,\ (\lambda^2+1)^2,\ \lambda-1$

(*d*) $\lambda,\ \lambda,\ \lambda,\ \lambda,\ \lambda^2+1,\ (\lambda^2+1)^2,\ (\lambda^2+1)^2,\ \lambda+1$

14. The following polynomials are the elementary divisors of a matrix whose rank is six. What are its invariant factors?

(*a*) $\lambda,\ \lambda,\ \lambda+1,\ \lambda+2,\ \lambda+3,\ \lambda+4$

(*b*) $\lambda^3,\ \lambda^2,\ \lambda,\ (\lambda-1)^2,\ \lambda-1$

(*c*) $(\lambda-1)^3,\ (\lambda-1)^2,\ (\lambda-1)^2,\ \lambda-1,\ (\lambda+1)^2$

(*d*) $\lambda^5,\ \lambda^3,\ \lambda,\ (\lambda+2)^5,\ (\lambda+2)^4,\ (\lambda+2)^2$

Ans. (*a*) $1,\ 1,\ 1,\ 1,\ \lambda,\ \lambda(\lambda+1)(\lambda+2)(\lambda+3)(\lambda+4)$

(*b*) $1,\ 1,\ 1,\ \lambda,\ \lambda^2(\lambda-1),\ \lambda^3(\lambda-1)^2$

(*c*) $1,\ 1,\ \lambda-1,\ (\lambda-1)^2,\ (\lambda-1)^2,\ (\lambda-1)^3(\lambda+1)^2$

(*d*) $1,\ 1,\ 1,\ \lambda(\lambda+2)^2,\ \lambda^3(\lambda+2)^4,\ \lambda^5(\lambda+2)^5$

15. Solve the system of ordinary linear differential equations

$$\begin{cases} Dx_1 + (D+1)x_2 = 0 \\ (D+2)x_1 - (D-1)x_3 = t \\ (D+1)x_2 + (D+2)x_3 = e^t \end{cases}$$

where x_1, x_2, x_3 are unknown real functions of a real variable t and $D = \frac{d}{dt}$.

Hint. In matrix notation, the system is

$$AX = \begin{bmatrix} D & D+1 & 0 \\ D+2 & 0 & -D+1 \\ 0 & D+1 & D+2 \end{bmatrix}\begin{bmatrix} x_1 \\ x_2 \\ x_3 \end{bmatrix} = \begin{bmatrix} 0 \\ t \\ e^t \end{bmatrix} = H$$

Now the polynomials in D of A combine as do the polynomials in λ of a λ-matrix; hence, beginning with a computing form similar to that of Problem 6, Chapter 5, and using in order the elementary transformations: $K_{12}(-1),\ H_1(-1),\ K_{21}(D+1),\ H_{21}(-D-2),\ H_{31}(D+1),\ K_{23}(D),\ H_{23}(-4),\ K_2(\tfrac{1}{2}),\ K_{32}(5D+7),\ H_{32}(-\tfrac{1}{2}D),$ $H_3(2),\ K_3(1/5)$ obtain

$$PAQ = \begin{bmatrix} -1 & 0 & 0 \\ 5D+6 & 1 & -4 \\ -5D^2-8D-2 & -D & 4D+2 \end{bmatrix} A \begin{bmatrix} 1 & \frac{1}{2}(D+1) & \frac{1}{10}(5D^2+12D+7) \\ -1 & -\frac{1}{2}D & -\frac{1}{10}(5D^2+7D) \\ 0 & \frac{1}{2}D & \frac{1}{10}(5D^2+7D+2) \end{bmatrix} = \begin{bmatrix} 1 & 0 & 0 \\ 0 & 1 & 0 \\ 0 & 0 & D^2+\frac{9}{5}D+\frac{4}{5} \end{bmatrix} = N_1$$

the Smith normal form of A.

Use the linear transformation $X = QY$ to carry $AX = H$ into $AQY = H$ and from $PAQY = N_1Y = PH$ get

$$y_1 = 0,\quad y_2 = t-4e^t,\quad (D^2+\tfrac{9}{5}D+\tfrac{4}{5})y_3 = 6e^t-1 \quad\text{and}\quad y_3 = K_1e^{-4t/5}+K_2e^{-t}+\tfrac{5}{3}e^t-\tfrac{5}{4}$$

Finally, use $X = QY$ to obtain the required solution

$$x_1 = 3C_1e^{-4t/5}+\tfrac{1}{2}t-\tfrac{3}{8},\qquad x_2 = 12C_1e^{-4t/5}+C_2e^{-t}-\tfrac{1}{2},\qquad x_3 = -2C_1e^{-4t/5}+\tfrac{1}{3}e^t+\tfrac{1}{4}$$

Chapter 25

The Minimum Polynomial of a Matrix

THE CHARACTERISTIC MATRIX $\lambda I - A$ of an n-square matrix A over F is a non-singular λ-matrix having invariant factors and elementary divisors. Using **(24.4)** it is easy to show

I. If D is a diagonal matrix, the elementary divisors of $\lambda I - D$ are its diagonal elements.

In Problem 1, we prove

II. Two n-square matrices A and B over F are similar over F if and only if their characteristic matrices have the same invariant factors or the same rank and the same elementary divisors in $F[\lambda]$.

From Theorems **I** and **II**, we have

III. An n-square matrix A over F is similar to a diagonal matrix if and only if $\lambda I - A$ has linear elementary divisors in $F[\lambda]$.

SIMILARITY INVARIANTS. The invariant factors of $\lambda I - A$ are called **similarity invariants** of A.

Let $P(\lambda)$ and $Q(\lambda)$ be non-singular matrices such that $P(\lambda)\cdot(\lambda I - A)\cdot Q(\lambda)$ is the Smith normal form

$$\text{diag}\,(f_1(\lambda),\ f_2(\lambda),\ \ldots,\ f_n(\lambda))$$

Now $\quad |P(\lambda)\cdot(\lambda I - A)\cdot Q(\lambda)| \;=\; |P(\lambda)|\cdot|Q(\lambda)|\,\phi(\lambda) \;=\; f_1(\lambda)\cdot f_2(\lambda)\cdot\ldots\, f_n(\lambda).$

Since $\phi(\lambda)$ and $f_i(\lambda)$ are monic, $|P(\lambda)|\cdot|Q(\lambda)| = 1$ and we have

IV. The characteristic polynomial of an n-square matrix A is the product of the invariant factors of $\lambda I - A$ or of the similarity invariants of A.

THE MINIMUM POLYNOMIAL. By the Cayley-Hamilton Theorem (Chapter 23), every n-square matrix A satisfies its characteristic equation $\phi(\lambda) = 0$ of degree n. That monic polynomial $m(\lambda)$ of minimum degree such that $m(A) = 0$ is called the **minimum polynomial** of A and $m(\lambda) = 0$ is called the **minimum equation** of A. ($m(\lambda)$ is also called the **minimum function** of A.)

The most elementary procedure for finding the minimum polynomial of $A \neq 0$ involves the following routine:

(i) If $A = a_0 I$, then $m(\lambda) = \lambda - a_0$;

(ii) If $A \neq aI$ for all a but $A^2 = a_1 A + a_0 I$, then $m(\lambda) = \lambda^2 - a_1\lambda - a_0$;

(iii) If $A^2 \neq aA + bI$ for all a and b but $A^3 = a_2 A^2 + a_1 A + a_0 I$, then

$$m(\lambda) = \lambda^3 - a_2\lambda^2 - a_1\lambda - a_0$$

and so on.

Example 1. Find the minimum polynomial of $A = \begin{bmatrix} 1 & 2 & 2 \\ 2 & 1 & 2 \\ 2 & 2 & 1 \end{bmatrix}$.

Clearly $A - a_0 I = 0$ is impossible. Set

$$A^2 = \begin{bmatrix} 9 & 8 & 8 \\ 8 & 9 & 8 \\ 8 & 8 & 9 \end{bmatrix} = a_1 \begin{bmatrix} 1 & 2 & 2 \\ 2 & 1 & 2 \\ 2 & 2 & 1 \end{bmatrix} + a_0 \begin{bmatrix} 1 & 0 & 0 \\ 0 & 1 & 0 \\ 0 & 0 & 1 \end{bmatrix}$$

Using the first two elements of the first row of each matrix, we have $\begin{cases} 9 = a_1 + a_0 \\ 8 = 2a_1 \end{cases}$; then $a_1 = 4$ and $a_0 = 5$. After (and not before) checking for every element of A^2, we conclude that $A^2 = 4A + 5I$ and the required minimum polynomial is $\lambda^2 - 4\lambda - 5$.

In Problem 2, we prove

V. If A is any n-square matrix over F and $f(\lambda)$ is any polynomial over F, then $f(A) = 0$ if and only if the minimum polynomial $m(\lambda)$ of A divides $f(\lambda)$.

In Problem 3, we prove

VI. The minimum polynomial $m(\lambda)$ of an n-square matrix A is that similarity invariant $f_n(\lambda)$ of A which has the highest degree.

Since the similarity invariants $f_1(\lambda), f_2(\lambda), \ldots, f_{n-1}(\lambda)$ all divide $f_n(\lambda)$, we have

VII. The characteristic polynomial $\phi(\lambda)$ of A is the product of the minimum polynomial of A and certain monic factors of $m(\lambda)$.

and

VIII. The characteristic matrix of an n-square matrix A has distinct linear elementary divisors if and only if $m(\lambda)$, the minimum polynomial of A, has only distinct linear factors.

NON-DEROGATORY MATRICES. An n-square matrix A whose characteristic polynomial and minimum polynomial are identical is called **non-derogatory**; otherwise, **derogatory**. We have

IX. An n-square matrix A is non-derogatory if and only if A has just one non-trivial similarity invariant.

It is also easy to show

X. If B_1 and B_2 have minimum polynomials $m_1(\lambda)$ and $m_2(\lambda)$ respectively, the minimum polynomial $m(\lambda)$ of the direct sum $D = \text{diag}(B_1, B_2)$ is the least common multiple of $m_1(\lambda)$ and $m_2(\lambda)$.

This result may be extended to the direct sum of m matrices.

XI. Let $g_1(\lambda), g_2(\lambda), \ldots, g_m(\lambda)$ be distinct, monic, irreducible polynomials in $F[\lambda]$ and let A_j be a non-derogatory matrix such that $|\lambda I - A_j| = \{g_j(\lambda)\}^{a_j}$, $(j = 1, 2, \ldots, m)$. Then $B = \text{diag}(A_1, A_2, \ldots, A_m)$ has $\phi(\lambda) = \{g_1(\lambda)\}^{a_1} \cdot \{g_2(\lambda)\}^{a_2} \cdots \{g_m(\lambda)\}^{a_m}$ as both characteristic and minimum polynomial.

COMPANION MATRIX. Let A be non-derogatory with non-trivial similarity invariant

(25.1) $$g(\lambda) = f_n(\lambda) = \lambda^n + a_{n-1}\lambda^{n-1} + \cdots + a_1\lambda + a_0$$

We define as the **companion matrix** of $g(\lambda)$,

(25.2) $$C(g) = [-a], \quad \text{if } g(\lambda) = \lambda + a$$

and for $n > 1$

$$(25.3) \qquad C(g) = \begin{bmatrix} 0 & 1 & 0 & & 0 & 0 & 0 \\ 0 & 0 & 1 & & 0 & 0 & 0 \\ \cdots & \cdots & \cdots & \cdots & \cdots & \cdots & \cdots \\ 0 & 0 & 0 & & 0 & 1 & 0 \\ 0 & 0 & 0 & & 0 & 0 & 1 \\ -a_0 & -a_1 & -a_2 & \cdots & -a_{n-3} & -a_{n-2} & -a_{n-1} \end{bmatrix}$$

In Problem 4, we prove

XII. The companion matrix $C(g)$ of a polynomial $g(\lambda)$ has $g(\lambda)$ as both its characteristic and minimum polynomial.

(Some authors prefer to define $C(g)$ as the transpose of the matrix given in **(25.3)**. Both forms will be used here.)

See Problem 5.

It is easy to show

XIII. If A is non-derogatory with non-trivial similarity invariant $f_n(\lambda) = (\lambda - a)^n$, then

$$(25.4) \qquad J = [a], \quad \text{if } n = 1, \qquad \text{and} \qquad J = \begin{bmatrix} a & 1 & 0 & \cdots & 0 & 0 \\ 0 & a & 1 & \cdots & 0 & 0 \\ \cdots & \cdots & \cdots & \cdots & \cdots & \cdots \\ 0 & 0 & 0 & \cdots & a & 1 \\ 0 & 0 & 0 & \cdots & 0 & a \end{bmatrix}, \quad \text{if } n > 1$$

has $f_n(\lambda)$ as its characteristic and minimum polynomial.

SOLVED PROBLEMS

1. Prove: Two n-square matrices A and B over F are similar over F if and only if their characteristic matrices have the same invariant factors or the same elementary divisors in $F[\lambda]$.

Suppose A and B are similar. From (i) of Problem 1, Chapter 20, it follows that $\lambda I - A$ and $\lambda I - B$ are equivalent. Then by Theorems **VIII** and **IX** of Chapter 24, they have the same invariant factors and the same elementary divisors.

Conversely, let $\lambda I - A$ and $\lambda I - B$ have the same invariant factors or elementary divisors. Then by Theorem **VIII**, Chapter 24 there exist non-singular λ-matrices $P(\lambda)$ and $Q(\lambda)$ such that

$$P(\lambda)\cdot(\lambda I - A)\cdot Q(\lambda) = \lambda I - B$$

or

$$\text{(i)} \qquad P(\lambda)\cdot(\lambda I - A) = (\lambda I - B)\cdot Q^{-1}(\lambda)$$

Let

$$\text{(ii)} \qquad P(\lambda) = (\lambda I - B)\cdot S_1(\lambda) + R_1$$

$$\text{(iii)} \qquad Q(\lambda) = S_2(\lambda)\cdot(\lambda I - B) + R_2$$

$$\text{(iv)} \qquad Q^{-1}(\lambda) = S_3(\lambda)\cdot(\lambda I - A) + R_3$$

where R_1, R_2, and R_3 are free of λ. Substituting in (i), we have

$$(\lambda I - B)\cdot S_1(\lambda)\cdot(\lambda I - A) + R_1(\lambda I - A) = (\lambda I - B)\cdot S_3(\lambda)\cdot(\lambda I - A) + (\lambda I - B)R_3$$

or

$$\text{(v)} \qquad (\lambda I - B)\{S_1(\lambda) - S_3(\lambda)\}(\lambda I - A) = (\lambda I - B)R_3 - R_1(\lambda I - A)$$

Then $S_1(\lambda) - S_3(\lambda) = 0$ and

(vi) $$(\lambda I - B)R_3 = R_1(\lambda I - A)$$

since otherwise the left member of (v) is of degree at least two while the right member is of degree at most one.

Using (iii), (iv), and (vi)

$$\begin{aligned} I &= Q(\lambda)\cdot Q^{-1}(\lambda) \\ &= Q(\lambda)\{S_3(\lambda)\cdot(\lambda I - A) + R_3\} \\ &= Q(\lambda)\cdot S_3(\lambda)\cdot(\lambda I - A) + \{S_2(\lambda)\cdot(\lambda I - B) + R_2\}R_3 \\ &= Q(\lambda)\cdot S_3(\lambda)\cdot(\lambda I - A) + S_2(\lambda)\cdot(\lambda I - B)R_3 + R_2R_3 \\ &= Q(\lambda)\cdot S_3(\lambda)\cdot(\lambda I - A) + S_2(\lambda)\cdot R_1\cdot(\lambda I - A) + R_2R_3 \end{aligned}$$

or

(vii) $$I - R_2R_3 = \{Q(\lambda)\cdot S_3(\lambda) + S_2(\lambda)\cdot R_1\}(\lambda I - A)$$

Now $Q(\lambda)\cdot S_3(\lambda) + S_2(\lambda)R_1 = 0$ and $I = R_2R_3$ since otherwise the left member of (vii) is of degree zero in λ while the right member is of degree at least one. Thus, $R_3 = R_2^{-1}$ and, from (vi)

$$\lambda I - B = R_1(\lambda I - A)R_2 = \lambda R_1R_2 - R_1AR_2$$

Since A, B, R_1, and R_2 are free of λ, $R_1 = R_2^{-1}$; then $\lambda I - B = \lambda I - R_2^{-1}AR_2$ and A and B are similar, as was to be proved.

2. Prove: If A is any n-square matrix over F and $f(\lambda)$ is any polynomial in $F[\lambda]$, then $f(A) = 0$ if and only if the minimum polynomial $m(\lambda)$ of A divides $f(\lambda)$.

By the division algorithm, Chapter 22,

$$f(\lambda) = q(\lambda)\cdot m(\lambda) + r(\lambda)$$

and then

$$f(A) = q(A)\cdot m(A) + r(A) = r(A)$$

Suppose $f(A) = 0$; then $r(A) = 0$. Now if $r(\lambda) \neq 0$, its degree is less than that of $m(\lambda)$, contrary to the hypothesis that $m(\lambda)$ is the minimum polynomial of A. Thus, $r(\lambda) = 0$ and $m(\lambda)$ divides $f(\lambda)$.

Conversely, suppose $f(\lambda) = q(\lambda)\cdot m(\lambda)$. Then $f(A) = q(A)\cdot m(A) = 0$.

3. Prove: The minimum polynomial $m(\lambda)$ of an n-square matrix A is that similarity invariant $f_n(\lambda)$ of A which has the highest degree.

Let $g_{n-1}(\lambda)$ denote the greatest common divisor of the $(n-1)$-square minors of $\lambda I - A$. Then

$$|\lambda I - A| = \phi(\lambda) = g_{n-1}(\lambda)\cdot f_n(\lambda)$$

and

$$\text{adj}(\lambda I - A) = g_{n-1}(\lambda)\cdot B(\lambda)$$

where the greatest common divisor of the elements of $B(\lambda)$ is 1.

Now $(\lambda I - A)\cdot\text{adj}(\lambda I - A) = \phi(\lambda)\cdot I$ so that

$$(\lambda I - A)\cdot g_{n-1}(\lambda)\cdot B(\lambda) = g_{n-1}(\lambda)\cdot f_n(\lambda)\cdot I$$

or

(i) $$(\lambda I - A)\cdot B(\lambda) = f_n(\lambda)\cdot I$$

Then $\lambda I - A$ is a divisor of $f_n(\lambda)\cdot I$ and by Theorem **V**, Chapter 23, $f_n(A) = 0$.

By Theorem **V**, $m(\lambda)$ divides $f_n(\lambda)$. Suppose

$$\text{(ii)} \qquad f_n(\lambda) = q(\lambda)\cdot m(\lambda)$$

Since $m(A) = 0$, $\lambda I - A$ is a divisor of $m(\lambda)\cdot I$, say

$$m(\lambda)\cdot I = (\lambda I - A)\cdot C(\lambda)$$

Then, using (i) and (ii),

$$(\lambda I - A)\cdot B(\lambda) = f_n(\lambda)\cdot I = q(\lambda)\cdot m(\lambda)\cdot I = q(\lambda)\cdot(\lambda I - A)\cdot C(\lambda)$$

and

$$B(\lambda) = q(\lambda)\cdot C(\lambda)$$

Now $q(\lambda)$ divides every element of $B(\lambda)$; hence $q(\lambda) = 1$ and, by (ii),

$$f_n(\lambda) = m(\lambda)$$

as was to be proved.

4. Prove: The companion matrix $C(g)$ of a polynomial $g(\lambda)$ has $g(\lambda)$ as both its characteristic and minimum polynomial.

The characteristic matrix of (25.3) is

$$\begin{bmatrix} \lambda & -1 & 0 & \cdots & 0 & 0 \\ 0 & \lambda & -1 & \cdots & 0 & 0 \\ \cdots & \cdots & \cdots & \cdots & \cdots & \cdots \\ 0 & 0 & 0 & \cdots & \lambda & -1 \\ a_0 & a_1 & a_2 & \cdots & a_{n-2} & \lambda + a_{n-1} \end{bmatrix}$$

To the first column add λ times the second column, λ^2 times the third column, ..., λ^{n-1} times the last column to obtain

$$G(\lambda) = \begin{bmatrix} 0 & -1 & 0 & \cdots & 0 & 0 \\ 0 & \lambda & -1 & \cdots & 0 & 0 \\ \cdots & \cdots & \cdots & \cdots & \cdots & \cdots \\ 0 & 0 & 0 & \cdots & \lambda & -1 \\ g(\lambda) & a_1 & a_2 & \cdots & a_{n-2} & \lambda + a_{n-1} \end{bmatrix}$$

Since $|G(\lambda)| = g(\lambda)$, the characteristic polynomial of $C(g)$ is $g(\lambda)$. Since the minor of the element $g(\lambda)$ in $G(\lambda)$ is ± 1, the greatest common divisor of all $(n-1)$-square minors of $G(\lambda)$ is 1. Thus, $C(g)$ is non-derogatory and its minimum polynomial is $g(\lambda)$.

5. The companion matrix of $g(\lambda) = \lambda^5 + 2\lambda^3 - \lambda^2 + 6\lambda - 5$ is

$$\begin{bmatrix} 0 & 1 & 0 & 0 & 0 \\ 0 & 0 & 1 & 0 & 0 \\ 0 & 0 & 0 & 1 & 0 \\ 0 & 0 & 0 & 0 & 1 \\ 5 & -6 & 1 & -2 & 0 \end{bmatrix} \quad \text{or, if preferred,} \quad \begin{bmatrix} 0 & 0 & 0 & 0 & 5 \\ 1 & 0 & 0 & 0 & -6 \\ 0 & 1 & 0 & 0 & 1 \\ 0 & 0 & 1 & 0 & -2 \\ 0 & 0 & 0 & 1 & 0 \end{bmatrix}$$

SUPPLEMENTARY PROBLEMS

6. Write the companion matrix of each of the following polynomials:

(*a*) $\lambda^3 + \lambda^2 - 2\lambda - 1$ (*d*) $\lambda^4 - 2\lambda^3 - \lambda^2 + 2\lambda$

(*b*) $(\lambda^2 - 4)(\lambda + 2)$ (*e*) $\lambda^2(\lambda^2 + 1)$

(*c*) $(\lambda - 1)^3$ (*f*) $(\lambda + 2)(\lambda^3 - 2\lambda^2 + 4\lambda - 8)$

Ans. (*a*) $\begin{bmatrix} 0 & 1 & 0 \\ 0 & 0 & 1 \\ 1 & 2 & -1 \end{bmatrix}$ (*b*) $\begin{bmatrix} 0 & 1 & 0 \\ 0 & 0 & 1 \\ 8 & 4 & -2 \end{bmatrix}$ (*c*) $\begin{bmatrix} 0 & 1 & 0 \\ 0 & 0 & 1 \\ 1 & -3 & 3 \end{bmatrix}$

(*d*) $\begin{bmatrix} 0 & 1 & 0 & 0 \\ 0 & 0 & 1 & 0 \\ 0 & 0 & 0 & 1 \\ 0 & -2 & 1 & 2 \end{bmatrix}$ (*e*) $\begin{bmatrix} 0 & 1 & 0 & 0 \\ 0 & 0 & 1 & 0 \\ 0 & 0 & 0 & 1 \\ 0 & 0 & -1 & 0 \end{bmatrix}$ (*f*) $\begin{bmatrix} 0 & 1 & 0 & 0 \\ 0 & 0 & 1 & 0 \\ 0 & 0 & 0 & 1 \\ 16 & 0 & 0 & 0 \end{bmatrix}$

7. Prove: Every 2-square matrix $A = [a_{ij}]$ for which $(a_{11} - a_{22})^2 + 4a_{12}a_{21} \neq 0$ is non-derogatory.

8. Reduce $G(\lambda)$ of Problem 4 to $\text{diag}(1, 1, \ldots, 1, g(\lambda))$.

9. For each of the following matrices A, **(i)** find the characteristic and minimum polynomial and **(ii)** list the non-trivial invariant factors and the elementary divisors in the rational field.

(*a*) $\begin{bmatrix} 1 & 0 & 0 \\ 0 & 2 & 0 \\ 0 & 0 & 3 \end{bmatrix}$ (*b*) $\begin{bmatrix} 1 & 1 & 3 \\ 5 & 2 & 6 \\ -2 & -1 & -3 \end{bmatrix}$ (*c*) $\begin{bmatrix} 2 & 0 & 0 \\ 0 & 1 & 0 \\ 0 & 0 & 1 \end{bmatrix}$ (*d*) $\begin{bmatrix} 1 & 2 & 2 \\ 2 & 1 & 2 \\ 2 & 2 & 1 \end{bmatrix}$ (*e*) $\begin{bmatrix} 1 & 1 & 2 \\ 1 & 1 & 2 \\ 1 & 1 & 2 \end{bmatrix}$

(*f*) $\begin{bmatrix} 2 & 1 & 1 & 1 \\ 4 & 2 & 3 & 0 \\ -6 & -2 & -3 & -2 \\ -3 & -1 & -1 & -2 \end{bmatrix}$ (*g*) $\begin{bmatrix} 2 & -3 & 1 & -3 \\ -1 & -6 & -3 & -6 \\ -3 & -3 & -4 & -3 \\ 2 & 6 & 4 & 6 \end{bmatrix}$ (*h*) $\begin{bmatrix} -5 & 4 & -6 & 3 & 8 \\ -2 & 3 & -2 & 1 & 2 \\ 4 & -3 & 4 & -1 & -6 \\ 4 & -2 & 4 & 0 & -4 \\ -1 & 0 & -2 & 1 & 2 \end{bmatrix}$

Ans. (*a*) $\phi(\lambda) = m(\lambda) = (\lambda-1)(\lambda-2)(\lambda-3)$; i.f. $(\lambda-1)(\lambda-2)(\lambda-3)$; e.d. $(\lambda-1), (\lambda-2), (\lambda-3)$

(*b*) $\phi(\lambda) = m(\lambda) = \lambda^3$; i.f. = e.d. = λ^3

(*c*) $\phi(\lambda) = (\lambda-1)^2(\lambda-2)$; i.f. $\lambda-1, (\lambda-1)(\lambda-2)$
$m(\lambda) = (\lambda-1)(\lambda-2)$; e.d. $\lambda-1, \lambda-1, \lambda-2$

(*d*) $\phi(\lambda) = (\lambda+1)^2(\lambda-5)$; i.f. $\lambda+1, (\lambda+1)(\lambda-5)$
$m(\lambda) = (\lambda+1)(\lambda-5)$; e.d. $\lambda+1, \lambda+1, \lambda-5$

(*e*) $\phi(\lambda) = \lambda^3 - 4\lambda^2$; i.f. $\lambda, \lambda^2 - 4\lambda$
$m(\lambda) = \lambda^2 - 4\lambda$; e.d. $\lambda, \lambda, \lambda-4$

(*f*) $\phi(\lambda) = \lambda(\lambda+1)^2(\lambda-1)$; i.f. $\lambda+1, \lambda^3 - \lambda$
$m(\lambda) = \lambda(\lambda^2-1)$; e.d. $\lambda, \lambda+1, \lambda+1, \lambda-1$

(*g*) $\phi(\lambda) = \lambda^2(\lambda+1)^2$; i.f. $\lambda, \lambda(\lambda+1)^2$
$m(\lambda) = \lambda(\lambda+1)^2$; e.d. $\lambda, \lambda, (\lambda+1)^2$

(*h*) $\phi(\lambda) = (\lambda-2)(\lambda^2-\lambda-2)^2$; i.f. $\lambda-2, \lambda^2-\lambda-2, \lambda^2-\lambda-2$
$m(\lambda) = \lambda^2-\lambda-2$; e.d. $\lambda-2, \lambda-2, \lambda-2, \lambda+1, \lambda+1$

10. Prove Theorems **VII** and **VIII**.

11. Prove Theorem **X**.

Hint. $m(D) = \text{diag}(m(B_1), m(B_2)) = 0$ requires $m(B_1) = m(B_2) = 0$; thus, $m_1(\lambda)$ and $m_2(\lambda)$ divide $m(\lambda)$.

12. Prove Theorem **XI**.

13. If A is n-square and if k is the least positive integer such that $A^k = 0$, A is called *nilpotent* of index k. Show that A is nilpotent of index k if and only if its characteristic roots are all zero.

14. Prove: (*a*) The characteristic roots of an n-square idempotent matrix A are either 0 or 1.
    (*b*) The rank of A is the number of characteristic roots which are 1.

15. Prove: Let A, B, C, D be n-square matrices over F with C and D non-singular. There exist non-singular matrices P and Q such that $PCQ = A$, $PDQ = B$ if and only if $R(\lambda) = \lambda C - A$ and $S(\lambda) = \lambda D - B$ have the same invariant factors or the same elementary divisors.
    Hint. Follow the proof in Problem 1, noting that similarity is replaced by equivalence.

16. Prove: If the minimum polynomial $m(\lambda)$ of a non-singular matrix A is of degree s, then A^{-1} is expressible as a scalar polynomial of degree $s-1$ in A.

17. Use the minimum polynomial to find the inverse of the matrix A of Problem 9(*h*).

18. Prove: Every linear factor $\lambda - \lambda_i$ of $\phi(\lambda)$ is a factor of $m(\lambda)$.
    Hint. The theorem follows from Theorem **VII** or assume the contrary and write $m(\lambda) = (\lambda - \lambda_i)q(\lambda) + r$, $r \neq 0$. Then $(A - \lambda_i I)\,q(A) + rI = 0$ and $A - \lambda_i I$ has an inverse.

19. Use $A = \begin{bmatrix} 1 & 1 \\ 0 & 1 \end{bmatrix}$ to show that the minimum polynomial is not the product of the distinct factors of $\phi(\lambda)$.

20. Prove: If $g(\lambda)$ is any scalar polynomial in λ, then $g(A)$ is singular if and only if the greatest common divisor of $g(\lambda)$ and $m(\lambda)$, the minimum polynomial of A, is $d(\lambda) \neq 1$.
    Hint. **(i)** Suppose $d(\lambda) \neq 1$ and use Theorem **V**, Chapter 22.
    (ii) Suppose $d(\lambda) = 1$ and use Theorem **IV**, Chapter 22.

21. Infer from Problem 20 that when $g(A)$ is non-singular, then $[g(A)]^{-1}$ is expressible as a polynomial in A of degree less than that of $m(\lambda)$.

22. Prove: If the minimum polynomial $m(\lambda)$ of A over F is irreducible in $F[\lambda]$ and is of degree s in λ, then the set of all scalar polynomials in A with coefficients in F of degree $< s$ constitutes a field.

23. Let A and B be square matrices and denote by $m(\lambda)$ and $n(\lambda)$ respectively the minimum polynomials of AB and BA. Prove:
    (*a*) $m(\lambda) = n(\lambda)$ when not both A and B are singular.
    (*b*) $m(\lambda)$ and $n(\lambda)$ differ at most by a factor λ when both A and B are singular.
    Hint. $B \cdot m(AB) \cdot A = (BA) \cdot m(BA) = 0$ and $A \cdot n(BA) \cdot B = (AB) \cdot n(AB) = 0$.

24. Let A be of dimension $m \times n$ and B be of dimension $n \times m$, $m > n$, and denote by $\phi(\lambda)$ and $\psi(\lambda)$ respectively the characteristic polynomials of AB and BA. Show $\phi(\lambda) = \lambda^{m-n}\psi(\lambda)$.

25. Let X_i be an invariant vector associated with a simple characteristic root of A. Prove: If A and B commute, then X_i is an invariant vector of B.

26. If the matrices A and B commute, state a theorem concerning the invariant vectors of B when A has only simple characteristic roots.

Chapter 26

Canonical Forms Under Similarity

THE PROBLEM. In Chapter 25 it was shown that the characteristic matrices of two similar n-square matrices A and $R^{-1}AR$ over F have the same invariant factors and the same elementary divisors. In this chapter, we establish representatives of the set of all matrices $R^{-1}AR$ which are **(i)** simple in structure and **(ii)** put into view either the invariant factors or the elementary divisors. These matrices, four in number, are called **canonical forms** of A. They correspond to the canonical matrix $N = \begin{bmatrix} I_r & 0 \\ 0 & 0 \end{bmatrix}$ introduced earlier for all $m \times n$ matrices of rank r under equivalence.

THE RATIONAL CANONICAL FORM. Let A be an n-square matrix over F and suppose first that its characteristic matrix has just one non-trivial invariant factor $f_n(\lambda)$. The companion matrix $C(f_n)$ of $f_n(\lambda)$ was shown in Chapter 25 to be similar to A. We define it to be the **rational canonical form** S of all matrices similar to A.

Suppose next that the Smith normal form of $\lambda I - A$ is

(26.1) $$\operatorname{diag}\left(1, 1, \ldots, 1, f_j(\lambda), f_{j+1}(\lambda), \ldots, f_n(\lambda)\right)$$

with the non-trivial invariant factor $f_i(\lambda)$ of degree s_i, $(i = j, j+1, \ldots, n)$. We define as the **rational canonical form** of all matrices similar to A

(26.2) $$S = \operatorname{diag}\left(C(f_j), C(f_{j+1}), \ldots, C(f_n)\right)$$

To show that A and S have the same similarity invariants we note that $C(f_i)$ is similar to $D_i = \operatorname{diag}(1, 1, \ldots, 1, f_i(\lambda))$ and, thus, S is similar to $\operatorname{diag}(D_j, D_{j+1}, \ldots, D_n)$. By a sequence of interchanges of two rows and the same two columns, we have S similar to

$$\operatorname{diag}\left(1, 1, \ldots, 1, f_j(\lambda), f_{j+1}(\lambda), \ldots, f_n(\lambda)\right)$$

We have proved

I. Every square matrix A is similar to the direct sum **(26.2)** of the companion matrices of the non-trivial invariant factors of $\lambda I - A$.

Example 1. Let the non-trivial similarity invariants of A over the rational field be

$$f_8(\lambda) = \lambda + 1, \qquad f_9(\lambda) = \lambda^3 + 1, \qquad f_{10}(\lambda) = \lambda^6 + 2\lambda^3 + 1$$

Then

$$C(f_8) = [-1], \qquad C(f_9) = \begin{bmatrix} 0 & 1 & 0 \\ 0 & 0 & 1 \\ -1 & 0 & 0 \end{bmatrix}, \qquad C(f_{10}) = \begin{bmatrix} 0 & 1 & 0 & 0 & 0 & 0 \\ 0 & 0 & 1 & 0 & 0 & 0 \\ 0 & 0 & 0 & 1 & 0 & 0 \\ 0 & 0 & 0 & 0 & 1 & 0 \\ 0 & 0 & 0 & 0 & 0 & 1 \\ -1 & 0 & 0 & -2 & 0 & 0 \end{bmatrix}$$

and

$$S \quad = \quad \operatorname{diag}\left(C(f_8),\, C(f_9),\, C(f_{10})\right) \quad = \quad \begin{bmatrix} -1 & 0 & 0 & 0 & 0 & 0 & 0 & 0 & 0 & 0 \\ 0 & 0 & 1 & 0 & 0 & 0 & 0 & 0 & 0 & 0 \\ 0 & 0 & 0 & 1 & 0 & 0 & 0 & 0 & 0 & 0 \\ 0 & -1 & 0 & 0 & 0 & 0 & 0 & 0 & 0 & 0 \\ 0 & 0 & 0 & 0 & 0 & 1 & 0 & 0 & 0 & 0 \\ 0 & 0 & 0 & 0 & 0 & 0 & 1 & 0 & 0 & 0 \\ 0 & 0 & 0 & 0 & 0 & 0 & 0 & 1 & 0 & 0 \\ 0 & 0 & 0 & 0 & 0 & 0 & 0 & 0 & 1 & 0 \\ 0 & 0 & 0 & 0 & 0 & 0 & 0 & 0 & 0 & 1 \\ 0 & 0 & 0 & 0 & -1 & 0 & 0 & -2 & 0 & 0 \end{bmatrix}$$

is the required form of Theorem **I**.

Note. The order in which the companion matrices are arranged along the diagonal is immaterial. Also

$$\begin{bmatrix} -1 & 0 & 0 & 0 & 0 & 0 & 0 & 0 & 0 & 0 \\ 0 & 0 & 0 & -1 & 0 & 0 & 0 & 0 & 0 & 0 \\ 0 & 1 & 0 & 0 & 0 & 0 & 0 & 0 & 0 & 0 \\ 0 & 0 & 1 & 0 & 0 & 0 & 0 & 0 & 0 & 0 \\ 0 & 0 & 0 & 0 & 0 & 0 & 0 & 0 & 0 & -1 \\ 0 & 0 & 0 & 0 & 1 & 0 & 0 & 0 & 0 & 0 \\ 0 & 0 & 0 & 0 & 0 & 1 & 0 & 0 & 0 & 0 \\ 0 & 0 & 0 & 0 & 0 & 0 & 1 & 0 & 0 & -2 \\ 0 & 0 & 0 & 0 & 0 & 0 & 0 & 1 & 0 & 0 \\ 0 & 0 & 0 & 0 & 0 & 0 & 0 & 0 & 1 & 0 \end{bmatrix}$$

using the transpose of each of the companion matrices above is an alternate form.

A SECOND CANONICAL FORM. Let the characteristic matrix of A have as non-trivial invariant factors the polynomials $f_i(\lambda)$ of **(26.1)**. Suppose that the elementary divisors are powers of t distinct irreducible polynomials in $F[\lambda]$: $p_1(\lambda), p_2(\lambda), \ldots, p_t(\lambda)$. Let

$$\textbf{(26.3)} \qquad f_i(\lambda) \quad = \quad \{p_1(\lambda)\}^{q_{1i}}\{p_2(\lambda)\}^{q_{2i}} \ldots \{p_t(\lambda)\}^{q_{ti}}; \qquad (i = j, j+1, \ldots, n)$$

where not every factor need appear since some of the q's may be zero. The companion matrix $C(p_k^{q_{ki}})$ of any factor present has $\{p_k(\lambda)\}^{q_{ki}}$ as the only non-trivial similarity invariant; hence, $C(f_i)$ is similar to

$$\operatorname{diag}\left(C(p_1^{q_{1i}}),\, C(p_2^{q_{2i}}),\, \ldots,\, C(p_t^{q_{ti}})\right)$$

We have

II. Every square matrix A over F is similar to the direct sum of the companion matrices of the elementary divisors over F of $\lambda I - A$.

Example 2. For the matrix A of Example 1, the elementary divisors over the rational field are $\lambda+1$, $\lambda+1$, $(\lambda+1)^2$, $\lambda^2-\lambda+1$, $(\lambda^2-\lambda+1)^2$. Their respective companion matrices are

$$[-1], \quad [-1], \quad \begin{bmatrix} 0 & 1 \\ -1 & -2 \end{bmatrix}, \quad \begin{bmatrix} 0 & 1 \\ -1 & 1 \end{bmatrix}, \quad \begin{bmatrix} 0 & 1 & 0 & 0 \\ 0 & 0 & 1 & 0 \\ 0 & 0 & 0 & 1 \\ -1 & 2 & -3 & 2 \end{bmatrix}$$

and the canonical form of Theorem **II** is

$$\begin{bmatrix} -1 & 0 & 0 & 0 & 0 & 0 & 0 & 0 & 0 & 0 \\ 0 & -1 & 0 & 0 & 0 & 0 & 0 & 0 & 0 & 0 \\ 0 & 0 & 0 & 1 & 0 & 0 & 0 & 0 & 0 & 0 \\ 0 & 0 & -1 & -2 & 0 & 0 & 0 & 0 & 0 & 0 \\ 0 & 0 & 0 & 0 & 0 & 1 & 0 & 0 & 0 & 0 \\ 0 & 0 & 0 & 0 & -1 & 1 & 0 & 0 & 0 & 0 \\ 0 & 0 & 0 & 0 & 0 & 0 & 0 & 1 & 0 & 0 \\ 0 & 0 & 0 & 0 & 0 & 0 & 0 & 0 & 1 & 0 \\ 0 & 0 & 0 & 0 & 0 & 0 & 0 & 0 & 0 & 1 \\ 0 & 0 & 0 & 0 & 0 & 0 & -1 & 2 & -3 & 2 \end{bmatrix}$$

THE JACOBSON CANONICAL FORM. Let A be the matrix of the section above with the elementary divisors of its characteristic matrix expressed as powers of irreducible polynomials in $F[\lambda]$. Consider an elementary divisor $\{p(\lambda)\}^q$. If $q = 1$, use $C(p)$, the companion matrix; if $q > 1$, build

$$(26.4) \qquad C_q(p) = \begin{bmatrix} C(p) & M & 0 & \cdots & 0 & 0 \\ 0 & C(p) & M & \cdots & 0 & 0 \\ \cdots & \cdots & \cdots & \cdots & \cdots & \cdots \\ \cdots & \cdots & \cdots & \cdots & \cdots & \cdots \\ 0 & 0 & 0 & \cdots & C(p) & M \\ 0 & 0 & 0 & \cdots & 0 & C(p) \end{bmatrix}$$

where M is a matrix of the same order as $C(p)$ having the element 1 in the lower left-hand corner and zeros elsewhere. The matrix $C_q(p)$ of **(26.4)**, with the understanding that $C_1(p) = C(p)$ is called the **hypercompanion matrix** of $\{p(\lambda)\}^q$. Note that in **(26.4)**, there is a continuous line of 1's just above the diagonal.

When the alternate companion matrix $C'(p)$ is used, the hypercompanion matrix of $\{p(\lambda)\}^q$ is

$$C_q(p) = \begin{bmatrix} C'(p) & 0 & 0 & \cdots & 0 & 0 \\ N & C'(p) & 0 & \cdots & 0 & 0 \\ 0 & N & C'(p) & \cdots & 0 & 0 \\ \cdots & \cdots & \cdots & \cdots & \cdots & \cdots \\ 0 & 0 & 0 & \cdots & C'(p) & 0 \\ 0 & 0 & 0 & \cdots & N & C'(p) \end{bmatrix}$$

where N is a matrix of the same order as $C'(p)$ having the element 1 in the upper right-hand corner and zeros elsewhere. In this form there is a continuous line of 1's just below the diagonal.

Example 3. Let $\{p(\lambda)\}^q = (\lambda^2 + 2\lambda - 1)^4$. Then $C(p) = \begin{bmatrix} 0 & 1 \\ 1 & -2 \end{bmatrix}$, $M = \begin{bmatrix} 0 & 0 \\ 1 & 0 \end{bmatrix}$, and

$$C_q(p) = \begin{bmatrix} 0 & 1 & 0 & 0 & 0 & 0 & 0 & 0 \\ 1 & -2 & 1 & 0 & 0 & 0 & 0 & 0 \\ 0 & 0 & 0 & 1 & 0 & 0 & 0 & 0 \\ 0 & 0 & 1 & -2 & 1 & 0 & 0 & 0 \\ 0 & 0 & 0 & 0 & 0 & 1 & 0 & 0 \\ 0 & 0 & 0 & 0 & 1 & -2 & 1 & 0 \\ 0 & 0 & 0 & 0 & 0 & 0 & 0 & 1 \\ 0 & 0 & 0 & 0 & 0 & 0 & 1 & -2 \end{bmatrix}$$

In Problem 1, it is shown that $C_q(p)$ has $\{p(\lambda)\}^q$ as its only non-trivial similarity invariant. Thus, $C_q(p)$ is similar to $C(p^q)$ and may be substituted for it in the canonical form of Theorem **II**. We have

III. Every square matrix A over F is similar to the direct sum of the hypercompanion matrices of the elementary divisors over F of $\lambda I - A$.

Example 4. For the matrix A of Example 2, the hypercompanion matrices of the elementary divisors $\lambda+1$, $\lambda+1$, and $\lambda^2-\lambda+1$ are their companion matrices, the hypercompanion matrix of $(\lambda+1)^2$ is $\begin{bmatrix} -1 & 1 \\ 0 & -1 \end{bmatrix}$ and that of $(\lambda^2-\lambda+1)^2$ is $\begin{bmatrix} 0 & 1 & 0 & 0 \\ -1 & 1 & 1 & 0 \\ 0 & 0 & 0 & 1 \\ 0 & 0 & -1 & 1 \end{bmatrix}$. Thus, the canonical form of Theorem **III** is

$$\begin{bmatrix} -1 & 0 & 0 & 0 & 0 & 0 & 0 & 0 & 0 & 0 \\ 0 & -1 & 0 & 0 & 0 & 0 & 0 & 0 & 0 & 0 \\ 0 & 0 & -1 & 1 & 0 & 0 & 0 & 0 & 0 & 0 \\ 0 & 0 & 0 & -1 & 0 & 0 & 0 & 0 & 0 & 0 \\ 0 & 0 & 0 & 0 & 0 & 1 & 0 & 0 & 0 & 0 \\ 0 & 0 & 0 & 0 & -1 & 1 & 0 & 0 & 0 & 0 \\ 0 & 0 & 0 & 0 & 0 & 0 & 0 & 1 & 0 & 0 \\ 0 & 0 & 0 & 0 & 0 & 0 & -1 & 1 & 1 & 0 \\ 0 & 0 & 0 & 0 & 0 & 0 & 0 & 0 & 0 & 1 \\ 0 & 0 & 0 & 0 & 0 & 0 & 0 & 0 & -1 & 1 \end{bmatrix}$$

The use of the term "rational" in connection with the canonical form of Theorem **I** is somewhat misleading. It was used originally to indicate that in obtaining the canonical form only rational operations in the field of the elements of A are necessary. But this is, of course, true also of the canonical forms (introduced later) of Theorems **II** and **III**. To further add to the confusion, the canonical form of Theorem **III** is sometimes called the rational canonical form.

THE CLASSICAL CANONICAL FORM. Let the elementary divisors of the characteristic matrix of A be powers of linear polynomials. The canonical form of Theorem **III** is then the direct sum of hypercompanion matrices of the form

$$C_q(p) = \begin{bmatrix} a_i & 1 & 0 & \cdots & 0 & 0 \\ 0 & a_i & 1 & \cdots & 0 & 0 \\ \cdots & \cdots & \cdots & \cdots & \cdots & \cdots \\ 0 & 0 & 0 & \cdots & a_i & 1 \\ 0 & 0 & 0 & \cdots & 0 & a_i \end{bmatrix} \tag{26.5}$$

corresponding to the elementary divisor $\{p(\lambda)\}^q = (\lambda - a_i)^q$. For an example, see Problem 2.

This special case of the canonical form of Theorem **III** is known as the **Jordan** or **classical canonical form**. [Note that $C_q(p)$ of **(26.5)** is of the type J of **(25.4)**.] We have

IV. Let $\mathfrak{F}$ be the field in which the characteristic polynomial of a matrix A factors into linear polynomials. Then A is similar over $\mathfrak{F}$ to the direct sum of hypercompanion matrices of the form **(26.5)**, each matrix corresponding to an elementary divisor $(\lambda - a_i)^q$.

Example 5. Let the elementary divisors over the complex field of $\lambda I - A$ be: $\lambda - i$, $\lambda + i$, $(\lambda - i)^2$, $(\lambda+i)^2$.

The classical canonical form of A is

$$\begin{bmatrix} i & 0 & 0 & 0 & 0 & 0 \\ 0 & -i & 0 & 0 & 0 & 0 \\ 0 & 0 & i & 1 & 0 & 0 \\ 0 & 0 & 0 & i & 0 & 0 \\ 0 & 0 & 0 & 0 & -i & 1 \\ 0 & 0 & 0 & 0 & 0 & -i \end{bmatrix}$$

From Theorem **IV** follows

V. An n-square matrix A is similar to a diagonal matrix if and only if the elementary divisors of $\lambda I - A$ are linear polynomials, that is, if and only if the minimum polynomial of A is the product of distinct linear polynomials.

See Problems 2-4.

A REDUCTION TO RATIONAL CANONICAL FORM. In concluding this discussion of canonical forms, it will be shown that a reduction of any n-square matrix to its rational canonical form can be made, at least theoretically, without having prior knowledge of the invariant factors of $\lambda I - A$. A somewhat different treatment of this can be found in Dickson, L. E., *Modern Algebraic Theories,* Benj. H. Sanborn, 1926. Some improvement on purely computational aspects is made in Browne, E. T., *American Mathematical Monthly*, vol. 48 (1940).

We shall need the following definitions:

If A is an n-square matrix and X is an n-vector over F and if $g(\lambda)$ is the monic polynomial in $F[\lambda]$ of minimum degree such that $g(A)\cdot X = 0$, then with respect to A the vector X is said to **belong** to $g(\lambda)$.

If, with respect to A, the vector X belongs to $g(\lambda)$ of degree p, the linearly independent vectors $X, AX, A^2X, \ldots, A^{p-1}X$ are called a **chain** having X as its **leader**.

Example 6. Let $A = \begin{bmatrix} 2 & -6 & 3 \\ 1 & -3 & 1 \\ 1 & -2 & 0 \end{bmatrix}$. The vectors $X = [1,0,0]'$ and $AX = [2,1,1]'$ are linearly independent while $A^2X = X$. Then $(A^2 - I)X = 0$ and X belongs to the polynomial $\lambda^2 - 1$. For $Y = [1,0,-1]'$, $AY = [-1,0,1]' = -Y$; thus, $(A+I)Y = 0$ and Y belongs to the polynomial $\lambda + 1$.

If $m(\lambda)$ is the minimum polynomial of an n-square matrix A, then $m(A)\cdot X = 0$ for every n-vector X. Thus, there can be no chain of length greater than the degree of $m(\lambda)$. For the matrix of Example 6, the minimum polynomial is $\lambda^2 - 1$.

Let S be the rational canonical form of the n-square matrix A over F. Then, there exists a non-singular matrix R over F such that

$$R^{-1}AR \quad = \quad S \quad = \quad \operatorname{diag}(C_j, C_{j+1}, \ldots, C_n) \tag{26.6}$$

where, for convenience, $C(f_i)$ in **(26.2)** has been replaced by C_i. We shall assume that C_i, the companion matrix of the invariant factor

$$f_i(\lambda) \quad = \quad \lambda^{s_i} + c_{i,s_i}\lambda^{s_i - 1} + \cdots + c_{i2}\lambda + c_{i1}$$

has the form

$$C_i \quad = \quad \begin{bmatrix} 0 & 0 & 0 & \cdots & 0 & -c_{i1} \\ 1 & 0 & 0 & \cdots & 0 & -c_{i2} \\ 0 & 1 & 0 & \cdots & 0 & -c_{i3} \\ \cdots & \cdots & \cdots & \cdots & \cdots & \cdots \\ 0 & 0 & 0 & \cdots & 0 & -c_{i,s_i-1} \\ 0 & 0 & 0 & \cdots & 1 & -c_{is_i} \end{bmatrix}$$

From **(26.6)**, we have

(26.7) $$AR = RS = R\,\text{diag}(C_j, C_{j+1}, \ldots, C_n)$$

Let R be separated into column blocks $R_j, R_{j+1}, \ldots, R_n$ so that R_i and C_i, $(i = j, j+1, \ldots, n)$ have the same number of columns. From **(26.7)**,

$$AR = A[R_j, R_{j+1}, \ldots, R_n] = [R_j, R_{j+1}, \ldots, R_n]\,\text{diag}(C_j, C_{j+1}, \ldots, C_n)$$
$$= [R_j C_j, R_{j+1} C_{j+1}, \ldots, R_n C_n]$$

and

$$AR_i = R_i C_i, \quad (i = j, j+1, \ldots, n)$$

Denote the s_i column vectors of R_i by $R_{i1}, R_{i2}, \ldots, R_{is_i}$ and form the product

$$R_i C_i = [R_{i1}, R_{i2}, \ldots, R_{is_i}]\,C_i = [R_{i2}, R_{i3}, \ldots, R_{is_i}, -\sum_{k=1}^{s_i} R_{ik} c_{ik}]$$

Since

$$AR_i = A[R_{i1}, R_{i2}, \ldots, R_{is_i}] = [AR_{i1}, AR_{i2}, \ldots, AR_{is_i}] = R_i C_i$$

we have

(26.8) $$R_{i2} = AR_{i1}, \quad R_{i3} = AR_{i2} = A^2 R_{i1}, \quad \ldots, \quad R_{is_i} = A^{s_i - 1} R_{i1}$$

and

(26.9) $$-\sum_{k=1}^{s_i} c_{ik} R_{ik} = AR_{is_i}$$

Substituting into **(26.9)** from **(26.8)**, we obtain

$$-\sum_{k=1}^{s_i} c_{ik} A^{k-1} R_{i1} = A^{s_i} R_{i1}$$

or

(26.10) $$(A^{s_i} + c_{is_i} A^{s_i - 1} + \ldots + c_{i2} A + c_{i1} I) R_{i1} = 0$$

From the definition of C_i above, **(26.10)** may be written as

(26.11) $$f_i(A) \cdot R_{i1} = 0$$

Let R_{i1} be denoted by X_i so that **(26.11)** becomes $f_i(A) \cdot X_i = 0$; then, since $X_i, AX_i, A^2 X_i, \ldots, A^{s_i - 1} X_i$ are linearly independent, the vector X_i belongs to the invariant factor $f_i(\lambda)$. Thus, the column vectors of R_i consist of the vectors of the chain having X_i, belonging to $f_i(\lambda)$, as leader.

To summarize: the n linearly independent columns of R, satisfying **(26.2)**, consist of $n-j+1$ chains

$$X_i, AX_i, \ldots, A^{s_i - 1} X_i \qquad (i = j, j+1, \ldots, n)$$

whose leaders belong to the respective invariant factors $f_j(\lambda), f_{j+1}(\lambda), \ldots, f_n(\lambda)$ and whose lengths satisfy the condition $0 < s_j \leqq s_{j+1} \leqq \ldots \leqq s_n$.

We have

VI. For a given n-square matrix A over F:

(i) let X_n be the leader of a chain $\mathcal{E}_n$ of maximum length for all n-vectors over F;

(ii) let X_{n-1} be the leader of a chain $\mathcal{E}_{n-1}$ of maximum length (any member of which is linearly independent of the preceding members and those of $\mathcal{E}_n$) for all n-vectors over F which are linearly independent of the vectors of $\mathcal{E}_n$;

(iii) let X_{n-2} be the leader of a chain $\mathcal{E}_{n-2}$ of maximum length (any member of which is linearly independent of the preceding members and those of $\mathcal{E}_n$ and $\mathcal{E}_{n-1}$) for all n-vectors over F which are linearly independent of the vectors of $\mathcal{E}_n$ and $\mathcal{E}_{n-1}$;

and so on. Then, for

$$R = [X_j, AX_j, \ldots, A^{s_j-1}X_j;\ X_{j+1}, AX_{j+1}, \ldots, A^{s_{j+1}-1}X_{j+1};\ \ldots;\ X_n, AX_n, \ldots, A^{s_n-1}X_n]$$

$R^{-1}AR$ is the rational canonical form of A.

Example 7. Let $A = \begin{bmatrix} 1 & 1 & 1 \\ 1 & 2 & 2 \\ 1 & 3 & 2 \end{bmatrix}$. Take $X = [1, 0, 0]'$; then $X, AX = [1, 1, 1]', A^2X = [3, 5, 6]'$ are linearly independent while $A^3X = [14, 25, 30]' = 5A^2X - X$. Thus, $(A^3 - 5A^2 + I)X = 0$ and X belongs to $f_3(\lambda) = m(\lambda) = \lambda^3 - 5\lambda^2 + 1 = \phi(\lambda)$. Taking

$$R = [X, AX, A^2X] = \begin{bmatrix} 1 & 1 & 3 \\ 0 & 1 & 5 \\ 0 & 1 & 6 \end{bmatrix}$$

we find

$$R^{-1} = \begin{bmatrix} 1 & -3 & 2 \\ 0 & 6 & -5 \\ 0 & -1 & 1 \end{bmatrix}, \qquad AR = [AX, A^2X, A^3X] = \begin{bmatrix} 1 & 3 & 14 \\ 1 & 5 & 25 \\ 1 & 6 & 30 \end{bmatrix}$$

and

$$R^{-1}AR = \begin{bmatrix} 0 & 0 & -1 \\ 1 & 0 & 0 \\ 0 & 1 & 5 \end{bmatrix} = S$$

Here A is non-derogatory with minimum polynomial $m(\lambda)$ irreducible over the rational field. Every 3-vector over this field belongs to $m(\lambda)$, (see Problem 11), and leads a chain of length three. The matrix R having the vectors of any chain as column vectors is such that $R^{-1}AR = S$.

Example 8. Let $A = \begin{bmatrix} 2 & 1 & 3 \\ 1 & 2 & 2 \\ 2 & 2 & 1 \end{bmatrix}$. Take $X = [1, -1, 0]'$; then $AX = X$ and X belongs to $\lambda - 1$. Now $\lambda - 1$ cannot be the minimum polynomial $m(\lambda)$ of A. It is, however, a divisor of $m(\lambda)$, (see Problem 11), and **could** be a similarity invariant of A.

Next, take $Y = [1, 0, 0]'$. The vectors $Y, AY = [2, 1, 2]', A^2Y = [11, 8, 8]'$ are linearly independent while $A^3Y = [54, 43, 46]' = 5A^2Y + 3AY - 7Y$. Thus, Y belongs to $m(\lambda) = \lambda^3 - 5\lambda^2 - 3\lambda + 7 = \phi(\lambda)$. The polynomial $\lambda - 1$ is not a similarity invariant; in fact, unless the first choice of vector belongs to a polynomial which could reasonably be the minimum function, it should be considered a false start. The reader may verify that

$$R^{-1}AR = \begin{bmatrix} 0 & 0 & -7 \\ 1 & 0 & 3 \\ 0 & 1 & 5 \end{bmatrix}$$

when $R = [Y, AY, A^2Y] = \begin{bmatrix} 1 & 2 & 11 \\ 0 & 1 & 8 \\ 0 & 2 & 8 \end{bmatrix}$.

See Problems 5-6.

SOLVED PROBLEMS

1. Prove: The matrix $C_q(p)$ of **(26.4)** has $\{p(\lambda)\}^q$ as its only non-trivial similarity invariant.

Let $C_q(p)$ be of order s. The minor of the element in the last row and first column of $\lambda I - C_q(p)$ is ± 1 so that the greatest common divisor of all $(s-1)$-square minors of $\lambda I - C_q(p)$ is 1. Then the invariant factors of $\lambda I - C_q(p)$ are $1, 1, \ldots, 1, f_s(\lambda)$. But $f_s(\lambda) = \{p(\lambda)\}^q$ since

$$\phi(\lambda) = |\lambda I - C_q(p)| = |\lambda I - C(p)|^q = \{p(\lambda)\}^q$$

2. The canonical form (a) is that of Theorems **I** and **II**, the non-trivial invariant factor and elementary divisor being $\lambda^4+4\lambda^3+6\lambda^2+4\lambda+1$. The canonical form of Theorem **III** is (b).

$$(a)\quad \begin{bmatrix} 0 & 1 & 0 & 0 \\ 0 & 0 & 1 & 0 \\ 0 & 0 & 0 & 1 \\ -1 & -4 & -6 & -4 \end{bmatrix} \qquad (b)\quad \begin{bmatrix} -1 & 1 & 0 & 0 \\ 0 & -1 & 1 & 0 \\ 0 & 0 & -1 & 1 \\ 0 & 0 & 0 & -1 \end{bmatrix}$$

3. The canonical form (a) is that of Theorem **I**, the invariant factors being $\lambda+2$, λ^2-4, $\lambda^3+3\lambda^2-4\lambda-12$ and the elementary divisors being $\lambda+2$, $\lambda+2$, $\lambda+2$, $\lambda-2$, $\lambda-2$, $\lambda+3$. The canonical form of both Theorems **II** and **III** is (b).

$$(a)\quad \begin{bmatrix} -2 & 0 & 0 & 0 & 0 & 0 \\ 0 & \mathbf{0} & \mathbf{1} & 0 & 0 & 0 \\ 0 & \mathbf{4} & \mathbf{0} & 0 & 0 & 0 \\ 0 & 0 & 0 & \mathbf{0} & \mathbf{1} & \mathbf{0} \\ 0 & 0 & 0 & \mathbf{0} & \mathbf{0} & \mathbf{1} \\ 0 & 0 & 0 & \mathbf{12} & \mathbf{4} & \mathbf{-3} \end{bmatrix} \qquad (b)\quad \begin{bmatrix} -2 & 0 & 0 & 0 & 0 & 0 \\ 0 & -2 & 0 & 0 & 0 & 0 \\ 0 & 0 & -2 & 0 & 0 & 0 \\ 0 & 0 & 0 & 2 & 0 & 0 \\ 0 & 0 & 0 & 0 & 2 & 0 \\ 0 & 0 & 0 & 0 & 0 & -3 \end{bmatrix}$$

4. The canonical form (a) is that of Theorem **III**. Over the rational field the elementary divisors are $\lambda+2$, $\lambda+2$, $(\lambda^2+2\lambda-1)^2$, $(\lambda^2+2\lambda-1)^3$ and the invariant factors are

$$(\lambda+2)(\lambda^2+2\lambda-1)^2, \qquad (\lambda+2)(\lambda^2+2\lambda-1)^3$$

The canonical form of Theorem **I** is (b) and that of Theorem **II** is (c).

$$(a)\quad \begin{bmatrix} -2 & 0 & 0 & 0 & 0 & 0 & 0 & 0 & 0 & 0 & 0 & 0 \\ 0 & -2 & 0 & 0 & 0 & 0 & 0 & 0 & 0 & 0 & 0 & 0 \\ 0 & 0 & \mathbf{0} & \mathbf{1} & \mathbf{0} & \mathbf{0} & 0 & 0 & 0 & 0 & 0 & 0 \\ 0 & 0 & \mathbf{1} & \mathbf{-2} & \mathbf{1} & \mathbf{0} & 0 & 0 & 0 & 0 & 0 & 0 \\ 0 & 0 & \mathbf{0} & \mathbf{0} & \mathbf{0} & \mathbf{1} & 0 & 0 & 0 & 0 & 0 & 0 \\ 0 & 0 & \mathbf{0} & \mathbf{0} & \mathbf{1} & \mathbf{-2} & 0 & 0 & 0 & 0 & 0 & 0 \\ 0 & 0 & 0 & 0 & 0 & 0 & \mathbf{0} & \mathbf{1} & \mathbf{0} & \mathbf{0} & \mathbf{0} & \mathbf{0} \\ 0 & 0 & 0 & 0 & 0 & 0 & \mathbf{1} & \mathbf{-2} & \mathbf{1} & \mathbf{0} & \mathbf{0} & \mathbf{0} \\ 0 & 0 & 0 & 0 & 0 & 0 & \mathbf{0} & \mathbf{0} & \mathbf{0} & \mathbf{1} & \mathbf{0} & \mathbf{0} \\ 0 & 0 & 0 & 0 & 0 & 0 & \mathbf{0} & \mathbf{0} & \mathbf{1} & \mathbf{-2} & \mathbf{1} & \mathbf{0} \\ 0 & 0 & 0 & 0 & 0 & 0 & \mathbf{0} & \mathbf{0} & \mathbf{0} & \mathbf{0} & \mathbf{0} & \mathbf{1} \\ 0 & 0 & 0 & 0 & 0 & 0 & \mathbf{0} & \mathbf{0} & \mathbf{0} & \mathbf{0} & \mathbf{1} & \mathbf{-2} \end{bmatrix}$$

$$(b)\quad \begin{bmatrix} \mathbf{0} & \mathbf{1} & \mathbf{0} & \mathbf{0} & \mathbf{0} & 0 & 0 & 0 & 0 & 0 & 0 & 0 \\ \mathbf{0} & \mathbf{0} & \mathbf{1} & \mathbf{0} & \mathbf{0} & 0 & 0 & 0 & 0 & 0 & 0 & 0 \\ \mathbf{0} & \mathbf{0} & \mathbf{0} & \mathbf{1} & \mathbf{0} & 0 & 0 & 0 & 0 & 0 & 0 & 0 \\ \mathbf{0} & \mathbf{0} & \mathbf{0} & \mathbf{0} & \mathbf{1} & 0 & 0 & 0 & 0 & 0 & 0 & 0 \\ \mathbf{-2} & \mathbf{7} & \mathbf{0} & \mathbf{-10} & \mathbf{-6} & 0 & 0 & 0 & 0 & 0 & 0 & 0 \\ 0 & 0 & 0 & 0 & 0 & \mathbf{0} & \mathbf{1} & \mathbf{0} & \mathbf{0} & \mathbf{0} & \mathbf{0} & \mathbf{0} \\ 0 & 0 & 0 & 0 & 0 & \mathbf{0} & \mathbf{0} & \mathbf{1} & \mathbf{0} & \mathbf{0} & \mathbf{0} & \mathbf{0} \\ 0 & 0 & 0 & 0 & 0 & \mathbf{0} & \mathbf{0} & \mathbf{0} & \mathbf{1} & \mathbf{0} & \mathbf{0} & \mathbf{0} \\ 0 & 0 & 0 & 0 & 0 & \mathbf{0} & \mathbf{0} & \mathbf{0} & \mathbf{0} & \mathbf{1} & \mathbf{0} & \mathbf{0} \\ 0 & 0 & 0 & 0 & 0 & \mathbf{0} & \mathbf{0} & \mathbf{0} & \mathbf{0} & \mathbf{0} & \mathbf{1} & \mathbf{0} \\ 0 & 0 & 0 & 0 & 0 & \mathbf{0} & \mathbf{0} & \mathbf{0} & \mathbf{0} & \mathbf{0} & \mathbf{0} & \mathbf{1} \\ 0 & 0 & 0 & 0 & 0 & \mathbf{2} & \mathbf{-11} & \mathbf{12} & \mathbf{17} & \mathbf{-14} & \mathbf{-21} & \mathbf{-8} \end{bmatrix}$$

$$(c)\quad \begin{bmatrix} -2 & 0 & 0 & 0 & 0 & 0 & 0 & 0 & 0 & 0 & 0 & 0 \\ 0 & -2 & 0 & 0 & 0 & 0 & 0 & 0 & 0 & 0 & 0 & 0 \\ 0 & 0 & 0 & 1 & 0 & 0 & 0 & 0 & 0 & 0 & 0 & 0 \\ 0 & 0 & 0 & 0 & 1 & 0 & 0 & 0 & 0 & 0 & 0 & 0 \\ 0 & 0 & 0 & 0 & 0 & 1 & 0 & 0 & 0 & 0 & 0 & 0 \\ 0 & 0 & -1 & 4 & -2 & -4 & 0 & 0 & 0 & 0 & 0 & 0 \\ 0 & 0 & 0 & 0 & 0 & 0 & 0 & 1 & 0 & 0 & 0 & 0 \\ 0 & 0 & 0 & 0 & 0 & 0 & 0 & 0 & 1 & 0 & 0 & 0 \\ 0 & 0 & 0 & 0 & 0 & 0 & 0 & 0 & 0 & 1 & 0 & 0 \\ 0 & 0 & 0 & 0 & 0 & 0 & 0 & 0 & 0 & 0 & 1 & 0 \\ 0 & 0 & 0 & 0 & 0 & 0 & 0 & 0 & 0 & 0 & 0 & 1 \\ 0 & 0 & 0 & 0 & 0 & 0 & 1 & -6 & 9 & 4 & -9 & -6 \end{bmatrix}$$

5. Let $A = \begin{bmatrix} -2 & 3 & 3 & -1 & -6 & -2 \\ 1 & 0 & -1 & 0 & 2 & 1 \\ 1 & -2 & 0 & 1 & 2 & 0 \\ 1 & 1 & -1 & -1 & 2 & 1 \\ 1 & -2 & -1 & 1 & 3 & 1 \\ 1 & 0 & -1 & 0 & 2 & 0 \end{bmatrix}$. Take $X = [1, 0, 0, 0, 0, 0]'$.

Then $X, AX = [-2, 1, 1, 1, 1, 1]'$, $A^2X = [1, 0, -1, 0, 0, -1]'$, $A^3X = [-3, 1, 1, 1, 1, 2]'$ are linearly independent while $A^4X = [1, 0, -2, 0, 0, -2]' = 2A^2X - X$; X belongs to $\lambda^4 - 2\lambda^2 + 1$. We tentatively assume $m(\lambda) = \lambda^4 - 2\lambda^2 + 1$ and write X_6 for X.

The vector $Y = [0, 0, 0, 1, 0, 0]'$ is linearly independent of the members of the chain led by X_6 and $AY = [-1, 0, 1, -1, 1, 0]'$ is linearly independent of Y and the members of the chain. Now $A^2Y = Y$ so that Y belongs to $\lambda^2 - 1$. Since the two polynomials complete the set of non-trivial invariant factors, we write X_5 for Y. When

$$R = [X_5, AX_5, X_6, AX_6, A^2X_6, A^3X_6] = \begin{bmatrix} 0 & -1 & 1 & -2 & 1 & -3 \\ 0 & 0 & 0 & 1 & 0 & 1 \\ 0 & 1 & 0 & 1 & -1 & 1 \\ 1 & -1 & 0 & 1 & 0 & 1 \\ 0 & 1 & 0 & 1 & 0 & 1 \\ 0 & 0 & 0 & 1 & -1 & 2 \end{bmatrix}, \qquad R^{-1}AR = \begin{bmatrix} 0 & 1 & 0 & 0 & 0 & 0 \\ 1 & 0 & 0 & 0 & 0 & 0 \\ 0 & 0 & 0 & 0 & 0 & -1 \\ 0 & 0 & 1 & 0 & 0 & 0 \\ 0 & 0 & 0 & 1 & 0 & 2 \\ 0 & 0 & 0 & 0 & 1 & 0 \end{bmatrix}$$

the rational canonical form of A.

Note. The vector $Z = [0, 1, 0, 0, 0, 0]'$ is linearly independent of the members of the chain led by X_6 and $AZ = [3, 0, -2, 1, -2, 0]'$ is linearly independent of Z and the members of the chain. However, $A^2Z = [-1, 1, 0, 0, 0, 1]' = -AX_6 + A^3X_6 + Z$; then $(A^2 - 1)(Z - AX_6) = 0$ and $W = Z - AX_6 = [2, 0, -1, -1, -1, -1]'$ belongs to $\lambda^2 - 1$. Using this as X_5, we may form another R with which to obtain the rational canonical form.

6. Let $A = \begin{bmatrix} -2 & -1 & -1 & -1 & 2 \\ 1 & 3 & 1 & 1 & -1 \\ -1 & -4 & -2 & -1 & 1 \\ -1 & -4 & -1 & -2 & 1 \\ -2 & -2 & -2 & -2 & 3 \end{bmatrix}$. Take $X = [1, 0, 0, 0, 0]'$.

Then $X, AX = [-2, 1, -1, -1, -2]'$, $A^2X = [1, 1, -1, -1, 0]'$ are linearly independent while $A^3X = [-1, 2, -2, -2, 0]' = 2A^2X - 3X$ and X belongs to $\lambda^3 - 2\lambda^2 + 3$. We tentatively assume this to be the minimum polynomial $m(\lambda)$ and label X as X_5.

When, in A, the fourth column is subtracted from the first, we have $[-1, 0, 0, 1, 0]'$; hence, if $Y = [1, 0, 0, -1, 0]'$, $AY = -Y$ and Y belongs to $\lambda+1$. Again, when the fourth column of A is subtracted from the third, we have $[0, 0, -1, 1, 0]'$; hence, if $Z = [0, 0, 1, -1, 0]'$, $AZ = -Z$ and Z belongs to $\lambda+1$. Since Y, Z, and the members of the chain led by X_5 are **linearly independent**, we label Y as X_4 and Z as X_3. When

$$R = [X_3, X_4, X_5, AX_5, A^2X_5] = \begin{bmatrix} 0 & 1 & 1 & -2 & 1 \\ 0 & 0 & 0 & 1 & 1 \\ 1 & 0 & 0 & -1 & -1 \\ -1 & -1 & 0 & -1 & -1 \\ 0 & 0 & 0 & -2 & 0 \end{bmatrix} \qquad R^{-1}AR = \begin{bmatrix} \mathbf{-1} & 0 & 0 & 0 & 0 \\ 0 & \mathbf{-1} & 0 & 0 & 0 \\ 0 & 0 & \mathbf{0} & \mathbf{0} & \mathbf{-3} \\ 0 & 0 & \mathbf{1} & \mathbf{0} & \mathbf{0} \\ 0 & 0 & \mathbf{0} & \mathbf{1} & \mathbf{2} \end{bmatrix}$$

the rational canonical form of A.

SUPPLEMENTARY PROBLEMS

7. For each of the matrices (a)-(h) of Problem 9, Chapter 25, write the canonical matrix of Theorems **I, II, III** over the rational field, Can any of these matrices be changed by enlarging the number field?

Partial Ans. (a) **I**, $\begin{bmatrix} 0 & 1 & 0 \\ 0 & 0 & 1 \\ 6 & -11 & 6 \end{bmatrix}$; **II, III**, $\text{diag}(1, 2, 3)$

(b) **I, II, III**, $\begin{bmatrix} 0 & 1 & 0 \\ 0 & 0 & 1 \\ 0 & 0 & 0 \end{bmatrix}$

(e) **I**, $\begin{bmatrix} \mathbf{0} & 0 & 0 \\ 0 & \mathbf{0} & \mathbf{1} \\ 0 & \mathbf{0} & \mathbf{4} \end{bmatrix}$; **II, III**, $\begin{bmatrix} 0 & 0 & 0 \\ 0 & 0 & 0 \\ 0 & 0 & 4 \end{bmatrix}$

(f) **I**, $\begin{bmatrix} \mathbf{-1} & 0 & 0 & 0 \\ 0 & \mathbf{0} & \mathbf{1} & \mathbf{0} \\ 0 & \mathbf{0} & \mathbf{0} & \mathbf{1} \\ 0 & \mathbf{0} & \mathbf{1} & \mathbf{0} \end{bmatrix}$; **II, III**, $\begin{bmatrix} \mathbf{0} & 0 & 0 & 0 \\ 0 & \mathbf{-1} & 0 & 0 \\ 0 & 0 & \mathbf{-1} & 0 \\ 0 & 0 & 0 & \mathbf{1} \end{bmatrix}$

(g) **I**, $\begin{bmatrix} \mathbf{0} & 0 & 0 & 0 \\ 0 & \mathbf{0} & \mathbf{1} & \mathbf{0} \\ 0 & \mathbf{0} & \mathbf{0} & \mathbf{1} \\ 0 & \mathbf{0} & \mathbf{-1} & \mathbf{-2} \end{bmatrix}$; **II**, $\begin{bmatrix} \mathbf{0} & 0 & 0 & 0 \\ 0 & \mathbf{0} & 0 & 0 \\ 0 & 0 & \mathbf{0} & \mathbf{1} \\ 0 & 0 & \mathbf{-1} & \mathbf{-2} \end{bmatrix}$; **III**, $\begin{bmatrix} \mathbf{0} & 0 & 0 & 0 \\ 0 & \mathbf{0} & 0 & 0 \\ 0 & 0 & \mathbf{-1} & \mathbf{1} \\ 0 & 0 & \mathbf{0} & \mathbf{-1} \end{bmatrix}$

(h) **I**, $\begin{bmatrix} \mathbf{2} & 0 & 0 & 0 & 0 \\ 0 & \mathbf{0} & \mathbf{1} & 0 & 0 \\ 0 & \mathbf{2} & \mathbf{1} & 0 & 0 \\ 0 & 0 & 0 & \mathbf{0} & \mathbf{1} \\ 0 & 0 & 0 & \mathbf{2} & \mathbf{1} \end{bmatrix}$; **II, III**, $\text{diag}(2, 2, 2, -1, -1)$

8. Under what conditions will (a) the canonical forms of Theorems **I** and **II** be identical? (b) the canonical forms of Theorems **II** and **III** be identical? (c) the canonical form of Theorem **II** be diagonal?

9. Identify the canonical form $\begin{bmatrix} 0 & 0 & 0 \\ 0 & 0 & 1 \\ 0 & 0 & 0 \end{bmatrix}$. Check with the answer to Problem 8(*b*).

10. Let the non-singular matrix A have non-trivial invariant factors (*a*) $\lambda + 1$, $\lambda^3 + 1$, $(\lambda^3+1)^2$, (*b*) $\lambda^2 + 1$, $\lambda^4 + 5\lambda^2 + 4$, $\lambda^6 + 6\lambda^4 + 9\lambda^2 + 4$. Write the canonical forms of Theorems **I**, **II**, **III** over the rational field and that of Theorem **IV**.

Ans. (*a*)

$$\text{I.}\quad \begin{bmatrix} -1 & 0 & 0 & 0 & 0 & 0 & 0 & 0 & 0 & 0 \\ 0 & 0 & 1 & 0 & 0 & 0 & 0 & 0 & 0 & 0 \\ 0 & 0 & 0 & 1 & 0 & 0 & 0 & 0 & 0 & 0 \\ 0 & -1 & 0 & 0 & 0 & 0 & 0 & 0 & 0 & 0 \\ 0 & 0 & 0 & 0 & 0 & 1 & 0 & 0 & 0 & 0 \\ 0 & 0 & 0 & 0 & 0 & 0 & 1 & 0 & 0 & 0 \\ 0 & 0 & 0 & 0 & 0 & 0 & 0 & 1 & 0 & 0 \\ 0 & 0 & 0 & 0 & 0 & 0 & 0 & 0 & 1 & 0 \\ 0 & 0 & 0 & 0 & 0 & 0 & 0 & 0 & 0 & 1 \\ 0 & 0 & 0 & 0 & -1 & 0 & 0 & -2 & 0 & 0 \end{bmatrix}$$

$$\text{II.}\quad \begin{bmatrix} -1 & 0 & 0 & 0 & 0 & 0 & 0 & 0 & 0 & 0 \\ 0 & -1 & 0 & 0 & 0 & 0 & 0 & 0 & 0 & 0 \\ 0 & 0 & 0 & 1 & 0 & 0 & 0 & 0 & 0 & 0 \\ 0 & 0 & -1 & -2 & 0 & 0 & 0 & 0 & 0 & 0 \\ 0 & 0 & 0 & 0 & 0 & 1 & 0 & 0 & 0 & 0 \\ 0 & 0 & 0 & 0 & -1 & 1 & 0 & 0 & 0 & 0 \\ 0 & 0 & 0 & 0 & 0 & 0 & 0 & 1 & 0 & 0 \\ 0 & 0 & 0 & 0 & 0 & 0 & 0 & 0 & 1 & 0 \\ 0 & 0 & 0 & 0 & 0 & 0 & 0 & 0 & 0 & 1 \\ 0 & 0 & 0 & 0 & 0 & 0 & -1 & 2 & -3 & 2 \end{bmatrix}$$

$$\text{III.}\quad \begin{bmatrix} -1 & 0 & 0 & 0 & 0 & 0 & 0 & 0 & 0 & 0 \\ 0 & -1 & 0 & 0 & 0 & 0 & 0 & 0 & 0 & 0 \\ 0 & 0 & -1 & 1 & 0 & 0 & 0 & 0 & 0 & 0 \\ 0 & 0 & 0 & -1 & 0 & 0 & 0 & 0 & 0 & 0 \\ 0 & 0 & 0 & 0 & 0 & 1 & 0 & 0 & 0 & 0 \\ 0 & 0 & 0 & 0 & -1 & 1 & 0 & 0 & 0 & 0 \\ 0 & 0 & 0 & 0 & 0 & 0 & 0 & 1 & 0 & 0 \\ 0 & 0 & 0 & 0 & 0 & 0 & -1 & 1 & 1 & 0 \\ 0 & 0 & 0 & 0 & 0 & 0 & 0 & 0 & 0 & 1 \\ 0 & 0 & 0 & 0 & 0 & 0 & 0 & 0 & -1 & 1 \end{bmatrix}$$

$$\text{IV.}\quad \begin{bmatrix} -1 & 0 & 0 & 0 & 0 & 0 & 0 & 0 & 0 & 0 \\ 0 & -1 & 0 & 0 & 0 & 0 & 0 & 0 & 0 & 0 \\ 0 & 0 & -1 & 1 & 0 & 0 & 0 & 0 & 0 & 0 \\ 0 & 0 & 0 & -1 & 0 & 0 & 0 & 0 & 0 & 0 \\ 0 & 0 & 0 & 0 & \alpha & 0 & 0 & 0 & 0 & 0 \\ 0 & 0 & 0 & 0 & 0 & \alpha & 1 & 0 & 0 & 0 \\ 0 & 0 & 0 & 0 & 0 & 0 & \alpha & 0 & 0 & 0 \\ 0 & 0 & 0 & 0 & 0 & 0 & 0 & \beta & 0 & 0 \\ 0 & 0 & 0 & 0 & 0 & 0 & 0 & 0 & \beta & 1 \\ 0 & 0 & 0 & 0 & 0 & 0 & 0 & 0 & 0 & \beta \end{bmatrix}$$

where $\alpha, \beta = \frac{1}{2}(1 \pm i\sqrt{3})$.

11. Prove: If with respect to an n-square matrix A, the vector X belongs to $g(\lambda)$ then $g(\lambda)$ divides the minimum polynomial $m(\lambda)$ of A.
Hint. Suppose the contrary and consider $m(\lambda) = h(\lambda)\cdot g(\lambda) + r(\lambda)$.

12. In Example 6, show that X, AX, and Y are linearly independent and then reduce A to its rational canonical form.

13. In Problem 6:

(a) Take $Y = [0, 1, 0, 0, 0]'$, linearly independent of the chain led by X_5, and obtain $X_4 = Y - (3A - 2I)X_5$ belonging to $\lambda+1$.

(b) Take $Z = [0, 0, 1, 0, 0]'$, linearly independent of X_4 and the chain led by X_5, and obtain $X_3 = Z - X_5$ belonging to $\lambda+1$.

(c) Compute $R^{-1}AR$ using the vectors X_3 and X_4 of (b) and (a) to build R.

14. For each of the matrices A of Problem 9(a)-(h), Chapter 25, find R such that $R^{-1}AR$ is the rational canonical form of A.

15. Solve the system of linear differential equations

$$\begin{cases} \dfrac{dx_1}{dt} = 2x_1 + x_2 + x_3 + x_4 + t \\ \dfrac{dx_2}{dt} = 4x_1 + 2x_2 + 3x_3 \\ \dfrac{dx_3}{dt} = -6x_1 - 2x_2 - 3x_3 - 2x_4 \\ \dfrac{dx_4}{dt} = -3x_1 - x_2 - x_3 - 2x_4 \end{cases}$$

where the x_i are unknown functions of the real variable t.

Hint. Let $X = [x_1, x_2, x_3, x_4]'$, define $\dfrac{dX}{dt} = \left[\dfrac{dx_1}{dt}, \dfrac{dx_2}{dt}, \dfrac{dx_3}{dt}, \dfrac{dx_4}{dt}\right]'$, and rewrite the system as

$$\textbf{(i)} \qquad \frac{dX}{dt} = \begin{bmatrix} 2 & 1 & 1 & 1 \\ 4 & 2 & 3 & 0 \\ -6 & -2 & -3 & -2 \\ -3 & -1 & -1 & -2 \end{bmatrix} X + \begin{bmatrix} t \\ 0 \\ 0 \\ 0 \end{bmatrix} = AX + H$$

Since the non-singular linear transformation $X = RY$ carries **(i)** into

$$\frac{dY}{dt} = R^{-1}ARY + R^{-1}H$$

choose R so that $R^{-1}AR$ is the rational canonical form of A. The elementary 4-vector E_1 belonging to $\lambda^3 - \lambda$ is leader of the chain $X_1 = E_1, AX_1, A^2X_1$ while E_4 yields $X_2 = E_4 - X_1 + 2AX_1$ belonging to $\lambda+1$. Now with

$$R = [X_1, AX_1, A^2X_1, X_2] = \begin{bmatrix} 1 & 2 & -1 & 3 \\ 0 & 4 & -2 & 8 \\ 0 & -6 & 4 & -12 \\ 0 & -3 & 2 & -5 \end{bmatrix}$$

$$\frac{dY}{dt} = \begin{bmatrix} 0 & 0 & 0 & 0 \\ 1 & 0 & 1 & 0 \\ 0 & 1 & 0 & 0 \\ 0 & 0 & 0 & -1 \end{bmatrix} Y + \begin{bmatrix} t \\ 0 \\ 0 \\ 0 \end{bmatrix} = \begin{bmatrix} t \\ y_1 + y_3 \\ y_2 \\ -y_4 \end{bmatrix}$$

Then

$$Y = \begin{bmatrix} C_1 + \frac{1}{2}t^2 \\ C_2e^t + C_3e^{-t} - t \\ -C_1 + C_2e^t - C_3e^{-t} - \frac{1}{2}t^2 - 1 \\ C_4e^{-t} \end{bmatrix} \quad \text{and} \quad X = RY = \begin{bmatrix} 2C_1 + C_2e^t + 3(C_3+C_4)e^{-t} + t^2 - 2t + 1 \\ 2C_1 + 2C_2e^t + 2(3C_3+4C_4)e^{-t} + t^2 - 4t + 2 \\ -4C_1 - 2C_2e^t - 2(5C_3+6C_4)e^{-t} - 2t^2 + 6t - 4 \\ -2C_1 - C_2e^t - 5(C_3+C_4)e^{-t} - t^2 + 3t - 2 \end{bmatrix}$$

INDEX

Absolute value of a complex number, 110
Addition
  of matrices, 2, 4
  of vectors, 67
Adjoint of a square matrix
  definition of, 49
  determinant of, 49
  inverse from, 55
  rank of, 50
Algebraic complement, 24
Anti-commutative matrices, 11
Associative laws for
  addition of matrices, 2
  fields, 64
  multiplication of matrices, 2
Augmented matrix, 75

Basis
  change of, 95
  of a vector space, 86
  orthonormal, 102, 111
Bilinear form(s)
  canonical form of, 126
  definition of, 125
  equivalent, 126
  factorization of, 128
  rank of, 125
  reduction of, 126

Canonical form
  classical (Jordan), 206
  Jacobson, 205
  of bilinear form, 126
  of Hermitian form, 146
  of matrix, 41, 42
  of quadratic form, 133
  rational, 203
  row equivalent, 40
Canonical set
  under congruence, 116, 117
  under equivalence, 43, 189
  under similarity, 203
Cayley-Hamilton Theorem, 181
Chain of vectors, 207
Characteristic
  equation, 149
  polynomial, 149
Characteristic roots
  definition of, 149
  of adj A, 151
  of a diagonal matrix, 155
  of a direct sum, 155
Characteristic roots (cont.)
  of Hermitian matrices, 164
  of inverse A, 155
  of real orthogonal matrices, 155
  of real skew-symmetric matrices, 170
  of real symmetric matrices, 163
  of unitary matrices, 155
Characteristic vectors
  (*see* Invariant vectors)
Classical canonical form, 206
Closed, 85
Coefficient matrix, 75
Cofactor, 23
Cogredient transformation, 127
Column
  space of a matrix, 93
  transformation, 39
Commutative law for
  addition of matrices, 2
  fields, 64
  multiplication of matrices, 3
Commutative matrices, 11
Companion matrix, 197
Complementary minors, 24
Complex numbers, 12, 110
Conformable matrices
  for addition, 2
  for multiplication, 3
Congruent matrices, 115
Conjugate
  of a complex number, 12
  of a matrix, 12
  of a product, 13
  of a sum, 13
Conjugate transpose, 13
Conjunctive matrices, 117
Contragredient transformation, 127
Coordinates of a vector, 88
Cramer's rule, 77

Decomposition of a matrix into
  Hermitian and skew-Hermitian parts, 13
  symmetric and skew-symmetric parts, 12
Degree
  of a matrix polynomial, 179
  of a (scalar) polynomial, 172
Dependent
  forms, 69
  matrices, 73
  polynomials, 73
  vectors, 68
Derogatory matrix, 197

Determinant
definition of, 20
derivative of, 33
expansion of
along first row and column, 33
along a row (column), 23
by Laplace method, 33
multiplication by scalar, 22
of conjugate of a matrix, 30
of conjugate transpose of a matrix, 30
of elementary transformation matrix, 42
of non-singular matrix, 39
of product of matrices, 33
of singular matrix, 39
of transpose of a matrix, 21
Diagonal
elements of a square matrix, 1
matrix, 10, 156
Diagonable matrices, 157
Diagonalization
by orthogonal transformation, 163
by unitary transformation, 164
Dimension of a vector space, 86
Direct sum, 13
Distributive law for
fields, 64
matrices, 3
Divisors of zero, 19
Dot product, 100

Eigenvalue, 149
Eigenvector, 149
Elementary
matrices, 41
n-vectors, 88
transformations, 39
Equality of
matrices, 2
matrix polynomials, 179
(scalar) polynomials, 172
Equations, linear
equivalent systems of, 75
solution of, 75
system of homogeneous, 78
system of non-homogeneous, 77
Equivalence relation, 9
Equivalent
bilinear forms, 126
Hermitian forms, 146
matrices, 40, 188
quadratic forms, 131, 133, 134
systems of linear equations, 76

Factorization into elementary matrices, 43, 188
Field, 64
Field of values, 171
First minor, 22

Gramian, 103, 111
Gram-Schmidt process, 102, 111
Greatest common divisor, 173

Hermitian form
canonical form of, 146
definite, 147
index of, 147
rank of, 146
semi-definite, 147
signature of, 147
Hermitian forms
equivalence of, 146
Hermitian matrix, 13, 117, 164
Hypercompanion matrix, 205

Idempotent matrix, 11
Identity matrix, 10
Image
of a vector, 94
of a vector space, 95
Index
of an Hermitian form, 147
of a real quadratic form, 133
Inner product, 100, 110
Intersection space, 87
Invariant vector(s)
definition of, 149
of a diagonal matrix, 156
of an Hermitian matrix, 164
of a normal matrix, 164
of a real symmetric matrix, 163
of similar matrices, 156
Inverse of a (an)
diagonal matrix, 55
direct sum, 55
elementary transformation, 39
matrix, 11, 55
product of matrices, 11
symmetric matrix, 58
Involutory matrix, 11

Jacobson canonical form, 205
Jordan (classical) canonical form, 206

Kronecker's reduction, 136

Lagrange's reduction, 132
Lambda matrix, 179
Laplace's expansion, 33
Latent roots (vectors), 149
Leader of a chain, 207
Leading principal minors, 135
Left divisor, 180
Left inverse, 63
Linear combination of vectors, 68
Linear dependence (independence)
of forms, 70
of matrices, 73
of vectors, 68
Lower triangular matrix, 10

Matrices
congruent, 115
equal, 2
equivalent, 40
over a field, 65

Matrices (cont.)
product of, 3
scalar multiple of, 2
similar, 95, 156
square, 1
sum of, 2
Matrix
definition of, 1
derogatory, 197
diagonable, 157
diagonal, 10
elementary row (column), 41
elementary transformation of, 39
Hermitian, 13, 117, 164
idempotent, 11
inverse of, 11, 55
lambda, 179
nilpotent, 11
non-derogatory, 197
non-singular, 39
normal, 164
normal form of, 41
nullity of, 87
of a bilinear form, 125
of an Hermitian form, 146
of a quadratic form, 131
order of, 1
orthogonal, 103, 163
periodic, 11
permutation, 99
polynomial, 179
positive definite (semi-definite), 134, 147
rank of, 39
scalar, 10
singular, 39
skew-Hermitian, 13, 118
skew-symmetric, 12, 117
symmetric, 12, 115, 163
triangular, 10, 157
unitary, 112, 164
Matrix polynomial (s)
definition of, 179
degree of, 179
product of, 179
proper (improper), 179
scalar, 180
singular (non-singular), 179
sum of, 179
Minimum polynomial, 196
Multiplication
in partitioned form, 4
of matrices, 3

Negative
definite form (matrix), 134, 147
of a matrix, 2
semi-definite form (matrix), 134, 147
Nilpotent matrix, 11
Non-derogatory matrix, 197
Non-singular matrix, 39
Normal form of a matrix, 41
Normal matrix, 164
Null space, 87
Nullity, 87
n-Vector, 85

Order of a matrix, 1
Orthogonal
congruence, 163
equivalence, 163
matrix, 103
similarity, 157, 163
transformation, 103
vectors, 100, 110
Orthonormal basis, 102, 111

Partitioning of matrices, 4
Periodic matrix, 11
Permutation matrix, 99
Polynomial
domain, 172
matrix, 179
monic, 172
scalar, 172
scalar matrix, 180
Positive definite (semi-definite)
Hermitian forms, 147
matrices, 134, 147
quadratic forms, 134
Principal minor
definition of, 134
leading, 135
Product of matrices
adjoint of, 50
conjugate of, 13
determinant of, 33
inverse of, 11
rank of, 43
transpose of, 12

Quadratic form
canonical form of, 133, 134
definition of, 131
factorization of, 138
rank of, 131
reduction of
Kronecker, 136
Lagrange, 132
regular, 135
Quadratic form, real
definite, 134
index of, 133
semi-definite, 134
signature of, 133
Quadratic forms
equivalence of, 131, 133, 134

Rank
of adjoint, 50
of bilinear form, 125
of Hermitian form, 146
of matrix, 39
of product, 43
of quadratic form, 131
of sum, 48

Right divisor, 180
Right inverse, 63
Root
  of polynomial, 178
  of scalar matrix polynomial, 187
Row
  equivalent matrices, 40
  space of a matrix, 93
  transformation, 39

Scalar
  matrix, 10
  matrix polynomial, 180
  multiple of a matrix, 2
  polynomial, 172
  product of two vectors (*see* Inner product)
Schwarz Inequality, 101, 110
Secular equation (*see*
    Characteristic equation)
Signature
  of Hermitian form, 147
  of Hermitian matrix, 118
  of real quadratic form, 133
  of real symmetric matrix, 116
Similar matrices, 95, 196
Similarity invariants, 196
Singular matrix, 39
Skew-Hermitian matrix, 13, 118
Skew-symmetric matrix, 12, 117
Smith normal form, 188
Span, 85
Spectral decomposition, 170
Spur (*see* Trace)
Sub-matrix, 24
Sum of
  matrices, 2
  vector spaces, 87
Sylvester's law
  of inertia, 133
  of nullity, 88
Symmetric matrix
  characteristic roots of, 163

Symmetric matrix (cont.)
  definition of, 12
  invariant vectors of, 163
System(s) of equations, 75

Trace, 1
Transformation
  elementary, 39
  linear, 94
  orthogonal, 103
  singular, 95
  unitary, 112
Transpose
  of a matrix, 11
  of a product, 12
  of a sum, 11
Triangular inequality, 101, 110
Triangular matrix, 10, 157

Unit vector, 101
Unitary
  matrix, 112
  similarity, 157
  transformation, 112
Upper triangular matrix, 10

Vector(s)
  belonging to a polynomial, 207
  coordinates of, 88
  definition of, 67
  inner product of, 100
  invariant, 149
  length of, 100, 110
  normalized, 102
  orthogonal, 100
  vector product of, 109
Vector space
  basis of, 86
  definition of, 85
  dimension of, 86
  over the complex field, 110
  over the real field, 100

Index of Symbols

Symbol	Page
a_{ij}	1
$[a_{ij}]$	1
A	1
Σ	3
I, I_n	10
A^{-1}; A^I	11
A'; A^T	11
$\overline{A}$; A^C	12
$\overline{A}'$; A^*; A^{CT}	13
$\|A\|$; det A	20
$\|M_{ij}\|$	22
$A_{i_1, i_2, \ldots, i_m}^{j_1, j_2, \ldots, j_m}$	23
α_{ij}	23
r	39
H_{ij}, K_{ij}	39
$H_i(k)$, $K_i(k)$	39
$H_{ij}(k)$, $K_{ij}(k)$	39
$\sim$	40
N	43
adj A	49
F	64
X, X_i	67
$V_n(F)$	85
$V_n^m(F)$	86
N_A	87

Symbol	Page
E_i, (vector)	88
$X \cdot Y$; $X\|Y$	100, 110
$\\|X\\|$	100, 110
G	103, 111
$X \times Y$	109
$\underset{\sim}{C}$	115
p	116
s	116
q	131
h	146
λ, λ_i	149
$\phi(\lambda)$	149
E_i, (matrix)	170
$f(\lambda)$	172
$F[\lambda]$	172
$A(\lambda)$	179
$A_R(C)$, $A_L(C)$	180
$N(\lambda)$	189
$f_i(\lambda)$	189
$m(\lambda)$	196
$C(g)$	198
J	198
S	203
$C_q(p)$	205